STANDARD HANDBOOK OF CONSULTING ENGINEERING PRACTICE

OTHER McGRAW-HILL HANDBOOKS OF INTEREST

American Institute of Physics · AMERICAN INSTITUTE OF PHYSICS HANDBOOK
Baumeister and Avallone · MARKS' STANDARD HANDBOOK FOR MECHANICAL
  ENGINEERS
Brady and Clauser · MATERIALS HANDBOOK
Callender · TIME-SAVER STANDARDS FOR ARCHITECTURAL DESIGN DATA
Chopey and Hicks · HANDBOOK OF CHEMICAL ENGINEERING CALCULATIONS
Condon and Odishaw · HANDBOOK OF PHYSICS
Croft, Watt, and Summers · AMERICAN ELECTRICIANS' HANDBOOK
Dean · LANGE'S HANDBOOK OF CHEMISTRY
Fink and Beaty · STANDARD HANDBOOK FOR ELECTRICAL ENGINEERS
Fink and Christiansen · ELECTRONICS ENGINEERS' HANDBOOK
Gaylord and Gaylord · STRUCTURAL ENGINEERING HANDBOOK
Harris and Crede · SHOCK AND VIBRATION HANDBOOK
Hicks · STANDARD HANDBOOK OF ENGINEERING CALCULATIONS
Hopp and Hennig · HANDBOOK OF APPLIED CHEMISTRY
Juran · QUALITY CONTROL HANDBOOK
Maynard · INDUSTRIAL ENGINEERING HANDBOOK
Merritt · STANDARD HANDBOOK FOR CIVIL ENGINEERS
Perry · ENGINEERING MANUAL
Perry and Green · PERRY'S CHEMICAL ENGINEERS' HANDBOOK
Rohsenow and Hartnett · HANDBOOK OF HEAT TRANSFER
Rosaler and Rice · STANDARD HANDBOOK OF PLANT ENGINEERING
Siedman and Mahrous · HANDBOOK OF ELECTRIC POWER CALCULATIONS
Tuma · ENGINEERING MATHEMATICS HANDBOOK
Tuma · HANDBOOK OF PHYSICAL CALCULATIONS

STANDARD HANDBOOK OF CONSULTING ENGINEERING PRACTICE

Starting, Staffing, Expanding, and Prospering in Your Own Consulting Business

TYLER G. HICKS, PE
International Engineering Associates

JEROME F. MUELLER, PE
Mueller Engineering Corporation

S. DAVID HICKS
Coordinating Editor

McGraw-Hill Book Company
New York St. Louis San Francisco Auckland Bogotá Hamburg
London Madrid Mexico Montreal New Delhi
Panama Paris São Paulo Singapore Sydney Tokyo Toronto

Library of Congress Cataloging in Publication Data
Hicks, Tyler Gregory, date.
 Standard handbook of consulting engineering practice.

 Bibliography: p.
 Includes index.
 1. Consulting engineers—Handbooks, manuals, etc.
I. Mueller, Jerome F.   II. Title.
TA216.H53   1985   620   84-29743
ISBN 0-07-028779-1

ISBN 0-07-028779-1

The editors for this book were Harold B. Crawford and Diane
Krumrey, the designer was Mark E. Safran, and the production
supervisor was Sally Fliess. It was set in Times Roman by
University Graphics, Inc.

Printed and bound by R. R. Donnelley & Sons Company.

Contents

ABOUT THE AUTHORS

Tyler G. Hicks, Professional Engineer, has had wide experience with consulting engineering firms of all sizes. He is author of *Standard Handbook of Engineering Calculations* and *Pump Application Engineering,* both published by McGraw-Hill.

Jerome F. Mueller, Professional Engineer, has had more than 30 years of experience as engineer of public record on more than 1000 engineering projects. His firm, Mueller Engineering Corporation, is well known in the northeast for its expertise in a variety of engineering disciplines. He is the author of *Standard Mechanical and Electrical Details, Standard Application of Mechanical Details,* and *Standard Application of Electrical Details,* all of which were published by McGraw-Hill.

Preface

Consulting engineering offers every engineer—young or not so young, male or female, introvert or extrovert, specialist or generalist—a chance to go into business for himself or herself. Being a consulting engineer in business for yourself is a viable alternative to stagnation on the much-discussed corporate "parallel ladder" to success.

As a consultant, the engineer daily does exciting and challenging work, free from many of the corporate bureaucratic rules of procedure and protocol. Consulting engineers are their own masters. They determine: (1) what kind of work they enjoy, (2) whom they want to work for, (3) how much they will charge for engineering services, and (4) when they will work. You might say that consulting engineering is the American dream (*and* the international dream) come true for engineers.

This Handbook shows all engineers how to start and prosper in their own consulting engineering businesses. Starting with the basic question, "Why become a consultant?" it moves on to a variety of important topics to guide the engineer in opening his or her own consulting business. Thus, the handbook shows you:

- How to evaluate your professional skills
- What your income outlook is as a consultant
- Types of consulting services you might offer

- Choosing consulting services to render (in view of your skills)
- Major fields of private professional practice
- Organizing your consulting engineering firm
- Financing your consulting practice
- Professional licensing requirements for consultants
- Budgeting your income and expenses
- Choosing an office for your business
- Professional insurance your practice should have
- Marketing your consulting skills
- Building an effective sales team
- Major markets you can earn money from
- Proposal preparation for profitable work
- Getting into (or staying out of) politics
- Staffing your consulting office
- Managing your firm and its projects
- Continuing education for your staff and yourself
- Office organization—files, catalogs, drawings, etc.
- Computer usage in your consulting office
- Protecting your firm against lawsuits
- Contract administration on large and small jobs
- How to make money in forensic engineering
- Expanding your professional practice
- Making money from joint ventures
- Many more money-making topics

Throughout the Handbook the emphasis is on practical and proven ways for you to earn a sizeable income in your own consulting business. With this Handbook on your desk, you can easily face the daily situations that occur in any consulting office. And you will be able to solve quickly the variety of small problems you can expect in any business—staffing, salary reviews, hiring, job descriptions, marketing techniques, office procedures, computer usage, joint ventures, and many others.

Both editors of this Handbook are licensed professional engineers with many years of consulting practice with large and small consulting firms doing domestic and overseas work. So the Handbook reflects actual work experience in the consulting engineering field today. Further, both editors have advised young and mature engineers who came to them with questions concerning the consulting engineering business. A number of these engineers took the editors' advice and today are successful consulting engineers.

To reflect the widest experience possible for the users of the Handbook, the editors have drawn on the recommendations of a number of outstanding consulting engineers in the areas of using computers in consulting, budgeting and scheduling projects, controlling costs, forensic engineering practice, etc. This approach offers the reader a wide view of the consulting field and the many opportunities it provides engineers. Further, the recommendations chosen are presented as "Consultant's Profit Keys," allowing the reader to focus on practical, useful, and profitable ways for the new consultant to start and run a successful consulting business.

The editors are indebted to a number of people who provided help or materials during preparation of the manuscript. These include Nicholas Chopey, Editor, *Chemical Engineering* magazine; Frank L. Evans, Editor, *Hydrocarbon Processing* magazine; Stanley Cohen, Editor, *Consulting Engineer* magazine; John Constance, PE, Engineering Registration Consultant; Joseph A. McQuillan, President, JMT Consultants, Inc,; Charleston C. K. Wang, Founding Partner, and Harold E. Parker, Law Office Administrator, of the law firm Groeber, Wang & Southard; Roy Peterson of Creative Research Services, Inc.; Lee F. Francis, Vice President and Manager of the International Department of James M. Montgomery, Consulting Engineers, Inc.; David Burstein, PE, Vice President and Manager of Operations, Engineering-Science; A. E. Kerridge, Project Manager, M. W. Kellogg Co.; Thomas G. Roberts, Information Systems Support Manager, Bechtel Petroleum, Inc.; Paul E. Pritzker, PE, Senior Engineer, George Slack Pritzker Forensic Engineering Consultants; Dr. Louis Berger, Vice Chairman of Louis Berger International, Inc.; International Federation of Consulting Engineers (FIDIC); Michael T. McGovern, PE, Paradigm Systems; William J. Meehan, PE, R. G. Vanderweil, Engineers; and Remarkable Products, Inc.

Finally, the editors thank the many active consulting engineers in all parts of the United States, Europe, South America, Asia, Australia, and Africa who gave advice, ideas, and criticisms of this work. We hope this Handbook lives up to their expectations.

TYLER G. HICKS
JEROME F. MUELLER

How to Become a Successful Consulting Engineer

WHY BECOME A CONSULTANT?

Reasons for becoming a consulting engineer are almost as numerous as the engineers in private practice today. Among the rewards most frequently mentioned are:

1. Your freedom to practice in professional areas of particular interest to you is increased, and you have a direct involvement in interesting work.
2. There is greater income potential than there is for a salaried engineer.
3. There is more time for hobbies and leisure-time pursuits.
4. There is more opportunity to receive professional recognition for your own achievements, i.e., discoveries, new patents, outstanding designs, and other contributions to your field.
5. A fuller life is possible since the usual retirement age ceases to loom as a shadow over career progress and creativity. You can build a solid business which will keep you employed as long as you wish.
6. There is growth and satisfaction through combining business and technical skills in your own exciting venture.

7. Security is enhanced because unlike the salaried engineer, you cannot be buffeted about by layoffs, job cancellations, firings, and other hazards of working for someone else.

8. The consultant has freedom from demanding bosses.

9. The flexibility in working hours means you work when you like.

10. You can choose where to work—at home, in your own office, or while traveling.

Undoubtedly you can add other reasons why you would like to become a consulting engineer. Try it—the list keeps growing when you realize that becoming a consultant:

• Gives the joy of small-business accomplishment
• Allows you to engage in the specific fields you like most
• Provides the tax relief and pension advantages of the professional corporation

HOW TO EVALUATE YOURSELF

Most engineers begin their careers in salaried positions. After 10 to 15 years, many begin to dream of starting their own businesses. For some, it is a time when their family responsibilities are heaviest. Lacking basic know-how about building a private practice, a good many engineers go on dreaming.

The dream can become reality, however, whether you have been an engineer for 5 years or 30 years. The first step uou must take is to look at your personal and professional assets. Clear self-evaluation should tell you if you have a solid foundation to enter this challenging and rewarding new career.

Let's start by looking at key points that distinguish the concept of the consulting engineer as it appears throughout this book. A professional engineering consultant, once he or she has established a private practice, usually owns and operates the business without earning a regular salary from an employer or works for a firm under the title "Consulting Engineer" for a regular salary. Consequently, most consulting engineers own or rent offices, laboratories, shops, or other facilities as places to conduct their practices. They also own or rent the equipment and supplies used there, such as drafting tables, desks, typewriters, computers, and drawing materials. Those who employ others in their practices maintain a payroll for office workers and professionals in the firm.

The professional consultant provides advice, design services, on-site evaluation of various projects, and other types of assistance in one or more of the several engineering disciplines: mechanical, civil, electrical, chemical, structural, environmental, etc. Consulting engineers rely on personal contact with clients and their representatives, plus visits to project locations during and after completion of the work for which they were hired.

The consulting engineer today is, perhaps more than anything else, a blend of two types of professionals: the experienced engineer and the active business person. When these two roles are combined in the proper mix, the results are likely to be outstanding personal satisfaction, financial success, and general well-being.

SELF-QUIZ FOR PROSPECTIVE CONSULTANTS

Here is a short quiz to help you evaluate your personal assets as you consider entering the world of consulting. All you need to do is answer each question and then tally your score according to the scale at the end of the quiz.

Check the appropriate line for each question.

	YES	NO
1. Are you the kind of person who enjoys working independently?		
2. Does the idea of having your own business strongly appeal to you?		
3. Do you find it frustrating working for others year after year?		
4. Do you feel that your current paychecks are barely meeting your living standards?		
5. Do you have an engineering specialty which is in demand in today's economy?		
6. Have you had experience as a project or group leader, manager, or supervisor?		
7. Do you enjoy "running a job" in which other people report to you?		
8. Are you a licensed professional engineer in your state?		
9. Can you write understandable reports on your professional work activities?		
10. Are you willing to take some risks if the potential rewards are sufficient?		

After answering all 10 questions, count the number of *yes* answers. If you checked 6 or more *yes,* you very likely have the basic work and personality characteristics shared by many successful consulting engineers. If you checked all 10 *yes,* the best advice is: Get your private practice started now! Use the guidelines throughout this book to build your new career.

The results of the self-quiz are not totally conclusive, of course. For those who have a score less than 6, a bit more evaluation may be necessary before entering private practice. The self-quiz is designed to sound out your feeling for some very general qualities of consulting work. The rest of this book presents more detailed guidelines you can use for further evaluation and, most importantly, for directing your present actions toward establishing a practice of your own in the future.

Before we go on to the details, let's review the chief attributes of the successful professional engineering consultant. As indicated by the quiz, the consultant has:

* The capacity to work alone (a common condition of work in this field)
* The desire to run her or his own business
* The goals to increase earnings to pay all personal and business expenses comfortably and to gain added income
* The ability to specialize in an area of engineering that is in demand (which helps the professional to start consulting work quickly and securely)
* The experience in job management, which provides sound business judgment and the ability to direct the work of others
* The skills for writing reports for successful bids and jobs—to communicate effectively
* The willingness to take calculated personal and professional risks—a must in almost any profession
* A valid professional engineering license (often required for consulting work)

THE CONSULTANTS INCOME OUTLOOK

Certainly, one of the most prominent reasons for becoming a consultant is to increase income. While any kind of business, including consulting, requires an initial start-up period in which cash may flow out faster than it comes in, a successful consultant's earnings generally exceed those of engineers employed by government agencies or private businesses at nearly all levels, except perhaps in top management.

A recent survey of engineers in private practice puts the gross annual income per consultant at over $40,000. The survey also indicates that a consultant's earnings remained relatively consistent regardless of the size of his or her firm. (More than 85 percent of all consulting firms have fewer than seven employees.) Another survey concluded that the annual salaries of engineers employed in industry peaked at around an average of $38,000. For those who were able to secure a supervisory position, the industry annual average peaked at $48,000.

Perhaps these figures do not seem to show much difference between earnings for the consultant and those of the engineer in industry. Consider the following actual example of what can happen in private practice, however, and the difference grows more apparent. Two engineers, one employed in industry and another a consultant, are both in top shape and willing to work long and hard to advance. The engineer in private industry has gained enough recognition for her work to be offered a supervisor's job and a possible $5000 raise.

The consultant, meanwhile, has a six-person firm which is bringing in $40,000, gross, per engineer. Now, let's say the consultant gets some additional work which requires at least half the time of a seventh employee. But rather than hiring someone new, the consulting engineer puts in a good deal of extra effort of his own and thereby substitutes for the time that a seventh employee would put in. The value of the extra effort could be the receipt of a $20,000 fee for the additional project. So the consultant's annual income leaps from $40,000 to $60,000 for his extra effort, an increase that could hardly be hoped for by the industry engineer—even over the course of several years!

The example illustrates a common problem faced by the industry engineer who must depend on a regular salary. For many very good salaried engineers, advancement is limited by the fact that there is not enough room in the company for everyone to rise to highly desirable managerial positions. But the consultant does not have this problem. She or he can work extra long and pocket the additional money or hire more people and expand the firm. Either way, the consultant wins.

Other benefits of private practice sometimes prove to be worth even more than the expanded income capabilities themselves. For example, the consulting firm enjoys many tax advantages not available to a salaried engineer. Travel, automobile use, entertainment, professional membership, and other fringe benefits are often available to consultants on a tax-deductible basis. Further, the engineer in private practice can set up his or her own corporation structure with liberal retirement benefits.

Moreover, the consultant gains inestimable amounts of "psychic income"—the personal, psychological satisfaction of being one's own boss and of seeing the results of one's professional skills applied directly to solve specific engineering problems. And there is gratification in being responsible for the successful completion of a project. You cannot use psychic income to pay your bills, but you can use it to give your life fulfillment and a sense of security and pride.

In the more concrete world of economics, the consultant also wins—by being able to adjust her or his fees to keep up with inflation. Salaried engineers, on the other hand, frequently do not manage to keep their earnings ahead of the inflation rate—even when they are given cost-of-living increases. The salaried employee commonly takes 15 to 20 years to reach the peak earning level in the profession whereas many consultants reach even higher levels within 4 to 8 years of starting their practice.

What Size Fees Do Consulting Engineering Firms Earn?

Little is published on the fees earned by consulting engineering firms. However, a recent survey by *Specifying Engineer* magazine gives some interesting numbers on annual fee earnings. Thus, for the top 50 fee-earning consulting engineering firms:

- The annual fees ranged from a high of $236 million to a low of $3.4 million.
- The average annual billings for fees was $19 million.
- Retrofit and modernization of plants and other facilities comprised nearly 51 percent of the fee earnings.
- Foreign work comprised nearly 18 percent of the fees received.
- Reports, investigations, studies, and surveys performed by consulting engineers comprise a major portion of the fee income of some firms.

While a new consulting engineering firm cannot expect fee earnings of this magnitude for a number of years, the numbers *do* show that significant revenues can be achieved. The fee earnings cited here are one more reason for the salaried engineer to consider opening his or her own consulting practice.

TYPES OF ENGINEERING CONSULTING SERVICES

Few if any limits exist in the range of engineering services that a consultant can provide. Once you consider becoming a consultant, one of the chief tasks is to draw up an informal outline of the types of services you will render to clients. Typical services could include:

1. Designing mechanical, electrical, and energy systems, transportation facilities, communications systems, and utilities
2. Analyzing client problems in specific technical areas, such as structural stress and strain, materials defects, equipment lubrication, transport scheduling, and control of energy and power processes
3. Calculating special structural or design features like piping flexibility, bridge and tower strength, foundation materials, and components for communications products and equipment
4. Providing computer services for professional firms, such as routine design calculations, specification preparation, payroll, project completion, and cost evaluation
5. Developing manufacturing designs and analyses; preparing assessments of the environmental impact of specific building projects.
6. Offering miscellaneous specialties in any one of many highly technical disciplines, such as magnetohydrodynamics, nuclear waste disposal, unique uses for electronic chips, all kinds of pollution control, etc.

Of these six major types of services, the first—design of mechanical and electrical systems and structures—is the area that involves the largest number of consulting engineers. This area probably accounts also for the largest proportion of the millions of dollars that consulting engineers earn each year. The usual arrangement for fees in this area of services is to have the consultant receive a percentage of the total job cost.

Next highest in popularity today is the analysis of client problems in specific technical areas. The most popular specialty is the analysis of stress and strain in machine and structural components. The reason is that many clients have serious problems in these areas, and solutions are often beyond the capabilities of the client's regular staff engineers. This area is probably more lucrative than machine and electrical design because the fees are often earned as a negotiated flat amount rather than as a percentage of the job cost.

Third in current popularity is the calculation and analysis of other features of an engineering job, such as special structural and design concerns. Thus, there are consultants who specialize in piping flexibility analysis, nuclear waste problems, merchant ship design, and other such areas. Here again, a flat fee arrangement for a one-time job is typical.

The relative prominence of the other three types of services listed is not as easy to estimate. Computer services are becoming more popular every day and may eventually rival analysis activities in overall popularity and revenue for consulting engineers.

CHOOSING THE SERVICES TO RENDER

Each engineer knows his or her own technical expertise better than anyone else. So the next step is to look over your experience in light of the current market for consulting services in your area. Most beginning consultants find the following to be reliable guidelines: Income can be accumulated fastest and market penetration achieved quickest in the mechanical, plant, or process design area, and solving clients' problems, such as those mentioned previously, also allows quick entry into consulting work, provided one has appropriate background experience. You should also consider your personal preferences for particular kinds of work, provided that you have appropriate background experience. Finally, you should take a sober look at the future market needs for the various types of services you might offer.

Fortunately, you do not necessarily have to embark on your consulting career as a total novice. More than likely, you had experience with the same types of services when you were an employee of a larger firm. And if you plan the transition from a salaried job to private practice appropriately, you may actually have an offer of a consulting project before you leave your old job. You probably already have several assets to help you get started: professional qualifications and license, a technical specialty, job contacts, and a reputation in your field. Some new consultants even begin by moonlighting—that is, they take on one or more small jobs in their spare time while working for the larger firm.

Once you develop a fairly solid idea of the kinds of services which would be best suited to your strongest technical skills, your personal preferences, and the marketplace, you can further develop your plans in the following way.

Write a brief description (about 250 words) setting out the basic details of your prospective services. This early step is very important because it will help clarify your thinking about the items you will be selling. The consulting engineer is indeed a salesperson of sorts; she or he sells her or his personality, firm, the services the firm offers, and the image that will be projected throughout the field. Here is a typical description a beginner can use as an example (your description can also come in handy for writing your firm's sales brochure later on):

ABC Design Services offer an experienced and highly capable engineering staff to design, specify, supervise construction of, and test equipment for heating, ventilating, energy conservation, and air conditioning and refrigeration. ABC's services are available for commercial, industrial, and institutional buildings anywhere in the world. ABC specializes in prompt, efficient, and cost-effective design of heating, ventilating, and air conditioning and refrigeration systems for a variety of clients, such as retail stores, manufacturing concerns, and operators of hospitals, jails, courthouses, and government office buildings. Architects, general contractors, individuals, industrial firms, and government agencies use ABC.

Among the specific services offered are: analysis of heating, ventilation, and cooling needs; selection of suitable equipment to meet those needs; specification preparation; supervision of the construction and installation; and testing and performance rating. Once an installation meets the client's specifications, ABC will monitor the ongoing performance of the work on a regular basis.

ABC Design Service personnel are professional engineers licensed in the states of California, Nevada, New Mexico, Illinois, Iowa, New York, and New Jersey. Where design work is to be performed in other states, ABC will seek licensing on a reciprocal basis.

To obtain more information on ABC's experience, capability, and availability, call or write N. L. Smith, ABC Design Services, 1000 Main Street, Anytown, 00001.

After you write your firm's description, put it aside for a day or two. Then read it over and polish the writing as thoroughly as possible to make it attractive and interesting to prospective clients.

As word spreads about your involvement in private practice, you may be surprised at how many people will ask for written information about your firm. If you have some copies of your description on hand which have been professionally printed or even duplicated on a copying machine, they might become the key to some of your first client contacts.

A WORD ON PROFESSIONALISM

Many people feel they know the meaning of the word "professional," but few find the definition easy to state. In engineering, one indication of the professional is that he or she usually possesses a license—a certificate and a wallet-sized card testifying that the engineer has earned a passing grade on the official state engineering exam.

The license is especially desirable—in fact, mandatory—if the engineer intends to provide consulting on matters relating to buildings, their construction or modification, or other work on structures which require building permits or such official approval.

Licensing is mandatory for work in many other areas as well. However, there is a trend in consulting today to provide collective work by several individuals of demonstrated competence in a variety of areas, some of which may not mandate a license.

So how does one distinguish the true professional? Basically, professionalism denotes the ability to make competent, unbiased judgments in a field requiring a high order of knowledge and specialization. The key word here is "unbiased." The true professional puts the bulk of her or his energy into doing the best possible job for the client, regardless of the engineer's feelings about any of the client's personal attributes, choice of supplies, associates, etc.

Professionalism in engineering has evolved perhaps in much the same way as it has evolved in medicine: More than a century ago, the physician's image in some sectors was one of being both "surgeon and barber," a person who dispensed advice and medical care as arbitrarily as the casually chatting barber. Engineering likewise was once fraught with experiment, personal opinion, and guesswork; but during the twentieth century it has emerged along with its life long partner, science, as a profession as refined, as accurate, and as beneficial to humanity as the specialties exemplified by the modern surgeon's use of such highly sophisticated tools as lasers to perform microsurgical operations.

The professional consultant—like today's surgeon—is the one who considers all perspectives of each project and submits thoroughly reasoned, objective conclusions at each step toward completion. As a consulting firm grows and hires more of these kinds of professionals, the professionalism of the firm grows too. Several examples of the Code of Ethics for Engineers are given at the end of Section 2 of this Handbook.

MAJOR FIELDS IN PRIVATE PRACTICE

Besides the six common types of consulting services reviewed earlier, private engineering firms also provide professional assistance in ways which could be said to comprise six major fields of endeavor. These are: (1) design of systems and products; (2) professional advice; (3) interprofessional assistance; (4) international consulting; (5) forensic engineering; and (6) research and development. These categories are not mutually exclusive. All are equally important to the engineering profession as a whole. They can overlap in many ways, although some firms perform the bulk of their work in one or two of the fields exclusively. For our purposes, however, the best way to examine the six major fields is to look at each one individually.

1. Design Firms

Firms which engage primarily in design—of a heating system for an apartment building, for example, or of other processes or products—are perhaps the most familiar kind of consulting organizations. Design firms prepare plans and designs for bridges, roads, sewers, environmental systems, buildings, and a variety of other structures and devices. They customarily earn commissions for their work from organizations such as other engineering firms, architects, industrial companies, developers, and government agencies.

The design firm may provide the plans and specifications for particular projects in whole or in part. Generally, the firm also secures bids from contractors on the completed design and specifications (often referred to as the "bid package") and provides field supervision during the actual construction phase.

2. Advisory Practice

Advisory consulting engineering organizations are both the best understood and the least understood of all consulting firms. Commonly, they are thought of as firms which give economic advice to a government representative or committee. Just as often, industrial companies seek out and retain consulting firms for engineering advice on topics such as production problems and facilities planning. What they offer, basically, is specialized expertise in areas of engineering with which a client may already be familiar but for which the client desires evaluation from an outside source.

Tasks the advisory firm might handle include feasibility studies in which the engineer examines the advisibility of performing particular engineering work. Feasibility studies can involve producing cost estimates of completed designs, work schedules, environmental impact statements, and similar types of analyses. The advisory firm usually does not provide a finished product but effectively performs professional inspections of both proposed and completed projects.

3. Interprofessional Practice

This term usually refers to engineering consulting firms that collaborate with other consulting firms in an advisory capacity or in producing parts of a project's design, specifications, construction, supervision, and so on. For example, an engineering consultant could be retained by an architect to determine and select the best building materials for a proposed high-rise building. Such interprofessional practice forms a substantial portion of the consulting work done by engineers trained in the structural, mechanical, and electrical disciplines.

4. International Consulting

The growth of international trade and increased awareness of energy needs throughout the world have spurred a boom in international engineering opportunities for the American engineer and others in the last few decades. International practice can hold surprises, however, for those who are unfamiliar with differences in business customs in other countries.

For instance, many firms and individuals outside the United States are accustomed to consulting arrangements wherein the consulting firm performs construction, management, and even financing operations for each project, in addition to its normal design or advisory services.

Two general rules that apply to international consulting ventures are: (1) Become doubly familiar with all aspects of your prospective duties on the project *before* agreeing to start the work; and (2) do not seek involvement in the project until you are certain that your firm has a staff which is large enough and diverse enough to perform all necessary duties. Generally, firms with fewer than 100 employees may expect to find international jobs difficult to secure, unless they enter into a joint venture with other firms that have complementary capabilities.

A large proportion of the engineering that goes into foreign work actually originates in the United States, where most of a project's financing arrangements are also based. The firm planning to engage in international practice therefore will benefit by cultivating good relationships with major banking and financial entities in the United States, as well as with key people involved in international trade activities for the federal government.

5. Forensic Engineering

This type of consulting practice may be unfamiliar to some people, but it is not new. Basically, "forensic engineering" refers to the professional engineer's involvement in legal matters—including cases related to accidents, fire, and engineering defects and omissions—for the purpose of providing expert testimony. The forensic engineer ("forensic" comes from a Latin word meaning "a public forum or place of debate") may be retained by law firms, insurance agencies, building owners, and individuals seeking redress for the results of possible engineering failures or other problems.

Some consultants view this field as a "dead-end street" which offers little room for creativity and personal accomplishment. On the contrary, forensic engineering provides the experienced consultant with almost unlimited opportunities for public service and recognition. The knowledge of engineers is frequently needed in courts of law, and there is little comparison to the feeling of fulfillment derived from providing data useful in obtaining justice.

Forensic practice may not be one of the most lucrative consulting fields, but a significant number of engineers who have worked in special subdisciplines as employees find that forensic work provides more than enough income to enable them to get a new private practice off to a solid start. The forensic field is especially well suited to the engineer with a wide range of experience and many years as an employee. For some, it can develop into a complete engineering career itself. For tips on being an effective expert witness, see Consultant's Profit Key 2, in this section.

6. Research and Development (R&D)

The tremendous sums of money allocated for research and development (R&D) in industrial and technical fields is a matter discussed in newspaper and magazines nearly every day. But the role of the professional consultant in this area is often overlooked. Actually, R&D accounts for a large and lucrative part of the consultant's practice.

The common obstacle to the new consultant in this field is lack of sufficient renown. Whether an R&D project is being sponsored by the federal government or by private industry, large amounts of money are likely at stake, and the sponsor may consequently be reluctant to retain consultants who do not have an established reputation as experts in the areas involved. Therefore, R&D consulting—like forensic engineering— tends to favor the older, more widely known engineer after he or she leaves a position with government or industry to set up a private firm. Such an engineer may have the benefit of well-respected contacts who already laud the engineer's expertise and who will recommend the consultant's services to others.

With diligence and determination, the consultant can build an R&D practice into a large and highly prosperous business. The most common mode of entry into this field for younger consultants is contracting for some of the extra design needs of companies or agencies on particular projects. While this is a more speculative approach, it too can lead to rewarding R&D experience.

CONSULTANT'S PROFIT KEY 1

Joseph A. McQuillan, President of JMT Consultants, Inc., a consulting engineering firm in San Francisco, California, is an attorney and former chairman on the expert witness committee of the Consulting Engineers Association of California; he wrote the following for *Consulting Engineer* magazine in an issue featuring forensic engineering.

A consulting engineer who assumes the role of an expert witness soon learns that a knowledge of his own profession is only part of the expertise necessary to fulfill his function. He will need, as well, to become something of an instant expert on certain aspects of the law and courtroom procedure. The testimony he is called upon to give at the trial may be only the final step in a series of procedures during which his services are required. The first is likely to be what is known as the process of discovery.

Discovery is a legal procedure by which a party to a lawsuit obtains data relating to the matters in contention. Since the purpose of litigation is to get to the truth of a matter, legislatures and the courts have instituted rules that promote early settlement and more rapid litigation procedures. Forthright disclosures of evidence in the possession of each party is one way to accomplish this. Under current rules, the days of the surprise witness and "fishing expeditions" by contending attorneys is a thing of the past. However, certain limitations are imposed on the rules of discovery to prevent attorneys from taking unfair advantage of the opportunity to examine evidence in the possession of their opponents.

One of these limitations is the doctrine of privilege. Certain confidential communications received by an attorney from his client are protected from discovery by the opposing party. If the consulting engineer is retained by the client, the communications between them are *not* privileged. However, if the engineer is retained by the attorney in the role of a consultant or advisor it is possible, under limited circumstances, that communications from the client made to the engineer for transmission to the attorney may be privileged.

Another limitation to the discovery process is the work product doctrine, which holds that the fruits of the attorney's labor constitute privileged information. This allows him to prepare his case in private and to investigate both the favorable and unfavorable aspects of the case without fear that such data might be used against him by opposing counsel. The extent to which this data is subject to discovery is a frequent source of controversy. The balance is not easy to maintain.

The decision as to whether a consulting engineer's efforts may be subject to discovery by opposing counsel depends largely upon whether the engineer has been retained simply as an advisor to the attorney or whether he has been retained as an expert witness. Usually the opinions and advice from a consulting engineer to the attorney who retains him solely as a consultant are protected by the work product doctrine.

The attorney may plan to convert the engineer from the role of an advisor to an expert witness only if the consultant's opinions prove to be favorable to the attorney's case. If a dispute arises as to whether the CE is an advisor or an expert witness, the court may make the final determination.

DEPOSITIONS

A deposition is a proceeding conducted before trial in which the attorneys representing the various parties may question persons who are identified as witnesses. The attorneys and the witness usually meet in the office of one of the attorneys. The witness is required to respond under oath which is administered by a certified shorthand reporter. The reporter then records the proceedings and later prepares a typewritten transcript for possible use at trial.

Once the consulting engineer is identified as an expert witness, the first notification received of the holding of a deposition is the receipt of a subpoena. This official document, frequently delivered by a process server, states that the presence of the engineer is required at a certain time and place. Failure to obey the subpoena may be treated as an offense punishable by a jail sentence and a fine. A subpoena *duces tecum* requires that the witness bring with him relevant records, documents, or other items within his custody or control. If the CE finds it difficult to comply with the terms of the subpoena, he should notify the attorney who retained him. Usually some accommodation will be worked out and another date set.

The purpose of the deposition is to permit opposing counsel to hear the expert's opinion before any scheduled trial. Exchange of information that attorneys have gathered in the form of expert opinions often results in rapid settlement of the issue.

The consulting engineer who is scheduled for deposition should meet with the attorney who retained him to go over the intended responses to possible questions. This prevents embarrassment and uncertainty during the deposition. A pre-deposition conference also assists the attorney in clarifying issues and preparing his case. The conduct of witnesses at deposition and trial—their demeanor, poise, and clarity of expression—can significantly affect the outcome of a case.

CONDUCT DURING DEPOSITIONS

It is essential that the expert witness remain composed while testifying, since professional demeanor lends credence to his testimony. While the substance of the engineer's expert testimony is of primary importance in the evaluation process, the behavior of the witness while giving the testimony cannot be ignored.

The deposition process gives the attorney a chance to see his engineer expert perform under fire. The strength or weakness of his testimony will not be lost upon the attorneys attending. Many cases do not go beyond the deposition stage because, having seen the direction in which the case is heading, the attorneys may see fit to begin serious settlement negotiations.

The ideal expert is one who is articulate but not verbose. He should not engage in verbal jousting with opposing counsel, but should calmly answer questions put to him by responding truthfully and clearly. When questions require a narrative response, the engineer should answer fully but resist the temptation to volunteer more information than is required. It also is important that the witness refrain from acting as an advocate and that he render an objective opinion.

The deposition is not the place for humorous or sarcastic comment. At times, opposing counsel may seem to exceed the bounds of common courtesy and appear to be hostile. The witness should nevertheless remain calm and concentrate on a clear and correct response. Should the question seem unclear or ambiguous, it is proper for the witness to ask that the question be repeated or rephrased.

The expert must keep in mind that his written report should serve as the complete and detailed answer to most questions. Answers that refer to the report for detail also help to keep the line of questioning on a consistent course and will help to avert inconsistencies in the expert's testimony. Opposing counsel will seize upon apparent inconsistencies for later use at trial in an effort to reduce the expert's credibility in the eyes of the jury.

If the engineer is uncertain of the answer to a question, it is better to decline to give an opinion than to conjure one on the spot. Presumably, most questions will have been anticipated, but occasionally a question may open a new door. In such an instance it may be best to state that the particular inquiry requires additional investigation. If the question is not relevant, the witness should say so. However, the fact that an expert witness is unable to supply an answer does not necessarily imply poor preparation.

On occasion, an attorney may ask the engineer a question that is improper. Usually the attorney who has retained the engineer will discern this and interrupt the proceeding to point out the impropriety. If not, the engineer may ask for a short recess to confer with the attorney. Such recesses should be kept to a minimum to avoid the impression that the engineer is unprepared or that the attorney is guiding him too carefully.

QUESTIONING AT DEPOSITION

There usually are two main areas of inquiry during the deposition. The first is the expert's qualifications; the second is the expert's opinion relative to the problem at hand. In most instances the attorneys will examine the expert's education, training, and experience in an effort to measure his background against those of opposing experts, since this may be important to the jury. He may be asked, for example, whether he has had experience with similar problems. After the attorneys have concluded this line of questioning, the expert usually is asked for his opinion relating to the matter in litigation and his reasons for holding the opinion. Pursuit of the details behind the expert's opinion may be of critical importance if the opinion is damaging to opposing counsel's case. In such an instance, opposing counsel will seek to uncover some invalid reasoning or an incorrect presumption on the part of the expert.

TESTIMONY AT TRIAL

Before the engineer is permitted to testify as an expert, the judge must be satisfied that he is competent at his profession. The attorney who proposes the CE as an expert witness will bring out his qualifications by direct examination. It is important to note that the expert's qualifications and experiences must be relevant to the particular problem about which he is to render an opinion. The fact that a witness has great qualifications in other related fields is of no consequence.

The judge has sole discretion to accept or reject someone as an expert witness. A lay witness is competent to testify only as to those things that are within his personal knowledge. The expert has no such limitation. While the lay, or ordinary, witness is forbidden to offer his opinion on a technical matter, an opinion is precisely what the expert witness is asked to give.

DIRECT EXAMINATION

Once the judge is satisfied that the engineer qualifies as an expert witness and there is no challenge by opposing counsel, the attorney may proceed with the direct examination. Normally, the attorney who has called the expert will lead up to the opinion by asking what he was retained to do and how he gathered the data that he required to reach his conclusion. The expert witness will then recite any assumptions he made and how they relate to the evidence.

In many cases, the actual testimony of the expert may have been preceded by many hours or days of careful study by both counsel and the engineer. Both may explore the alternative opinions that may suggest themselves and the reasons for rejecting them. There are few aspects of a case that are more important than the verbal picture painted for the jury by counsel and his expert during direct examination. Painstaking preparation includes a search for any flaws that could be attacked on cross-examination.

CROSS-EXAMINATION

The aim of opposing counsel is to probe for any flaws in the expert's testimony and his credibility, thus reducing the strength of his opponent's case. If the expert has any weaknesses in education or experience, opposing counsel is likely to seize upon them to enhance his own expert's credibility if his witness has superior credentials or training. Once the issue of the expert's qualifications is addressed, opposing counsel may move on to a detailed examination of the basis on which the expert formed his opinion.

At this stage, the opposing counsel knows the contents of the CE's report and has heard a brief recital of the conclusions reached in that report. He also has had the benefit of a review of the report by his own expert, whose opinions may differ considerably. If a key assumption the expert has expressed can be shown to be faulty, his credibility is in serious jeopardy.

Impeachment of the credibility of an expert witness also can be achieved by showing that he is biased in some way or that his general character or reputation for honesty and truthfulness is questionable. Inconsistent statements are a much more common method of impeaching credibility. Attorneys carefully scrutinize an expert's report and compare it with statements made during deposition and direct examination. If not in conflict with rules of evidence dealing with hearsay, the opposing counsel may offer the testimony of others, relating statements made to them by the engineer expert that are inconsistent with the engineer's report or testimony.

Frequently, statements made during deposition may initiate further investigations that require revised opinions. In such instances, the attorney who presents the expert must offer an explanation for any apparent or real inconsistency. This attempt to explain away an apparent impeachment of credibility is called rehabilitation of the witness by redirect examination.

Cross-examination by a skilled and aggressive attorney who is well-prepared can be a stimulating challenge to the expert witness, and it requires considerable preparation. Unlike most engineering experiences, it takes place in a legal forum with special rules and procedures that may be disconcerting to those who are not familiar with them.

As a rule, a consulting engineer has little in his background that prepares him for service as an expert witness. Therefore, a CE who chooses to assume such a role would do well to acquaint himself with the rules, the techniques, and the pitfalls of what will be, in many respects, the practice of a new discipline.

ORGANIZING THE FIRM

The engineer who decides to move into the world of private practice faces a number of challenges. A big question one must address is how to organize the business. With a wide variety of organizational arrangements to choose from—partnerships, corporations, joint ventures, associations, and others—the engineer who is new to private practice can, understandably, feel unsure about how to proceed. For those who start their consulting practice alone or with one or two others, the choices are somewhat narrowed; sole proprietorships and partnerships are the more common modes of organization for the beginning consultant.

But there is a broad range of possibilities you can explore to find the kind of firm that is right for you. Some of the factors to take into account as you begin this important step are your personal "style" and feelings. For example: Do you work best and feel most comfortable when you can be your own boss? Preferences like this are no less significant in the long run than some of the more practical decisions you will make about whom to hire, what kinds of paper to buy, and so on. Disregarding such feelings could lead to lack of satisfaction and potential conflicts later on, even though there are times, too, when it is more appropriate to place your personal desires in the background.

When organized as a sole proprietor, some consultants choose to display their name and title simply, as in "John Smith, Professional Engineer" or "Jane Smith, Consultant," thereby indicating that they own and operate the practice independently. There is nothing wrong with this, but sound business strategy dictates that one should give serious consideration to all aspects of the image projected to the public. In this case, the simple title might satisfy the consultant, but it could suggest to potential clients that the firm is too small to take on certain projects. The logical alternative might be to name the firm in a way that better reflects its capacity to handle all kinds of jobs: "Jones and Associates, Engineering Consultants," "Jones Engineering Enterprises," or "Jones and Company," among other possibilities.

Each type of organization has its own benefits and limitations. A partnership using the names and titles of the partners in the firm's name may project an attractive image to clients; but if the partners find it difficult to get along with each other, problems might arise. Thus it is important to consider your personal style when you set up your organization.

Besides yourself and potential partners, the most important person to consult while considering setting up any type of organization (except a sole proprietorship) is an attorney. The attorney can effectively point out the advantages and disadvantages of each arrangement and help you prepare for ways to protect the firm against potential problems. All partners or officers involved in starting a firm should develop a clear and detailed agreement, in writing, with the attorney's aid. A thorough legal agreement will help to take care of business difficulties before they arise.

The recommendation to consult an attorney may sound a bit foreboding, but it does not imply that business problems are likely to occur. Instead, it merely means that you will be building a more solid foundation for your success. While the individual proprietor is generally responsible for all aspects of his or her business, other types of organizations, such as the partnership, warrant the advice of an attorney by virtue of their more complex structures. For example, in a partnership an individual partner can create a liability for all the firm's partners, possibly making them liable for some of his or her debts. Likewise, unless special protective agreements are spelled out with an attorney in advance, a partnership may have to divide its total income evenly among the partners, even though some of them may have invested much more time and effort than others.

The consulting engineer who is considering forming a corporation faces additional questions. In consulting engineering, there are two basic types of corporations: the *professional* and the *nonprofessional* organization. The first type is usually owned and operated by licensed engineers and thus enjoys freedoms which may not be permitted to unlicensed practitioners, in accordance with state laws. Specific privileges and licensing laws vary from state to state, so an attorney should always be consulted. The basic distinction between the professional and nonprofessional engineering corporation is that the owner-operators' nonprofessional organizations usually do not possess a license. Consequently, the nonprofessional group may not be able to do business in certain areas of consulting engineering.

A nonprofessional corporation can build a successful business in some fields; but lacking the ability to "seal" a job with a registered engineer's official stamp, it may have no way to guarantee responsibility for the finished work, and its job prospects will consequently be limited.

Another organizational arrangement you might consider is the *joint venture*. In it, engineers maintain their individual practice and the ability to work alone on some projects while joining together with fellow consultants for particular jobs. One advantage of this arrangement is that it can effectively expand the talent and capacity of a small firm, enabling it to compete for work with firms which may already have large and diversified staffs. In this case as in the other types of organizations, an attorney should help in drawing up agreements between the parties involved. The joint venture sometimes has special restrictions with regard to liability insurance and other legal matters; for instance, insurers will sometimes require separate policies for each project or venture.

Besides the more common types of organizations, there are some specialized consulting firms which engage in R&D and at the same time invest money in the items under study. They may buy stock, use the services of venture-capital firms, or proceed along any of a number of other financing routes. The rewards can be substantial, yet so can the pains. You need not look far in the business press and newspapers to read of lawsuits brought by individuals and firms claiming they have been robbed of trade secrets or experimental processes. Few consultants enter this area at the beginning of their private practice; but regardless of the amount of experience one has, the best advice here, as elsewhere, is still: See a knowledgeable attorney, discuss the situation, and make certain that any business actions are covered by a legal agreement before you proceed.

A variety of opportunities exist for the consultant to optimize success. Many engineers are so busy concentrating on the tasks immediately in front of them that they fail to recognize their true success potential. The basis of a rewarding career in consulting may be with you while you work at a salaried job—or it could be across the room in the mind of a colleague and potential business partner.

The important point to remember is that only by opening your mind to the broad range of possibilities and exploring them with other knowledgeable professionals can you begin to take the steps needed to build a successful firm. The old adage, "Investigate before you invest," holds doubly true in engineering. Following it wisely can protect you financially and project you into a fulfilling professional life.

FINANCING YOUR PRACTICE

In the actual formation of the new firm, the first order of business is finance—i.e., money. Unfortunately, few engineers find themselves independently wealthy at the time they decide to enter private practice. Therefore, some thought must be given to how much of your savings or current earnings you can afford to spend and how much you will be required to spend to get the business started and keep it going.

Established consulting firms have many different ways of planning their budgets, some of which will be discussed later. All that you really need in order to start budget planning at the beginning, however, is a basic, rather abbreviated outline of financial requirements for the first year. Such an outline can be developed in a manner similar to the example shown in Exhibit 1-1.

The example may appear simple, but it covers the basic needs of a new consulting firm and gives reasonable percentages of the total business dol-

EXHIBIT 1-1   Income and expense, cash basis.

Gross fee income $50,000

Expenses

Percent of gross income	Item	Value
40	Direct labor	$20,000
5	Clerical payroll	2,500
5	Administrative payroll	2,500
10	Office expense	5,000
4	Car expense	2,000
3	Insurance	1,500
2	Dues, licenses, publications	1,000
10	Rent	5,000
4	Payroll taxes	2,000
4	Telephone	2,000
5	Travel and entertainment	2,500
	Total expenses	$46,000
8	Profit	$4,000
100		

lars one might allocate for each type of expense. The budget is based on an initial gross fee income of $50,000, which a consultant could be earning through projects in which she is already involved at the time she starts the practice. Alternatively, of course, the money could come from personal savings, loans, partners' savings, even government or foundation grants, or a combination of sources.

Although some of the figures in the example may not be as high or as low as you are accustomed to, they do reflect the percentage breakdowns of expenses for many engineers in private practice. In some cases the figure for the first year's salary might be considered inordinately low. The principal reason for this is that the example represents the new firm, which does not yet have a large number of clients and has focused its efforts on start-up operations.

Funds needed for start-up can sometimes seem as difficult to obtain as the impossible dream. Banks are often hesitant to provide loans for fledgling professional firms merely on the basis of a firm's income projections. Instead, they may lend money on real assets, such as cars, houses, or other properties in which one has equity. Of course, the loan must eventually be repaid with interest—a factor which discourages a significant number of potentially successful business people from securing bank loans, especially at high interest rates.

Some banks and lending institutions offer special plans, however, whereby you can borrow about 95 percent of the cash value of your life insurance at rates of between 5 and 8 percent. In addition, you might be eligible to borrow 95 percent of the value of your savings account at rates approximately 2½ percent higher than the normal savings account rates. These are just two of a number of financing plans which are worth looking into.

But just as important as obtaining the funds to start your consulting practice is the question of how *much* money you will need. Our example in Exhibit 1-1 shows how one could operate a successful beginning practice with $50,000 gross income. But if you were to ask the opinions of a number of consultants and other business people on this topic, you might get a different answer from each person.

Basically, your initial funds will have to provide for an office, travel, supplies, and telephone use. If you are lucky enough to be able to set up an office in your home, the rental portion of the budget will be negligible. For many engineers, however, the home office may not be a practical alternative (a point which will be addressed later in this section).

For consultants renting offices, the figures quoted most often at the time of this writing in informal surveys ranged from a low of $15,000 to a high of $25,000 for basic start-up expenses. This wide range is due mainly to differences in rental and other consumer prices in various locations, with the highest costs in or near major cities. The figures do not include estimates for offices in the most expensive rental areas of the largest cities, nor do they include allocations for the consultant's personal living expenses.

The basic implication of these figures is that if you can live for a year without substantial personal income, then you can probably get your practice started for a minimum investment of $15,000 to $25,000. Inflation will increase these start-up costs, of course, at the same time that it alters interest rates.

Forecasts like these make it appear all the more desirable to have mounds of money sitting in your savings account, to have a rich uncle, or to have a list of eager clients waiting at the door when you open the new office. Too many salaried engineers lack such advantages when they set out on their own, although nearly every engineer can begin to build contacts and a client list while working for others. There is a brighter side to the financing issue, however.

You can actually start your consulting practice with a nominal outlay for the first month's expenses and let client fees pay for your start-up. Here is an example of how it might be done:

Ideally, you could acquire a project for which you could bill the client on an hourly basis and receive payments monthly. Now, assuming that your total expenses would amount to around $25,000 for the first year, you would need to have, say, $2000 for each month. Consulting fees tend to average approximately 2½ times actual payroll costs; therefore, if you set payroll at $10 per hour, a reasonable client fee would be about $25 per hour.

If you work 175 hours during the first month, your bill will be $4375. This leaves you, after expenses, with a net of $2375. Now even if you pay yourself $1700 for your first month's effort, you will have $675 left over—a nice monthly nest egg to put toward strengthening and expanding your firm.

A significant number of consultants start their practices on this basis. The advantage of it is that it practically eliminates the need for large capital sources during the first year. The engineer in the example actually needed no start-up funds other than the $2000 to cover the first month's expenses. The only other ingredients needed were the client and his agreement on a monthly pay plan.

More details about how you can take this second crucial step—securing your first clients—will be given in the sections that follow. Now let's look at a few more examples of consultants who got themselves off to a good start in their own practice.

Self-Financing Can Work

A 30-year-old engineer—we will call him Mr. A—who had worked for several companies in the heating, ventilating, and air conditioning contract business, decided to go into private practice, using his $12,000 savings as seed money. He had already made a number of contacts with architects and engineers at his regular job and felt that it would not be terribly hard to get his first clients. Using his savings forced him to forego moving from an apartment to a house, however, which Mr. A and his wife had wanted to do for some time. At the beginning of the new practice, he rented desk space in an accounting office for a nominal cost.

It was not all that easy to get the first clients, though, and Mr. A found that most of the $12,000 savings was gone before his efforts began to pay off. The end of the first year saw him earning $21,000 for himself and enough profit to pay, with the help of the seed money, $14,000 in expenses. Mr. A's first-year income thus equaled roughly the amount he had been making at his former job (not including bonuses and other benefits).

When he began to expand the firm during the second and third years, buying new equipment, renting more office space, and hiring two new employees, the economic picture darkened. Total operating costs exceeded revenues, and Mr. A found himself gazing into the gloomy gap between income and expenses that can easily appear during times of expansion in a private practice. Recognizing that he could not expect to maintain the same level of earnings for himself that he enjoyed during the first year, Mr. A wisely decided to reduce his own salary by $5000 and $7000 in the second and third years, respectively. The savings helped the firm stay afloat and gain added momentum so that by the fourth year revenues actually exceeded Mr. A's earlier projections. He was now able to pay himself $35,000 for the whole fourth year—an increase of almost 65 percent over the engineer's salary.

From Part-Time to Full-Time Consulting

Another engineer, Ms. B, entertained the idea of starting her own practice for several years. But it was not until the age of 43—when she developed a group of solid contacts and prospective clients—that she felt ready to leave her salaried position. In the course of 2½ years on the regular job, Ms. B had cultivated her contacts so well that she actually developed a lucrative sideline: At nights and on weekends, she inspected homes for structural problems and other defects.

The part-time business soon became the foundation of Ms. B's private practice. Rather than use the extra income for luxuries and other personal amenities, she invested the part-time profits in money market funds at high interest rates. In a little more than 2 years, the accumulated savings amounted to $47,000 (with interest). This enabled Ms. B to pay her basic start-up expenses and embark swiftly on a private practice with steadily rising revenues.

Finding a Partner with Money

A third engineer, Mr. C, faced a very different situation. Like Ms. B, he developed a firm backlog of contacts. But his savings were meager, not nearly sufficient to start the business he had for so long dreamed of owning. One engineer with whom Mr. C was acquainted had inherited a large sum of money, though, and Mr. C discussed with him the possibility of forming a partnership. The acquaintance conceded that he was considering going into private practice, but he had hesitated because he was not sure about how to get started. In this case, the two engineers' personal assets complemented each other ideally: Mr. C had the contacts and know-how to start the business, while his colleague had the capital.

They both recognized the opportunity before them and consulted lawyers, who then assisted the engineers in developing formal agreements to establish a partnership. The result: $100,000 in seed money and a list of prospective clients. With some skillful marketing during the months after the engineers left their regular jobs, several of the prospective clients became eager customers; no more than $20,000 had to be spent on start-up before the partnership began turning a profit. Within 1 year, each partner was earning $23,000 in base salary.

Certainly there are other success stories in consulting. Those just recounted illustrate only the beginnings of a few rewarding consulting practices. But one might wonder: What about the failures? Not everyone who goes into private practice is a resounding success, surely. However, few, if any, real failures were found in the course of researching this book. Engineers who viewed their practices as unsuccessful most frequently had based their notions of success on the prospect that they would be earning more at the beginning of their practice than they had earned at earlier jobs. Such a measure can be misleading. The few consultants who believed they were failing felt this way merely because they declined to recognize that it is not unusual for many businesses to take up to several years to start showing profits. This may happen even more frequently in fields other than consulting.

So we will probably never know how successful some engineers could have been as consultants had they not turned away from private practice because they feared the challenge of that start-up period. The numerous successes speak for themselves, as do the rewards.

LICENSING REQUIREMENTS

One of the questions that invariably arises during the early phases of organizing an engineering practice is: What specific forms of licensing will my practice require? Which license will provide the most advantages to the firm?

If you intend to offer consulting services on management topics such as budgeting, sales, and finance, for example, your firm may not need a professional engineering license at all. But if you will be involved with work on public buildings, highways, or other structures directly related to public safety, you will certainly need to be licensed in particular areas. Generally, at least one member of the firm must be licensed in each state in which the firm intends to practice.

Procedures for obtaining the professional engineer's (PE) license vary from state to state, but a common route taken by many seeking registration is to enroll in an Engineer in Training (E.I.T.) internship program during college. This is a relatively simple way of securing the PE, though it is sometimes overlooked by the busy student. Once the student satisfies the E.I.T. requirements, she or he is eligible to apply for a PE license (laws in most states mandate an E.I.T. examination).

Some students take the first half of the PE exam after completing the E.I.T. internship program. In accordance with state laws, the second half of the exam is often taken after the engineer has engaged in 3 or more years of field work. Successful completion can then lead directly to licensing.

Later in this section you will find listed the different requirements of each state; they show that the engineer who does not follow the preceding route toward registration will be required to take both parts of the exam in sequence. In that case, it is advisable for the engineer to spend a few months reviewing some basic textbooks and other academic materials used by the engineering student—particularly in fields such as physics, mathematics, and structural and electrical engineering. Many state professional engineering societies and colleges offer special refresher programs like the "Professional Engineering Exam Course" for this purpose. Special refresher books which an engineer can use to prepare for the license exam are listed in bibliography form later in this chapter.

In the past, many states allowed people to obtain engineering licenses without taking an examination, provided they had a substantial number of years' experience working for professional firms. But more and more states are reducing or eliminating this option. Some states recognize licenses from other states or will grant additional registration based on the licensed engineer's passing an oral exam in the second state. In addition, licensing may not be required to work in some states when the engineer is affiliated with a government agency or other licensed superiors.

To be safe, you should contact the professional engineering societies or related groups in each state in which you intend to practice to get full details on specific licensing requirements. A few states even have special additional restrictions for work done within their borders. For example, California emphasizes a need for familiarity with earthquake-resistant structural systems, while Florida emphasizes knowledge of structures that can withstand hurricanes.

To clarify the whole question of licensing, the authors asked John D. Constance, PE, a well-known consultant on the licensing of PEs, to prepare a summary of licensing requirements, a bibliography of pertinent works in the field, and a glossary of frequently encountered terms. Here is the summary, followed by the bibliography and glossary, all prepared by John Constance to assist readers in obtaining their first PE license or a license to practice in another state.

PE LICENSURE REQUIREMENTS AT A GLANCE

If you are thinking of becoming licensed, this summary of requirements will tell you how to qualify. If you are already registered, here is what you may need to learn about practicing in states other than your own. Summarized in Exhibit 1-2 are the requirements for getting a PE license in your own and in every other state.

Requirements of Licensure

There are seven general requirements for licensure as a PE, although the details may vary from state to state. The procedure by which an engineer with average experience goes about getting licensed is essentially the same in all states.

1. **Age:** The minimum age for an Engineer in Training (Intern Engineer) varies from 19 in New York to 21 in most other states. The minimum age for full licensure is 25 in most states.

EXHIBIT 1-2

State	Minimum Age, E.I.T.	Minimum Age, PE	Degree Accreditation ABET / School or Board approved (BA)	Experience Credit for No Degree, Years	Experience Credit for PE with Degree, Years	Credit Teaching Experience	Credit Military Experience	Credit Contracting Experience	Board Provides Licensure by Endorsement	Month Written Exam Held: Apr = 4, Oct = 10	Written Examination: E.I.T. or Fundamentals, Hours	Written Examination: PE, Branch, or Specialty, Hours	Reciprocity or Comity Practices for PE	Certificate of Verification Recognized	Temporary Practice Permit Granted	Time Limit on Permit, Days	Register in Branches or PE	Seal Required on Drawings, Plans, Specs	Time Limit on E.I.T. Certificates, Years	Reinstatement Policy
Alabama	—	—	ABET	8	4	Asst prof	Eng	Design	No	4, 10	8	8	Yes	Yes	Yes	Bd rev	PE	No	10	—
Alaska	—	—	ABET	8	4	Eng	Eng	Eng	Mt req	4, 10	8	8	Yes	Yes	No	—	Branch	Yes	5	—
Arizona	—	25	ABET	8	4	5 yr max	Yes	Eng	Yes	4, 10	8	8	Yes	Yes	No	—	Branch	Yes	8	Yes
Arkansas	—	—	BA	8	2	Yes	Eng	Ind bas	Yes	4, 10	8	8	Yes	No	No	No limit	—	No	—	—
California	—	25	—	6	4	—	Eng	—	Yes	4, 10	8	8	Yes	Yes	No	—	Branch	No	11	Yes
Canal Zone	—	21	Deg	12	4	Yes	Eng	No	No	4, 10	8	8	Yes	Yes	Yes	1 yr	PE	No	None	Yes
Colorado	—	25	Deg	8	4	Yes	Eng	Eng	No	4, 10	8	8	Yes	Yes	No	—	PE	Yes	None	—
Connecticut	—	—	ABET	10	4	Yes	Eng	Eng	Yes	4, 10	8	8	Yes	Yes	Yes	—	PE	Yes	10	Yes
Delaware	—	—	ABET	8	4	Eng	None	None	Yes	4, 10	8	8	Yes	Yes	Yes	30	PE	Yes	None	None
District of Columbia	21	25	BA	8	4	Yes	No	Yes	—	4, 10	8	8	Yes	Yes	Yes	1 proj	PE	Yes	None	Yes
Florida	—	24	ABET	10	4	Yes	None	None	Yes	4, 10	8	8	Yes	Yes	Yes	1 yr	PE	Yes	12	Yes
Georgia	—	25	ABET	12	4	B.A.	—	—	No	4, 10	8	8	Yes	Yes	Yes	90	PE	Yes	10	None
Guam	—	21	BA	8	3	1 yr	B.A.	Eng	Yes	4, 10	8	8	Yes	Yes	Yes	—	Branch	Yes	12	—
Hawaii	—	—	Deg	12	4	Eng	Yes	Yes	No	4, 10	8	8	Yes	Yes	Yes	1 yr	Branch	Yes	None	—
Idaho	—	—	Deg	8	4	—	Yes	No	—	4, 10	8	8	Yes	Yes	Yes	30	PE	Yes	—	Yes
Illinois	—	—	BA	8	4	1 yr	—	—	Yes	4, 10	8	8	Yes	Yes	Yes	30	PE	Yes	12	Yes
Indiana	—	—	ABET	5-12	4	Yes	Eng	None	No	4, 10	8	8	Yes	Yes	Yes	30	PE	No	10	Yes
Iowa	—	—	ABET	12	4	Eng	Eng	No	Yes	4, 10	8	8	Yes	Yes	No	Bd rev	Branch	Yes	10	Yes
Kansas	—	—	BA	8	4	Eng	Eng	—	No	4, 10	8	8	Yes	Yes	No	—	—	Yes	None	Yes
Kentucky	—	—	ABET	8	4	Yes	Yes	—	—	4, 10	8	8	Yes	Yes	Yes	—	Branch	Yes	—	Yes
Louisiana	—	—	ABET	8	4	Yes	Maybe	Yes	Yes	4, 10	8	8	Yes	Yes	Yes	30	Branch	Yes	10	Yes
Maine	—	25	ABET	12	4	B.A.	B.A.	B.A.	Yes	4, 10	8	8	Yes	Yes	Yes	30	PE	Yes	12	Yes
Maryland	—	—	ABET	12	4	Yes	Eng	No	Yes	4, 10	8	8	Yes	Yes	Yes	1 yr	PE	Yes	None	Yes
Massachusetts	—	—	ABET	8	4	Asst prof	Eng	Eng	No	4, 10	8	8	Yes	Yes	Yes	30	PE	Yes	None	Yes
Michigan	—	21	ABET	8	4	Eng	Eng	No	—	4, 10	8	8	Yes	Yes	Yes	60	PE	Yes	None	Yes

State	1	2	3	4	5	6	7	8	9	10	11	12	13	14	15	16	17	18	19
Minnesota	—	25	ABET	13	4	1 yr Eng	Eng	Eng	Yes	4, 10	8	Yes	Yes	Yes	30	Branch	No	10	Yes
Mississippi	—	—	BA	8	4	Yes	Eng	Eng	Yes	4, 10	8	Recog	Yes	Yes	—	PE	Yes	10	Yes
Missouri	21	—	BA	8	4	Yes	No	No	Yes	4, 10	8	Yes	Yes	No	—	PE	Yes	8	Yes
Montana	—	—	ABET	8	4	Yes	Eval	Yes	Yes	4, 10	8	Yes	Yes	No	—	PE	Yes	10	Yes
Nebraska	—	25	ABET	8	4	Yes	Eng	Eng	Yes	4, 10	8	Yes	Yes	No	Bd rev	PE	Yes	None	Yes
Nevada	—	—	BA	8	4	Yes	Lim	Yes	Yes	4, 10	8	Yes	Yes	Yes	—	PE	Yes	8	Yes
New Hampshire	—	—	BA	8	4	Eng	No	No	Yes	4, 10	8	Yes	Yes	No	30	PE	Yes	10	Yes
New Jersey	21	25	ABET	8	4	Yes	Yes	Eng	Yes	4, 10	8	Yes	Yes	No	—	PE	Yes	10	Yes
New Mexico	19	—	BA	8	4	Yes	No	Policy	Yes	4, 10	8	Yes	Yes	No	30	PE	Yes	None	Yes
New York	19	19	ABET	12	4	Eng	Eng	Policy	Yes	4, 10	8	Yes	Yes	Yes	30	PE	Yes	10	Yes
North Carolina	—	21	BA	10	4	4 yr max	Policy	Policy	Yes	4, 10	8	Yes	Yes	Yes	30	PE	Yes	None	Yes
North Dakota	21	21	ABET	—	4	4 yr min	Eng	Eng	No	4, 10	8	Yes	Yes	Yes	1 yr	PE	Yes	12	Yes
Ohio	—	—	ABET	8	4	Eng	Eng	Eng	Yes	4, 10	8	Yes	Yes	Yes	60	PE	Yes	10	Yes
Oklahoma	—	—	ABET	8	4	Yes	Yes	Eng	Yes	4, 10	8	Yes	Yes	Yes	1 yr	PE	Yes	7	Yes
Oregon	—	—	BA	8	4	Eng	Eng	Eng	Yes	4, 10	8	Yes	Yes	Yes	30	PE	Yes	10	Yes
Pennsylvania	—	25	ABET	12	4	Yes	Yes	Some	Yes	4, 10	8	Yes	Yes	Yes	30	PE	Yes	None	Yes
Puerto Rico	—	—	ABET	—	4	B.A.	B.A.	B.A.	No	4, 10	8	Yes	Yes	Yes	Bd rev	PE	Yes	Indef	Yes
Rhode Island	—	—	ABET	12	4	1 yr	None	None	Yes	4, 10	8	Yes	Yes	Yes	30	PE	No	None	Yes
South Carolina	—	—	BA	8	4	Yes	Eng	None	No	4, 10	8	Yes	Yes	Yes	30	PE	Yes	10	Yes
South Dakota	21	21	ABET	8	4	Yes	Eng	Eng	Yes	4, 10	8	Yes	Yes	Yes	1 yr	PE	Yes	None	Yes
Tennessee	—	—	BA	8	4	—	2 yr	B.A.	Yes	4, 10	8	Yes	Yes	No	—	Eng	Yes	None	Yes
Texas	—	—	Deg	8	4	Asst prof	Eng	No	No	4, 10	8	Yes	Yes	No	Bd rev	PE	Yes	12	Yes
Utah	—	—	ABET	8	4	2 yr	No	No	No	4, 10	8	Yes	Yes	Yes	30	PE	Yes	10	Yes
Vermont	—	—	BA	12	4	Yes	Eng	Eng	Yes	4, 10	8	Yes	Yes	Yes	30	PE	Yes	12	Yes
Virginia	—	21	ABET	10	4	Yes	Eng	Lim	Yes	4, 10	8	Yes	Yes	No	No	PE	No	10	Yes
Washington	—	21	ABET	8	4	Yes	Eng	Eng	No	4, 10	8	No	No	No	—	PE	Yes	None	Yes
West Virginia	—	—	BA	8	8	Yes	Eng	Yes	No	4, 10	8	Yes	Yes	No	60	Branch	No	12	Yes
Wisconsin	—	—	ABET	12	4	Eng	Eng	Eng	Yes	4, 10	8	Yes	Yes	Yes	60	PE	Yes	10	Yes
Wyoming	—	—	ABET	8	4	Yes	Yes	Yes	Yes	4, 10	8	Yes	Yes	No	—	PE	No	None	Yes

*The exhibit was developed from data furnished by boards of registration of the states and represents the latest information available. Because boards retain the right to change conditions, and the exhibit contains space-saving simplifications, engineers are cautioned to verify answers to questions with their boards of record. Addresses of state boards may be obtained from the National Council of Engineering Examiners (NCEE), Box 1686, Clemson, South Carolina 29633-1686.

Most states have no minimum age requirement for Engineers in Training. Upon graduation from an accredited program in engineering or science or upon completion of the board-approved equivalent in engineering experience, the graduate is permitted to sit for the first of two parts (Fundamentals of Engineering) of the written examination. After serving his or her internship (4 years or more) in engineering practice, he or she may qualify to sit for the second part of the examination (Principles and Practice). In all states, each of the two parts of the written examination runs 8 hours long.

†Some states require an engineering degree from a school accredited by ABET, and some are satisfied with a degree from a board-approved school.

2. **Citizenship:**   In most states, being a U.S. citizen is no longer required.

3. **High School Graduation:**   The requirement that applicants hold a high school diploma, or the equivalent, is general, but movements are under way to make at least an engineering degree or science degree the minimum qualification.

4. **College Degree:**   The standard requirement is an undergraduate degree in engineering or science from a school program accredited by the Accreditation Board for Engineering and Technology (ABET), or the equivalent in board-approved engineering experience. Graduates of nonaccredited schools must show additional years of experience of an approved nature. All curricula must be approved by the boards of record.

5. **Experience:**   Evidence of sufficient qualifying experience as of the date of the application must be properly presented. States couple the degree and experience requirements for graduates. Nondegreed applicants must show longer engineering experience. In each case, all experience must be attested to.

6. **Character:**   References (usually five, three of which must be licensed engineers) attesting to the applicant's character and integrity are required.

7. **Examination:**   In most all cases, the applicant must take a written examination. Rarely, and only under unusual circumstances, boards may waive a part or parts of the written examination.

Details of state requirements are subject to change without notice because of revisions in state registration laws and board policy decisions.

Engineering Experience

An applicant's experience must be of a character and grade satisfactory to the board. Boards generally hold that calendar years of experience are not necessarily equivalent to years of board-approved experience, although "approved" experience is something that defies exact definition. All boards insist that experience be broad in scope and such as to have developed the candidate's knowledge and judgment. Whether credit is given for teaching, military, and contracting experience is indicated in Exhibit 1-2.

Licensure by Endorsement

The provision in the laws of some states for "long-established practice" permits boards to waive a part or parts of the written examination, and in a few cases all of it, for engineers who can qualify by education, by a longer period of lawful experience and practice (usually 15 to 25 years after graduation), and by showing appreciable experience in responsible charge of engineering work of high caliber. In practice, very few registrations are granted under this provision unless the applicant has been registered for a number of years in another state.

Boards often give special consideration to engineers who have had outstanding experience and who are of mature age. Boards sometimes grant interviews, possibly combined with short oral or written examinations, to determine an engineer's basic knowledge. Before a board grants a registration by endorsement, however, it will endeavor to ascertain that the applicant is fully qualified. Boards recognize that years of experience of an outstanding nature and achievement constitute evidence of qualifications superior to the results of written examinations. Factors that help an applicant in this respect include presentation of technical papers, advanced degrees and studies, technical society memberships and activities, patents, special developments, and the esteem of fellow engineers.

Experience in the armed services is given credit only if the service was not routine. Any experience is questionable if the applicant has not applied the tools of engineering: mathematics, physics, chemistry, strength of materials, and mechanics.

All states give some credit for teaching experience. Certain states place limitations on the years of credit, on the acceptability of the subjects taught, or on the rank of the teacher; for instance, rank of assistant professor or higher.

Reciprocity and the Mobile Engineer

Because professional engineers are likely to move from one state to another a number of times during their careers, they must see to it that geographic boundaries do not limit their practices.

Fortunately, all states now allow engineers licensed in another state to practice within them for a short period of time without requiring a second license. Some states permit this for the duration of a single project; others allow it for definite periods of time.

Written and oral examinations may be waived if the applicant has passed a 16-hour written examination, or the equivalent, in another state. Most states now require that the applicant be licensed in his or her home state before he or she will be considered for reciprocal registration.

Applicants may be considered for reciprocal registration under this category if they hold: (1) a license or certificate to practice professional engineering, issued after examination by a legally constituted board in any state or political subdivision of the United States, and (2) a Certificate of Qualification (National Engineering Certifications) issued by the National Council of Engineering Examiners (NCEE)—provided that, when the license or certificate was issued, the issuing state's examination was fully equivalent (16 hours of examination) to the current examination in the state in which the application is being filed and the applicant's record fully meets all other statutory requirements such as age, citizenship, education, character, and experience.

Committee on Records Verification

The Committee, a function of NCEE, was established to minimize effort and expense when registered engineers applied for out-of-state licenses. A fact-finding and verifying agency, it serves as a clearinghouse for the professional records of licensed engineers. The engineer's experience record and licensure status in his or her home state is verified by the NCEE, and upon submission of the appropriate documentation, a Council Record is compiled, and a certificate of verification is issued.

The certificate serves as a verified record of an engineer's experience. It is not a national license to practice, nor may it be used in lieu of a state license to practice. Its sole purpose and use is to help competent engineers cross state lines and to eliminate duplication of effort by state boards. All states recognize it. Because a search by the Committee takes considerable time, an engineer should not wait until she or he needs a certificate to apply for it. To apply, contact NCEE headquarters in Clemson, South Carolina.

Legal licensing in one state does not authorize the practice of engineering except in that state. Most states, however, have provisions in their laws that allow licensed engineers of other states to practice for a short period of time in those states if they request permission in writing to do so from the state board concerned.

Selected Books for the Professional Engineer and Engineering Fundamentals (Engineer-in-Training) Examination Preparation, Including Home Study Courses

Publications listed in this bibliography are available directly from the sources indicated. When inquiring about any of these publications, please indicate, "latest edition."

Engineering Fundamentals Examination

Apfelbaum, H. J., and Ottesen, W. O. *Basic Engineering Sciences and Structural Engineering for Engineer-in-Training Examinations.* Hayden Book Company, Rochelle Park, N.J. (1970).

Baldwin, A. J., and Hess, K. M. *A Programmed Review of Engineering Fundamentals.* Van Nostrand Reinhold, New York (1978).

Morrison, J. W. *Engineering Fundamentals.* Arco Publishing Company, New York (1978).

National Council of Engineering Examiners. *Typical Questions—Fundamentals of Engineering.* Clemson, S.C. (1983). No answers.

————. *Fundamentals of Engineering (FE) Sample Examinations.* Clemson, S.C. (1983). Includes answers.

Newnan, C. D. *Engineer-in-Training License Review.* Engineering Press, Inc., San Jose, Calif. (1981).

Newnan, D. G., and Larock, B. E. *Engineering Fundamentals Examination Review.* Wiley Interscience, New York (1978).

Polentz, L. M. *Engineering Fundamentals for Professional Engineers' Examinations.* McGraw-Hill Book Company, New York (1979).

Professional Engineering Institute. *Engineer-in-Training Review Manual.* San Carlos, Calif. (1982).

Professional Engineer (Principles and Practice) Examination. Bentley, J. H. *A Programmed Review for Electrical Engineering.* Van Nostrand Reinhold, New York (1978).

Constance, J. D. *Electrical Engineering for Professional Engineers' Examinations.* McGraw-Hill Book Company, New York (1975).

————. *How to Become a Professional Engineer.* McGraw-Hill Book Company, New York (1978).

Hafer, C. R. *Electronics Engineering for Professional Engineers' Examinations.* McGraw-Hill Book Company, New York (1980).

Illinois Society of Professional Engineers. *Typical Questions from Illinois Examinations for Professional Engineer Registration.* Springfield, Ill. (1970).

Jones, L. D., and Lima, J. A. *Electrical Engineering License Review.* Engineering Press, Inc. San Jose, Calif. (1980).

Kurtz, M. *Structural Engineering for Professional Engineers' Examinations.* McGraw-Hill Book Company, New York (1978).

————. *Engineering Economics for Professional Engineers' Examinations.* McGraw-Hill Book Company, New York (1975).

LaLonde, W. S., Jr., and Stack-Staikidis, W. J. *Professional Engineers' Examination Questions and Answers.* McGraw-Hill Book Company, New York (1976).

Lyons, J. S., and Dublin, S. W. *Electrical Engineering and Economics and Ethics for Professional Engineering Examinations.* Hayden Book Company, Rochelle Park, N.J. (1970).

Morrison, J. W. *Professional Engineer Registration: Problems and Solutions.* Arco Publishing Company, New York (1978).

National Council of Engineering Examiners. *Professional Engineer Examinations.* Vol. I (1965–1971). A compilation of questions from examinations

administered by state boards from E.I.T. and PE exams during period 1965–
1971 in chemical, civil, electrical, and mechanical engineering.

————. *Solutions: Professional Engineering Examinations.* Vol. II (1965–
1971). Verified solutions to those questions which appear in Vol. I above.

————. *Professional Engineer Examinations.* Vol. III (1972–1976). Study guide
for *Fundamentals of Engineering and Principles and Practice of Engineering
Examinations.* Contains sample fundamentals of engineering examination.
Principles and practice portion includes following disciplines: agricultural,
ceramic, industrial, manufacturing, nuclear, petroleum, sanitary, structural,
and aeronautical-aerospace.

————. *Typical Questions Pamphlets—Principles and Practice of Engineering
(1983).* Includes disciplines of aerospace-aeronautical, agricultural, ceramic,
industrial, manufacturing, nuclear, and petroleum engineering.

Newnan, D. G. *Civil Engineering License Review.* Engineering Press, Inc., San
Jose, Calif. (1980).

————. *Engineering Economic Analysis.* Engineering Press, Inc., San Jose,
Calif. (1980).

Packer, M. R., editor and contributor. *Professional Engineer (Civil) State Board
Examination Review.* Arco Publishing Company, New York (1977).

Pefley, R. K., and Newnan, D. G. *Mechanical Engineering License Review.* Engineering Press, Inc., San Jose, Calif. (1980).

Probhudesai, R. K., and Das, D. K. *Chemical Engineering for Professional Engineers' Examinations.* McGraw-Hill Book Company, New York (1983).

Professional Engineering Institute. *Civil Engineering Review Manual.* San Carlos,
Calif. (1981).

————. *Electrical Engineering Review Manual.* San Carlos, Calif. (1981).

————. *Mechanical Engineering Review Manual.* San Carlos, Calif. (1981).

Stamper, E., and Dublin, S. W. *Mechanical Engineering and Economics and Ethics for Professional Engineering Examinations.* Hayden Book Company,
Rochelle Park, N. J. (1971).

Home Study Courses

Constance, J. D. *A Guide to P.E. Refresher Courses, New Engineer,* Vol. 6, No. 9,
October 1977. Suggested reference as a general guide to refresher courses.

Engineering Enterprises, 12 Hugenot Street, East Hanover, N.J. 07936. Three
courses are offered: engineering fundamentals, mechanical engineering, electrical engineering.

Engineering Registration Studies (ERS), Post Office Box 24550, Los Angeles,
Calif. 90024. Four courses are offered: civil engineering 1 (structural) civil engineering 2 (nonstructural), electrical engineering, fundamentals of engineering.

PERC, Inc., Post Office Box 123, Northport, N.Y. 11768. Four study groupings
are offered: Engineering fundamentals (E.I.T. and Intern Engineer), Engineering

fundamentals mini reviews, professional engineer review in engineering, economic principles, and practice.

PE Readiness Courses. Conducted by the MGI Management Institute and available through the National Society of Professional Engineers (SPE), 2029 K Street, NW, Washington, D.C. 20006. Four courses are offered: Engineer in Training, civil engineering, electrical engineering, mechanical engineering.

University of Wisconsin—Extension, Department of Engineering and Applied Sciences, 432 North Lake Street, Madison, Wis. 53706.

Publications of Pertinent Interest

National Council of Engineering Examiners (NCEE). *Compendium of Registration Laws for the Design Professions.* Clemson, S.C. (1978).

————. *Supplement to the Compendium of Registration Laws.* Clemson, S.C. (1981). The *Supplement* and the *Compendium* (above) were published in cooperation with the National Council of Architectural Registration Boards (NCARB) and include the registration laws for engineers, architects, land surveyors, and landscape architects and planners for all 50 states and the U.S. territories.

————. *A Task Analysis of Licensed Engineers, Final Report (1981).* Contains the basis for NCEE's efforts to provide meaningful uniform and objective standards with which to measure minimum competency for professional licensure through the examination process. This analysis forms the basis for current examination specifications.

Addresses of Publishers

Arco Publishing Company, 219 Park Avenue South, New York, N.Y. 10003.

Engineering Press, Inc., Box 1, San Jose, Calif. 95103.

Hayden Book Company, 50 Essex Street, Rochelle Park, N.J. 07662.

Illinois Society of Professional Engineers, 1304 South Lowell Street, Springfield, Ill. 62704.

John Wiley & Sons, Inc., Publishers, Wiley Interscience Division, 605 Third Avenue, New York, N.Y. 10158.

McGraw-Hill Book Company, 1221 Avenue of the Americas, New York, N.Y. 10020.

National Council of Engineering Examiners, Box 1686, Clemson, S.C. 29633-1686.

Professional Engineering Institute, Post Office Box 639, Department 71, San Carlos, Calif. 94070.

Van Nostrand Reinhold, 135 West 50th Street, New York, N.Y. 10020.

ENGINEERING PRACTICE IN
ANOTHER STATE

There have been a significant number of cases in the various states dealing with individual engineers, or engineers practicing in the partnership form, where the client sought to avoid paying the engineer on the grounds that the design contract was "illegal" since the engineer was unlawfully practicing in the jurisdiction. In general, the engineer has been denied fees under such circumstances except in cases where the engineer was working for another design professional who was qualified to practice in the jurisdiction in which the project was located. The distinction in these cases involving individual engineers seems to be that the licensing statutes were enacted for the protection of the public and not for the protection of a fellow design professional.

Licensing of Individuals

The state licensing laws make it clear that it is unlawful for an individual to practice engineering within the state unless he or she is duly licensed and registered in accordance with state law. The laws are much less clear as to what constitutes the practice of engineering in the state.

Further complexities arise with respect to statutory provisions which provide for limited practice in state B by an individual who is licensed only in state A. Several states have exceptions in their licensing laws which grant limited permission to practice to nonlicensed engineers who have no established place of business within the state. For example, New York, New Jersey, North Carolina, and South Carolina will give a limited permit to practice for 30 days to a nonresident, provided that the nonresident is legally qualified to practice engineering in his or her home state (see Exhibit 1–2). Michigan's practice limitation extends to 60 days, with the additional requirement that the nonresident's own state grant reciprocity to Michigan engineers. Another variation is found in West Virginia's statute, which will grant a permit to practice on a specific project for the duration of the project. The limited-practice candidate can avoid licensure, but only if the time limit is not exceeded. To continue to practice in state B, the individual thereafter must file for a certificate of licensure or run afoul of the registration laws.

What Is Engineering Practice?

Legally, it is arguable that an individual engineer registered in state A, who does not physically render her or his services in state B (state in which the project is located) and who designs a project to be built in state B, is not practicing engineering in state B. Without some other provision in state B's statutes (e.g., requiring the sealing of plans or precluding public agencies from contracting to work performed by nonlicensed—in state B—engineers), state B has no jurisdiction over the state A engineer. In any event, most states do have provisions for a foreign engineer, not licensed in that state, to render limited services within the state.

Despite state B's apparent lack of jurisdiction over the state A engineer in the foregoing, a state B client nevertheless may seek to avoid paying the engineer on the grounds that the engineer was illegally performing services in state B. The nonpaying client often has been successful in such cases, which as a general rule must be tried in state B.

Affixing-of-Seal Requirement

Various state and local statutes require that plans for certain construction bear the seal of a professional engineer licensed in that state. Such requirements are straightforward and usually present no insurmountable problems. However, it should be noted that a state A engineer who designs a project in state B and arranges for a state B licensed engineer to review and seal the plans may still be in violation of the licensing statute.

Government-as-Client Limitation

The licensing statutes include provisions—other than the requirements of licensure to practice in the state and the sealing of plans—which have a direct bearing on the performance of work in connection with an out-of-state project. Most licensure statutes also require that governmental entities have their engineering services performed by engineers licensed in the state. Some states require only that the plans be prepared by the in-state licensed engineer; others prohibit the agency from contracting with non-in-state licensed engineers. But here arises a difference. Under the first restriction the state A engineer apparently could contract for work and then (if there is no contractual prohibition against subcontracting) subcontract to a licensed state B engineer. But such procedures must be carefully watched for infringement against state statutes.

Corporation Limitations

Some states permit domestic business corporations (i.e., other than professional corporations) to practice professional engineering, provided that the engineering work is performed under the direct control of licensed PEs, and often requiring that certain officers, directors, etc., of the business corporation be licensed. As a general rule, an out-of-state business corporation will be permitted to practice engineering in state B to the same extent as a domestic state B business corporation, provided that the out-of-state business corporation is also permitted to practice under its home-state laws.

Virtually all, if not all, states permit professional corporations to practice engineering. Generally, all officers, directors, and stockholders of a professional corporation must be licensed PEs.

In regard to practice by an out-of-state professional corporation, most of the licensing statutes are either silent or ambiguous. Laws relating to professional corporations have been enacted in most states but need to be subjected to a greater amount of interpretation. Even when these relatively new acts are read in conjunction with the business corporation and the licensing statutes, some uncertainty arises in determining if a foreign professional corporation can practice lawfully within a particular state.

Separate from the question as to the practice of engineering in a foreign state is the question of whether a corporation (whether business or professional) formed in state A may conduct any business in state B. This question usually will arise even though all the stockholders, directors, and officers of the state A corporation are individually licensed in state B. A corporation is an artificially instituted being created by the state of its incorporation and has no legal existence outside of the state of incorporation (unlike an individual, who legally exists anywhere).

The need for the state A corporation to qualify to do business in state B is independent of the need for the state A corporation to qualify under the professional corporate laws. Thus, for example, in the states which permit corporate practice of engineering as long as the persons in charge of the work are licensed, a state A corporation can readily comply with the licensing statute (if the proper individuals are licensed in state B). However, such qualification under the licensing laws does not mean that the state A corporation can lawfully conduct business in state B.

In the case of a business corporation, all states make provisions in their corporate laws for the authorization for a foreign corporation to do business in that state. The principal limitation that state B places on the state A corporation is that it may conduct business (such as it may be) only as is permitted a state B corporation. Other requirements to qualify in the foreign state are largely formalistic, i.e., the filing of certain information, payment of taxes, etc.

In most jurisdictions there are three sets of relevant statutes: the licensing statutes, the business corporation statutes, and the professional corporation statutes. These statutes are administered by at least two separate agencies of the state, and the statutes are generally uncoordinated with each other. Hence, there are many unresolved questions with respect to the practice of engineering in a foreign state by a professional corporation.

The following glossary presents an explanation of many of the terms encountered by PEs seeking licenses. The editors recommend that license candidates review these terms *before* applying for licenses.

GLOSSARY FOR PROFESSIONAL ENGINEERING LICENSE CANDIDATES AND PROFESSIONAL ENGINEERS

More than just a word list, this glossary answers many of the questions asked by interested individuals about engineer's licensure. It will help those who are interested to answer their questions about this important pursuit.[1]

This compendium offers quick answers, free of the legal jargon often found in the wording of the various state registration laws. Study of this glossary will provide an excellent introduction to the more formal literature on the subject.

Applicants without Engineering Degrees   These applicants must show they have become self-educated in the application of engineering works. Those who cannot attest to having obtained an engineering degree accredited by the Accreditation Board for Engineering and Technology, or approved equivalent (ABET) may not be accepted.

Applications   A candidate should use the forms prescribed and furnished by a board of registration. These applications contain statements made under oath, showing the applicant's education and a detailed summary of his or her technical work and experience. Applications should contain no less than five references, three of whom should have personal knowledge of the applicant's engineering experience and should be licensed engineers of any state. Some states have special provisions on this and should be questioned.

Basic Requirements for Licensure   See the preceding section, PE Licensure Requirements at a Glance.

Board-Approved Programs in Engineering   The ABET is responsible for maintaining an accreditation list of programs in engineering which is accepted by state boards of registration and the NCEE.

---

[1]Because the definitions have been condensed to save space and because boards of registration can change the rules without prior notice, interested applicants should contact their boards of concern for clarification.

Certificate of Registration   Upon successfully passing the written examination in the fundamentals of engineering (FE) and the principles and practice (PP) or the equivalent as determined by a board, having given satisfactory evidence of competency and fitness, and having completed all the requirements of the board, an individual is granted licensure to practice engineering in her or his state and is issued a certificate.

Closed-Book Examination   When taking this type of examination, the examinee is not permitted to use references, books, or notes.

Comity   A licensed engineer in one state may properly request licensure in another state. This request may be handled on one or two different bases: namely, comity or reciprocity. *Comity* is an established and legal practice, defined by the courts, whereby one state extends to citizens of another state the same privileges that it provides for its own citizens. In matters of licensure or registration, the widespread practice of state boards is to require the same qualifications of the applicant as it would have required of one of its own citizens seeking original licensure at the time of the applicant's reference licensure.

Committee on Records Verification   This is a service agency for engineers needing registration or licensure in more than one state (multiple licensure). It minimizes the effort, helps avoid embarrassment, and reduces the expense required of licensed engineers when seeking licensure in other states.

Compilation of Past or Sample Examinations   State boards no longer make available sample examinations. Write to the NCEE for complete information as to availability.

Concurrent Time   Time spent in engineering work while attending undergraduate engineering courses in evening sessions may be given credit.

Contracting Experience   Boards feel that the experience required of a graduate in an internship is for the purpose of developing and maturing his or her engineering knowledge and judgment. Design work is a very good way of achieving this, but there are many other devices by which it may be accomplished. In general, boards look for positions which require an engineering education to obtain and the amplification of it to do the work. Boards seek breadth as well as depth. It often follows that experience with a contractor may or may not provide this and that the interested individual should not always expect full credit for this type of work.

Eminent Practice (or Eminence)   Boards may waive specific requirements for qualifications, except those of age, character, and citizenship for applicants who possess long-established practice and recognized standing in the engineering profession and who have practiced engineering for more than 15 years after graduation from an accredited ABET course or program in engineering. The definition of *eminence* takes on different interpretation in different states and obtaining licensure by eminence is extremely difficult.

Endorsement   Licensure by endorsement by unanimous agreement of the full complement of board membership provides licensure of previously licensed engineers in one or more states. Although requirements differ to some degree due to individual state prerogative, written or oral examinations may be waived by boards if the applicant has passed all parts of the NCEE format examinations in another state of prior licensure through a 16-hour written examination in the FE and the PP of her or his discipline. Waiver may also be applied if the applicant is a licensed PE with an established practice of long standing in the engineering profession. Most states require the applicant hold prior licensure in her or his home state. In addition, the applicant must meet the statutory requirements of age, citizenship, education, and experience.

Engineer in Training (E.I.T.)   Also known as an Intern Engineer or Associate Engineer, an E.I.T. is a candidate who has qualified and who has passed the preliminary examination in engineering fundamentals (FE) and is approved and certified by his or her board. The E.I.T. is also a candidate for full licensure as a PE, and has been granted a certificate as an E.I.T. upon graduation from an ABET accredited program in engineering and passing a written examination in FE. Upon completion of the required number of years of experience in engineering, he or she will also be eligible for the final examination in PP prescribed for full licensure as a PE. An E.I.T. does not have the privilege of practicing engineering before the public.

Engineering Roster   Most states publish a roster of their registrants. Copies are furnished each registrant.

Engineering Teaching   Such experience may be credited if it was at the college level and if it is satisfactory to the board of reference. Experience must be in engineering teaching and often in advanced subjects. Assistant professorship is a minimum in some states.

Engineer's Seal   Upon obtaining licensure, each licensed engineer shall obtain a seal of a design authorized by the board, bearing the registrant's number and name and the legend "registered (or licensed) professional engineer." Plans, specifications, plates, calculations, and reports issued by a licensee shall be stamped with the seal during the life of the certificate, but it is unlawful for anyone to stamp or seal any document if his or her certificate has expired or has been revoked, unless his or her certificate has been renewed or reissued.

Experience   The nature and character of experience must be approved by the board of record, and it should be broad in scope. Experience must be accumulated in any state or country. The nature and caliber of the experience is the important consideration, not merely the number of calendar years of experience. Qualifying experience is the legal minimum number of years of creative engineering work requiring the application of the engineering sciences to the investigation, planning, design, and construction of major engineering projects. The mere layout of design details, acceptable performance of engineering calculations, writing of specifications, or making tests are not sufficient; instead, the experience must be a combination of these, plus the exercise of good judgment, taking into account economic and social factors, in arriving at decisions and giving advice to client or employer. The aspiring engineer should seek the widest possible responsibility of a broad and diversified nature. Experience should be progressive and of an increased standard of quality and responsibility. The applicant's record of experience should reflect this background and be attested to by responsible references.

Experience   As defined by ABET, an experienced engineer is characterized by his or her ability to apply creative scientific principles to design or develop structures, machines, apparatus, or manufacturing processes, to construct or operate them, or to forecast their behavior under specific operating conditions. *Process design* may be defined briefly as the best process as determined to accomplish a given end from the standpoint of economy, safety, raw materials, and available equipment.

Filing Instructions   List job titles for each engagement; give complete data, describing your function in detail; show how much and what kind of experience; look for areas of progression to more important work on the job; attest to more than a passing interest and initiative in your work; show evidence of design in its broadest sense and not confined solely to making drawings and computations.

First Day's Examination   The first day's examination covers the FE and tests the candidate's facility in mathematics and engineering theory. This is an 8-hour examination covering a full day, and boards want to know whether or not the candidate understands basic engineering principles.

Foreign Credentials   Applicants submitting educational credits obtained from a foreign engineering school must submit them in original. This applies also to degrees and other pertinent papers.

Graduates of Nonaccredited Engineering Programs (ABET)   Candidates who are not graduates of accredited engineering (ABET) programs or nonaccredited programs of study in accredited engineering schools may be admitted to the preliminary examination (FE) provided they satisfy the age, citizenship, secondary education (high school), and endorsement requirements, and, in addition, qualify for 8 years of credit for combined educational and experience records.

Grandfather Clause   When a state first passes a licensing law, it must recognize persons already practicing successfully in that field. That part of the law making this possible is known as a "grandfather clause." It usually expires one year after passage of the statute.

Inquiries on Licensure or Registration   Engineers who wish to secure legal registration (licensure) must communicate directly with the individual state board or department and request application forms and a copy of the state registration law. This is an important requirement.

Interstate Licensure   The very nomadic nature of engineering may require interstate licensure. An applicant may obtain licensure in another state by endorsement (see Endorsement) but her or his original registration is not transferable. Most states require previous licensure in the resident state as a preliminary requirement.

Military Experience   To be creditable, military experience must have been spent in engineering work. This experience must have been progressive and of an increasing standard of quality and responsibility. Final decision as to its qualifying nature rests with the board.

Minimum Evidence for Qualification for Practice of Engineering   The following facts established in the application shall be regarded as minimum evidence satisfactory to the board: (1) a specific record of 8 or more years of active practice in engineering work of a character satisfactory to the board and indicating that the applicant is competent to be placed in responsible charge of such work, or (2) graduation from a school or engineering school accredited by ABET or approved by the board as being satisfactory in standing and a specific record of an additional 4 years of active practice in engineering work of a character satisfactory to the board and indicating that the applicant is competent to be placed in responsible charge of such work, provided that no person shall be eligible for licensure who is not of good character and repute.

Multiple Registration   See Interstate registration.

National Council of Engineering Examiners   The purpose of the NCEE is to provide an organization through which state boards of registration may act and counsel together to better discharge their responsibilities in regulating the practice of engineering and land surveying as it relates to the welfare of the public in safeguarding life, health, and property. Address: National Council of Engineering Examiners, Box 1686, Clemson, S.C. 29633-1686.

NCEE Format Examinations   The NCEE, through its various working committees, supervises the preparation of and is responsible for the content and format of all uniform examinations in the FE and PP. All states, except Illinois, participate in the administration of the NCEE format examinations which are offered twice each year, usually in April and October.

National Registration of Engineers   Under the Constitution of the United States, there can be no national or federal licensing of any profession; the authority to legally register or license citizens to practice a given profession is given to the states.

Open-Book Examination   This is a written examination in which all reference books, notes, etc., that can help the candidate in any way are permitted inside the examination room and used by the candidate. All states use the open-book examination and provide each candidate with specific instructions as to what is permitted inside the examination room.

Oral or Written Examination   When required, this examination is conducted at such time and place as the board shall determine. The scope and method or procedure shall be as prescribed by the board, with special reference to the applicant's ability to design and/or supervise engineering works, ensuring the life and safety, health, and property of the public. A candidate failing in any examination may, at the discretion of the board, be examined again. Once a candidate passes, his or her passing mark is the only one retained for record, as all other grades are discarded.

Practice of Engineering   "Practice" refers to any service or creative work requiring education, training, and experience and the application of the special knowledge of mathematics and the physical sciences to such services or creative work, such as consultation, investigation, evaluation, planning, design, and the supervision of construction for the purpose of assuring compliance with specifications and design, in connection with any public or private utilities, structures, building, machines, equipment, processes, works, or projects. A person shall be construed to practice or offer to practice engineering who, by verbal claim, sign, advertisement, letterhead, card, telephone listing, or other way, represents herself or himself to be an engineer; or who holds herself or himself out as able to perform or does perform any engineering service or work. But the term shall not include persons who merely operate or maintain machinery or equipment. Any engineering work wherein the public welfare or the safeguarding of life, health, or property is concerned or involved (when related professional services require the application of engineering principles, data, and ethics) is practice of engineering.

Presentation of Additional Evidence In cases where the evidence presented by the candidate or applicant does not appear to the board to be conclusive or to warrant the issuing of a certificate of licensure, the applicant or candidate may be required to present additional evidence for board consideration.

Professional Engineer (PE) A PE is a person who, by reason of his or her knowledge of mathematics, the physical sciences, and the principles and practice of engineering, acquired by professional education and practical experience, is qualified to engage in engineering practice.

Qualifying Experience The basic requirement of qualifying experience is to assure that the applicant has acquired through practice of suitable caliber in engineering the professional judgment, capacity, and competence in the application of the engineering sciences to demonstrate that she or he is qualified to design engineering works, structures, or systems.

Reciprocity See also Comity. Reciprocity must be previously established by agreement between the legislatures of two states. Many states make use of the term "reciprocity," but there are very few reciprocal agreements. Thus, by "reciprocity" one state does to the applicants from the other states as it does to its own citizens.

Registration by Endorsement It is possible to exempt from examination an applicant who has satisfied requirements for admission to the examination but who possesses a license from another state received as a result of passing written examinations. If the examination he or she passed was the equivalent of that of a particular state at the time passed, most state laws provide for its acceptance in lieu thereof. This consideration applies equally as well to E.I.T.'s from another state. No applicant should consider that the requirements have automatically been met merely because the required number of years has elapsed. Most states require that the candidate pass a 16-hour written examination in the previous state.

Registration in State of Residence This is a requirement for licensure by reciprocity or comity in most states.

Registration Law Each state has its own statutory qualification requirements as reflected in its registration law. Educational requirements for licensure of engineers have been written into the laws. Qualification of good moral character, evidence of completion of academic and professional education, and field of experience of a character satisfactory to the boards have been included. Copies of registration laws are available from boards, whose addresses are available from the NCEE.

Responsible Charge   This term refers to the independent control and direction, by the use of initiative, skill, and independent judgment, of the investigation or design of professional engineering work or the supervision of such projects. It implies "of work" and/or "of men." Work assigned by a supervisor but not requiring her or his direct and continued supervision may be classified as "in responsible charge" on the application form. Boards look for "healthy" progress. The college graduate who starts out on the drafting board must show how she or he progresses through various states to a point where she or he is independently applying basic engineering principles in her or his everyday work in the engineering design office or in the field.

One of the most common mistakes engineers make when describing their work experience is to use vague generalities. For example, it is common for applicants to write something like: "My duties consisted of designing power plants, including economic evaluation of the sites and original investment costs." This description is merely a paraphrasing of the definition of engineer as given in the statutes. Or, an applicant will use ambiguous phrasing, such as "I performed," "I was engaged in," "I assisted in," or "I worked on," with no mention of specific work duties and responsibilities. Board members cannot tell from such descriptions whether an applicant was a drafter, designer, print coordinator, or project manager.

An engineer is in responsible charge if she or he signs engineering documents and is capable of answering questions asked by equally qualified engineers. These questions would be relevant to the engineering decisions made during the engineer's participation in the project and in sufficient detail to leave little question as to her or his technical knowledge of the work performed.

Sales Engineering Experience   For sales engineering experience to be creditable, it must be demonstrated conclusively that engineering principles and engineering knowledge were actually employed. The mere selection of data on equipment performance from a manufacturer's catalog or similar publication will not be considered engineering experience acceptable to a board. The term "sales engineer" in many states is not permitted if the holder is unlicensed.

Second Day's Examination   This test reflects the principles and practice of engineering (PP). After the first day's examination has tested the candidate's facility in mathematics and engineering theory, the second day's questions require him or her to show proper judgment in selecting correct formulas, making economic evaluations, and using practical approaches to problems.

State Board Duties   Boards determine what constitutes adequate education and experience and authorize an applicant to practice only when she or he has been found to have satisfied the full qualification requirements. The function of a board of registration is to administer the state registration act according to the intent of the state legislature.

Subprofessional Experience   This type of experience includes field work, drafting, construction, computations of a routine nature, cost data, testing routines, operations, detailing, installation, repairs, maintenance, apprenticeship, routine analysis, airport operation, test piloting, and engine testing, wherein the basic disciplines of engineering knowledge are not applied.

Superior Qualifications   This is a form of eminent practice. It includes, but is not limited to, advanced studies in the community of scholars, advanced degrees, technical society membership and committee participation, inventions and patents, patent pendings, contributions to the technical literature, copyrights, special developments, high regard and esteem by fellow engineers and contemporaries, contributions to engineering knowledge, support of engineering education, and community interest.

Temporary Practice   Most states have a provision in their registration law to allow a registrant of another state to practice for a short period of time in their state if he or she requests permission in writing to do so from the state board of record and, in some cases, pays a temporary filing fee. In some states, an application for temporary registration must be filed, and permission may be granted to practice while the application is being acted on if the applicant is licensed in the state of his or her place of business. Penalties for failure to advise the board regarding temporary practice can be severe and embarrassing to both the engineer and his or her client.

Transcripts of Records   When required, to comply with the requirements set forth in the application, transcripts must be mailed directly from the school or college to the board with the signature of the proper school officer and the embossed seal of the school impressed thereon. Transcripts of grades are normally required of applicants who are not graduates of ABET accredited programs in engineering or other board-approved curricula.

Uniformity of State Registration Laws   State registrations may not be uniform due to local conditions, date of enactment, and other conditions. An applicant should determine the latest requirements by writing directly to the state board office concerned.

OTHER FORMS OF REGISTRATION

Besides the professional license, there are other forms of registration which can also benefit the engineer in private practice. The National Engineers' Council issues a national certificate of registration; it is not a license per se, but it provides a detailed list of each professional's qualifications. Each registrant's certificate is kept on file by the National Board of Certification; it should be updated regularly by the engineer. Further, the board acts as a clearinghouse for information about engineers and is a valuable reference for the engineer applying for work in places where she or he may not yet have a license or an established reputation.

The PE can further enhance his or her credentials by joining one or more of the many professional organizations around the country. Professional societies are nearly always seeking to increase their memberships. If you are active in such groups, your affiliation with them serves as an added indication to clients that your firm is truly professional. You might even gain special recognition by joining a professional society. For example, your city may have a large number of consulting firms which specialize in air conditioning; but if yours is the only one headed by a Distinguished Fellow of the American Society of Heating, Air Conditioning and Refrigeration Engineers, it could well become one of the more renowned firms in the area.

Perhaps you are still wondering, "How many licenses will I need in how many states?" As was noted earlier, you—or at least one person in your firm—should be registered in the state in which your office is located. It may be helpful, however, to consider that you have a practical trading area for your talents in adjacent states as well. This does not necessarily mean that you will automatically be able to obtain licenses in every one of those states solely on the basis of your present license. But it is certainly worthwhile to open your mind to the full range of areas in which you may be able to provide some services.

The concept of broad trading areas is most applicable to regions in which the states are geographically small, such as in the mid-Atlantic and New England areas. This is not to say that licensing in any one state is superior to that in others; but it is true that some states stand out in regard to the prestige of their engineering licenses. Holding a license in such states may thus give your firm greater influence.

In the northeast, for example, New York is probably considered by many to be a dominant force in engineering. In the midwest, Illinois tends to stand out, while in the west, California is highly influential. In the south, Texas and Florida are especially well regarded. The actual quality of the work done by engineers in each state may not vary significantly, although the origin of the license can affect a firm's marketing clout and the self-image of its members.

ADVANTAGES AND LIMITATIONS IN LICENSING

Moonlighters can be found in nearly every kind of work; but in the world of professional engineering, "moonlighting" may mean more than merely working at two jobs on the same day. While it is not unusual for a registered engineer who heads a consulting firm to put her or his official seal on the work of an unlicensed employee of the firm, a few engineers actually moonlight their license to gain a bit of extra income. For a fee, these engineers permit their seal to be placed on work prepared by an unlicensed engineer who has no formal relationship to the licensed professional or the firm.

Moonlighting the license is not only ethically questionable but it is also dangerous. Even when the licensed engineer is assured that moonlighting the seal is safe and courteous in a particular instance, he or she runs a real risk of damaging his or her reputation—or worse, having the license revoked. By placing his or her seal on a job, the engineer formally declares that he or she is personally responsible for the work and its results. If problems arise in connection with the job, the licensed engineer must assume the consequent burdens. Moreover, the registered professional may not be able to shift the burden to the actual author of the work without added legal problems involving charges of fraud and threats of suit.

Possessing a license enables the engineer to open many doors to new opportunities. Occasionally, people might complain that licensing requirements result in unhealthy restrictions for the engineer's practice. But most professionals agree that registration requirements safeguard both the public and the profession itself. Like any other license, the engineer's should be used carefully and responsibly.

Technically, the license does not prevent its holder from acquiring additional skills in fields other than those in which she or he is specifically trained when first registered. However, many states require a separate license for each engineering discipline—mechanical, chemical, electrical, or other, more specific, areas. This places some common sense limitations on the individual's work: The wise consultant generally confines her or his services to fields in which the firm and its members have solid qualifications and documented expertise.

In this regard it is interesting to note that several states have in the last few years begun issuing a new type of engineering license. Usually referred to as the "corporation license," it may be required for all individuals or groups who incorporate their firms. Generally, all members, or at least the chief executive of the firm, must already have PE registrations in order to obtain the corporate license. There is some disagreement in professional circles about the value of this type of registration, but many people feel that the corporation license helps to demonstrate the merits of an entire firm rather than those of just one or two members. Also, some feel that the corporation license helps to distinguish actual engineering firms from companies which may attempt to do engineering consulting without having the appropriate qualifications. Of course, specific licensing requirements should be investigated in each state for each type of work under consideration.

PLANNING FOR EFFECTIVE MANAGEMENT

Once you decide what type of licensing you may need to begin work in your area, another vital question emerges: What arrangements should you seek in setting up the firm's management? Approaching this topic thoughtfully may put you ahead of competitors: too many people leap into their first project as soon as they have the necessary credentials, leaving management decisions until a time when observers might start asking, "How does this firm get its work done?" Needless to say, by *that* time the firm's reputation may already be hurt.

So it is a good idea to begin setting management priorities while you organize your new firm. Basically, the organizational structure of the firm itself helps to mold the managerial framework and its functions. If, for example, you start the business alone or with one employee, you naturally become the proprietor and manager; it is your responsibility to solicit and plan projects, to direct operations, and to control finances and other aspects of daily performance.

As the size and responsibilities of the firm grow, so does the complexity of management; but this need not be a hindrance. Authority that is wisely delegated leads to smoother operation. This is especially true in a partnership, where the firm's members should have skills which complement each other. However, it may be easier to imagine such a situation than to actually find the appropriate partner. So maximum effort should be geared toward determining a potential partner's skill and toward planning the management structure before the partnership is officially formed.

Corporations have a different set of needs. Those who form a corporation must decide on who will act as president, vice president, treasurer, secretary, and so on. Many firms combine the secretary's and treasurer's jobs into one office since the secretarial function is usually very limited in a small company. Others may appoint one executive as both president and secretary and another person as vice president and treasurer.

The arrangements can make for some interesting results. For instance, if the new corporation has three members, one of them will most likely become president and chief operating officer. The president will probably be supervised, to some extent, by a chairperson of the board of directors, who is elected by the firm's stockholders. But the stockholders will very likely consist of the three founding members of the firm. This leaves all three members having to choose one of their group as chairperson and another as president, while the third member may initially have no major title at all.

Such a selection process might put the members to a serious test of their ability to cooperate, especially if each desires a particular executive position. The problem can be mitigated somewhat if a new chairperson is elected each year. The best course to take, basically, is to recognize that the working environment should accommodate the self-images of the firm's members and should be geared toward personal satisfaction and economic and professional prosperity. And the most effective tool for accomplishing this is clear communication among the individuals involved.

There is a classic idea in business that says that in lieu of salary increases, first you put your name and title on your desk; only later do you put them on the door; then, finally, they go on the firm's stationery. This idea is as relevant to consulting as to any other kind of firm. Generally, management responsibilities should be delegated only through procedures which carefully match the individual's skills, personality, and needs with the needs of the firm. More specifically, in the case of certain types of firms such as joint ventures or associations, management authority—and the placing of executives' names on letterheads—may have certain legal restrictions. For example, in some states, one cannot legally list as "associates" names of members of the firm who are not registered engineers in the state.

Fortunately, most engineers gain familiarity with basic areas of management through their regular work. Whether they realize it or not, many enter into private practice after learning about management by having observed management teams at their salaried jobs; some have actually been part of such teams. Therefore, the greatest challenge to the new consultant may lie not in learning how to be a manager but rather in adapting to the change of perspective that comes from moving from a staff position to a directional one. Initially, the change can be disorienting; but with time and patience, the consultant can master the new perspective just as he or she mastered basic engineering skills.

An examination of some of the general foundations of business and management may help smooth this transition. Production, sales, administration, and service—no matter how big or small a firm is—are the nuts and bolts used to build a consulting engineer's practice.

Too many firms suffer from paying too little attention to marketing and sales. Others tend to neglect production while they concentrate on expanding sales and other routine functions.

In sole proprietorships, some engineers are tempted to immerse themselves in the details of each project while they neglect steps to establish themselves as leaders in sales and administration. It can be misleading to think that the business will prosper merely by doing good work and providing follow-up services. Certainly, these are among the most fundamental areas of concern in any business, but it can be self-defeating to try to attend to every aspect of the firm's performance at once. More than anything else, what the consultant needs to do—whether in a sole proprietorship, partnership, corporation, or joint venture—is establish a pattern of leadership.

Successful leadership stems from skill in delegating authority plus a knowledge of time management. While there is no fixed formula for the amount of time usually spent in various areas of management in successful firms, there are some reliable rules of thumb. Most consulting firms devote about 40 percent of all employee hours to producing the work (drawings, designs, specifications, on-site technical assistance, etc.). From 5 to 10 percent of available time goes toward peripheral and follow-up services. The remainder, 45 to 50 percent, is taken up by marketing and administration.

A practical business approach would dictate that about 2 days' time each week should be devoted to sales and management activities, with 1 day per week spent in supplying peripheral services and planning, and 2 or 3 days going toward work on the actual projects at hand.

In a partnership or small corporation with several employees, one should tally up the total amount of time available. For example, if there are three partners or three members of the corporation and they are to spend about 20 percent of the total time in marketing and 20 percent in administration, then one of the partners or officers can devote at least 3 days each week to direct selling and another can put 3 days' work on administration.

In actual practice, these figures will vary according to the needs of each firm; but they are reasonable approximations for good time allocations. In some cases, the administrative and planning functions may take only 2 employee-days. If, in a firm of 10 people or so, there are 3 owners or partners, then ideally 1 of the 3 could devote all of her or his time each week to selling. Another could put 2 days toward administration and management, and the last could devote 3 days or more to supervising production and client services.

In joint ventures, where the member firms may each have their own work in addition to the joint project, it is common for one firm to handle the administrative aspects (usually for a proportionately larger share of the total fee) while the other firm handles production. Alternatively, members of each firm could jointly administer the larger project. Either way, the person or persons in charge of management for the individual firms would simply use part of the 2 days or so normally devoted to overseeing their own firm's operation and allocate that time to administrative aspects of the joint venture.

Administrative functions will be treated in greater detail in later sections of this Handbook; but for now, one should keep in mind that the administrator's primary functions include not only overseeing staff but also making certain that all forms, contracts, and reports required for each project are properly completed—from start to finish. In sales, some of the major work is done on paper too; advertising brochures and other publications produced as part of your marketing efforts help promote the firm's growth. But the most important activities in sales often are those involving face-to-face contact.

No statistics are available on how many firms have run out of work because they neglected to market themselves through in-person meetings. But it is safe to say that a significant proportion of firms experiencing financial setbacks do so, at least in part, because of poor marketing. There is no substitute for direct selling in consulting because professional consulting engineering is a people-service. How to get in-person interviews and what to do once you have obtained them is a topic that will be covered in detail in a subsequent section of this Handbook.

SETTING OBJECTIVES

If you were to ask a group of consultants what the chief objectives of their firms were, you would probably find that most people point to profits and business growth as their main goals. These are quite appropriate, but there are other kinds of objectives that may be just as important in helping to motivate a firm and its members toward optimum performance. Objectives, in the stricter sense, are much like the targets that a long-distance runner would set up while training. The runner's ultimate goal might be to win a marathon, but he or she does not begin training merely by running the marathon itself. The prudent athlete—like the smart business person—gradually builds strength by setting daily, weekly, and monthly objectives, diligently devoting nearly all his or her energy to meeting the targets.

For example, a consulting business might set objectives such as a 15 percent annual increase in sales and profits, or it could plan to add several new clients and two more staff members, during the current year. In setting objectives, the athlete analyzes his or her physical strength and endurance to decide how much he or she can reliably increase his or her workout each month. The professional consultant can analyze the characteristics of members of his or her firm to set effective growth projections. Generally speaking, objectives of 10 to 20 percent growth per year are reasonable, provided the proper energy is put into reaching the goals.

Certainly, few people work at 100 percent efficiency. In most cases it is safe to say that one can probably increase output by 10 to 15 percent if a good growth plan is clearly worked out and consistently followed. For example, an analysis of office procedures might indicate that certain changes could be made to improve office efficiency, perhaps reducing or eliminating the need for revisions of paperwork caused by typographic or mathematical errors. This could be accomplished, perhaps, by assigning one employee to proofread and check specifications on each draft for an hour or more each day. Progress can be made toward other types of growth by assigning a staff member to investigate additional marketing avenues, such as magazine advertising or trade exhibitions. One might very likely be able to add a few new clients by examining the market for specialized services which the firm has not yet engaged in. Of course, these are only some general ideas of a few of the many kinds of objectives you could set.

The point to remember is that setting and reviewing such objectives on an annual or semiannual basis can help you gain a keen perspective on the strengths and weaknesses of your firm's operation. If you find that certain objectives have not been reached, you can proceed with an informed discussion regarding the background and possible solutions. The minimum benefit you can obtain from objective setting is that you will build a forum for communication among the firm's members. This in itself can give you a "handle" on what is right or wrong with job performance. The maximum benefits include attaining and even exceeding your growth projections.

If you have an objective, for example, to increase sales 15 percent during the year and you find little improvement during that time period, a thorough examination of staff behavior and market conditions might disclose that production and sales appear to be proceeding well; but perhaps the market is experiencing a slump. During an economic recession, clients may be reluctant to start new projects, for instance. If this is the case, you might find that the best course is to put more of the firm's energy toward innovative means of market penetration, such as involvement in projects designed to reduce clients' expenses or in overseas work. The maxim "Plan your work and work your plan" is a tried-and-true guide to progress and prosperity in consulting.

BUDGETING: THE "2.5 FACTOR"

Earlier, a brief overview of financing procedures was presented for new consulting firms. The economics of engineering self-employment warrant more detailed examination as the consultant begins dealing with specific administrative issues such as marketing and growth forecasts. So let's look at how some of these issues might come into play in the budget context.

As has already been noted, a great many established engineering firms charge for their services at a rate roughly amounting to 2½ times actual payroll costs. This is referred to, simply, as the "2.5 factor." It merits special attention at this point because the factor provides the foundation for much budget planning throughout the consulting field (although it may be adjusted to suit particular situations).

Applying the 2.5 factor indicates, for example, that a $10 hourly pay rate for employees would be accompanied by a $25 hourly fee to clients.

The cost to retain an employee at this rate for a 2000-hour year would thus be $20,000, while the consequent charges to clients would amount to $50,000. A reasonable estimate of profits would be 10 to 20 percent, before taxes. This profit estimate is based on data derived from a number of consulting firms and on the following breakdown of expenses: Of the $25 hourly fee, $10 goes to the employee payroll, $10 is reserved for all operational expenses, and $5 remains as profit before taxes. Most often, the after-tax profit falls in the range of 4 to 6 percent.

Recent surveys by the American Engineers' Council and other professional organizations found that the average annual amount charged to clients in the United States per employee equals about $38,000. (It is important to remember that $38,000 is the average annual charge per staff member rather than an individual fee charged for each employee since pay rates may vary depending on seniority and other factors.) The average annual salary for all kinds of employees in consulting engineering firms—from drafters to project engineers—is approximately $17,000.

Now, based on our survey estimates and the 2.5 factor, the cost to the firm would break down to roughly $17,000 for the labor of one employee and $17,000 for overhead, with about $4,000 remaining as profit. (The survey's figures indicate a factor closer to 2.2, mostly as a result of using averages in calculation. The 100 percent overhead estimate, moreover, may be slightly exaggerated as a result of rounding. Actual overhead in most firms would probably be closer to 85 to 90 percent of employee salaries and thus would result in higher profit margins.)

The example presented earlier in this section, which gave a rough budget outline for a beginning firm, indicated a direct payroll expense of $20,000 and an overhead of $26,000, or about 130 percent of payroll. The reason for the apparently huge overhead in that case was that extra start-up costs (for such items as advertising and promotion) were added. A new firm may find it somewhat difficult to keep overhead under strict control until after the firm has passed through the start-up period, during which equipment buying costs and advertising expenses can take big bites out of revenues.

Two budget areas that should be watched especially carefully by both new and experienced firms are rent and travel expenses. These areas are prone to wide variations unless the firm is operating under a strict, fixed budget. Stationery stores and bookstores which specialize in business materials often sell various types of accounting forms which can help you organize your firm's preliminary accounting programs to help cope with such potential trouble spots. But the best authority is an accountant skilled in small-business budgeting, who can set up a series of on-the-line cost estimates for you based on a 100-percent-of-labor-cost overhead figure.

What do you do if you do not know how much labor will be involved during the start-up period? The answer to this question touches upon another maxim: "Cut the cloth to fit the garment"; that is, spend no more than you have, or not more than you can reasonably expect to earn, in a given period of time.

OTHER TYPES OF BUDGETS

There are other types of budgets than those for expenses. Most likely, you will eventually make use of budgets for time allocation, budgets for growth, and budgets for staffing the firm. Here are a few hints which can help you plan in these areas.

First, let us return to the earlier budget example, wherein the firm charged $25 per hour while paying $10 to an employee for 1 hour's work. Using these figures as our base, a $10,000 fee for a project which takes about 400 hours to complete would indicate a cost of about $4000 in direct labor. To produce a time budget for the project, many established consulting firms would weigh the hourly pay rate to compensate for individuals in their organization who earn more (or less) than the basic $10. For example, if you have a manager on a project who is earning a few dollars more than average, you might increase the basic pay rate to $12 per hour. Dividing the $4000 labor cost by $12 per hour indicates that you have about 333 employee-hours available to complete the project.

Budgeting for company growth and staffing are both quite similar. The basic question to ask in each case is, "How many dollars must the firm plan to take in before an additional employee can be hired?" If the firm is charging $25 per hour per employee and it has prospects of adding an additional $40,000 in gross income through new work, it will either need to hire another employee or will have to put forth extra effort on an overtime basis with existing staff resources. The same sort of figuring applies to growth projection, which is one of the elemental business objectives mentioned earlier.

Granted, these examples may appear simple. But simplicity can be a vital aid in developing an overall budget plan. Budgeting will benefit most in the long run from the designer's creativity and diligence. Lacking specifics on each individual firm, however, the best sort of guidelines emphasize the basic goals of performing the firm's work effectively and expanding the business profitably. The complex, detailed sort of budget planning which often appears in large corporations' quarterly or annual reports is rarely necessary in the beginning consulting practice during start-up. For the older engineering firm which employs 50 engineers and grosses several hundred thousand dollars a year, the need for highly detailed reports and control procedures is often as crucial as it is for a national manufacturing enterprise. The majority of engineering consulting firms that consist of fewer than 10 persons find that elaborate in-house reporting procedures become too costly and unwieldy to be worthwhile.

There is a level of record keeping which you can develop to help your firm run with optimum effectiveness, however. To start, the major item every firm needs to keep track of is *time*. The firm should always keep an accurate log of the amount of time devoted to each project by each individual involved. Hardly any other factor is more directly related to the proper gauging of revenues—especially when the project fee is a lump sum or fixed percentage of cost. If, for example, you and your client agree on a lump-sum fee of $10,000 for a job estimated to take 400 employee-hours but you do not properly control the hours your staff works, you could find in the end that you have used as much as 1000 hours. Your fixed fee would then be barely enough to cover your firm's payroll, not to mention that the fee would not even provide for overhead costs or profits.

Other items which can seriously affect the profitability of your operation if not watched carefully are such expenses as those generated by telephone use, document reproduction, and entertainment for clients and for members of the firm while traveling. Each of these areas should be budgeted and controlled as carefully as possible through ongoing accounting methods. Not only will this help you plan and direct an efficient practice but it should help to place your firm in a healthy position when and if a time arises to deal with accounting analyses or tax audits.

During times of increased fiscal insecurity throughout the nation's economy, engineering firms, agencies, and their clients are more frequently involved in auditing by independent accounting firms. During an audit or analysis, some of the first items an accountant or auditor will ask for are payroll and expense records. Records which are simple but thorough can obviously be valuable at such times, regardless of whether the audit is being done at the request of your firm itself or of some other individual or group.

A skilled accountant is almost invariably the most reliable source of beneficial information and advice about financial matters for any company. This point can hardly be overemphasized. The accountant may be able to help reduce overhead, point to trouble spots in time management, suggest tax shelters; he or she may even act as a sounding board for new management ideas in addition to carrying out some of the more routine accounting tasks.

A good accountant can help you set up a budget to prepare for the effects that inflation rates will have on wages. He or she may help you develop contingency plans to guard against changes in markets and the economic climate. But if for some reason you find that you cannot retain an accountant on a regular basis, you can at least bolster your position by reviewing your budget regularly to determine what sorts of expenses and staffing can be reduced or eliminated in the unhoped-for event that income begins to lag significantly behind the firm's projections. Emergency answers like "Fire half the staff!" simply will not heal an ailing fiscal profile—especially at the last minute.

The consulting firm should also maintain strong communication ties with its bank. While the bank and its branch manager may not always be able to provide answers for your money questions, they are usually able to provide an added perspective and to recommend other sources of help.

The other type of professional with whom the firm should maintain solid contacts is the attorney who is well versed in your specific field. The best time to start this relationship is while the firm is being formed. Most firms will eventually require an attorney's assistance in situations involving contracts and the use of standard contract clauses in various arrangements. Without a knowledgeable professional to guide you in your understanding of such items, you could become confused, or worse, misled. The skilled attorney can help the firm avoid needless hassles by reviewing prospective contracts and advising on legal details. Clients understand this as a part of professional consulting work.

GAUGING PERSONNEL NEEDS

The professional consulting engineer is constantly acquiring new knowledge in a variety of areas—legal, financial, managerial, and others—through personal contacts and business experiences. The enhancement of one's professionalism begins as soon as one decides to enter private practice; with time, the consultant gains greater confidence in budgeting, develops familiarity with accounting and legal procedures, and learns how best to manage daily operations.

One very important skill that tends to develop with experience is the ability to estimate personnel needs. As the consultant gains greater familiarity with particular types of projects, he or she can make more and more accurate projections of the number and kinds of personnel (and the amount of time) required for similiar jobs. For the newer firm and for times when you may be taking on new types of projects, however, a few basic guidelines can be a great help in this regard.

First, be sure to select your personnel on the basis of projected sales, and vice versa. For instance, if you are the sole member of your firm and you acquire a project which will put your gross income for the year at $50,000 above your original projections, you can probably hire one employee at $20,000 to handle the extra work and enjoy a welcome addition to your projected profits too. Conversely, sales should be sought to meet or exceed the 2.5 factor for payroll, as discussed earlier.

Second, take special care to base your fees on reliable estimates of the amount of time a given project will require. There is no substitute for experience in this regard, but if you find that your $50,000 project will require more time than you and your one additional employee could provide, you may want to hire that first employee at $17,000 and take on a part-time worker to cover the extra hours. Keeping rigorous time records now can help you immeasurably in making reliable estimates later.

One concern voiced by many firms arises from the need to hire and pay additional personnel before fees are paid by the client. This concern is, of course, related to one of the most serious issues facing the growing business: how to finance the company so that it can grow. To cope with this, many new and rapidly growing firms accept reduced salaries for their owners and partners so that they can hire more people.

As a firm becomes more established, some of the money needed for new staff can be borrowed on short-term loans (3-, 6-, or 9-month) from a local bank. This is often a satisfactory option for the established firm; but for the 1- or 2-year-old firm, there seem to be few solutions other than building sufficient funds at the start of the business or accepting a lower income until adequate cash flow is developed.

There is, however, another potential remedy for the problem of needing to hire more employees than you can afford to pay during an extended period: You can gradually increase or decrease the size of staff as project requirements vary.

Most contracts begin slowly, with fewer employees needed, then gather momentum. Figure 1-1 shows the typical *s* curve for contract completion (staffing) versus time. A common assumption is that if the contract calls for a total of five persons overall, it may in the beginning require the services of one or two; around the halfway point of the job, it might involve five or more, and toward the end it will probably require only one or two employees again. In other cases, a contract may call for one person's effort at the start and five at the end. The point is: Each contract should be analyzed, before work begins, for weekly and/or monthly personnel needs. Doing so will help you gauge the job accordingly.

Obviously, you will not want to be hiring and firing people to accommodate every little change in a project. But you can learn to gauge staffing optimally. The most vital ingredients of this gauging will be reliable projections of time requirements and income flow.

It is also wise to check whether a client is willing to make partial payments as work proceeds and whether or not it would be effective to retain certain employees on a part-time or freelance basis.

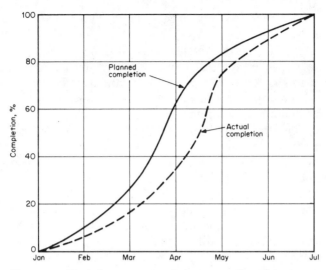

Figure 1-1   Typical contract completion-versus-time plot.

The emphasis should be *balance:* to develop a well-rounded staff with skills which complement each other and people who can supplement one another. Highly skilled employees may not always be easy to find, but sometimes you can build a stronger staff by hiring a younger engineer trained in a specialty which your firm frequently requires than by hiring an older, more experienced person whose skills merely duplicate those of others on the staff. It is also best to avoid hiring people who have thorough knowledge only in some areas in which the firm engages infrequently. A well-balanced staff promotes a well-balanced business!

CHOOSING YOUR OFFICES

One of the more difficult chores faced by the professional in private practice is that of selecting an office. Whether the office you seek will be your first or your fifth, it will become an integral part of your practice, a reflection of your personality and of the functions of the firm itself. The hunt for the right place can actually be enjoyable, if you follow a few simple guides and gauge finances properly.

Each type of consulting engineer has different office needs. A technical consultant specializing in emergency repairs of mine shaft supports may need hardly any permanent office space, for example, since most of the work is done on the client's location. A civil engineer, however, might need space for several drafters and for clients who will visit to view the firm's designs.

The most vital questions which arise in choosing an office are: (1) How much space is needed and (2) how much can you afford to pay? One option that might come up is that of opening an office in your home. Under ideal circumstances, a home office can save money in several ways. You will not have to commute or pay additional rent or utility bills; in addition, you can deduct your office expenses on your federal and state income tax return. But there are some serious drawbacks as well.

If you live in an apartment, it may not have enough space to accommodate an office. If you own a house, you still have to find sufficient space for the office, away from distractions that can be brought on by family members and friends. You will need an additional telephone line, or else you must contend with nonbusiness calls during work hours. Your home may be in a location that is difficult for clients to travel to and find. Further, you may not be able to expand the office area as your practice grows. So for many engineers, the home office is not a practical alternative. Hence, the bulk of our discussion is based on the prospect that you will be setting up an office outside your home.

TO LEASE OR BUY?

Today it is nearly impossible to find office space without rental limitations. Most offices are leased or bought by the firms using them. At the start, unless you have substantial capital and want to try your hand at a real estate investment by buying the office space, your best bet is to seek a relatively short-term lease—3 to 5 years at most. A short-term lease will help protect you against problems which could arise in the event that you later find the rent too high to pay. A longer-term lease can be sought after your private consulting practice becomes more established and secure. Some firms do choose to stay at the same location for many years, but others want to move and are restrained by a long lease or by the need to pay extra money to break the lease.

While it is best to seek a short-term lease at the start, attention should also be given to the idea that your firm might remain in the new location as it grows. Ideally, you should choose space that will be sufficient not only for your present and projected needs but which also provides access to additional nearby office areas that can be used in case your growth exceeds projections. In a large office building, an adjoining room or one a short walk down the hall could be perfect for adding space. The focus of the selection process should still be to find space with rental costs suitable to your income, however. For some consultants entering private practice, the amount of space is sharply limited by the funds available.

THE RIGHT LOCATION

Before we consider determining cost and space requirements, there are other factors worth considering. First, many people starting their private consulting practice look forward to having their office in an impressive-looking area. But how important is a prestigious location? The answer varies somewhat, depending on what kinds of visitors you might expect. Basically, though, it will not matter much to your clients whether the office is in the city's newest high-rise office tower or in a conveniently located shopping mall or professional complex. You are selling a professional service. As long as you have a clean, organized office, your prestige will come from the quality of the work you do, more than from any other sources.

Unless you expect your office to be visited frequently by major corporate or legal representatives, public officials, or news reporters, the best course of action is to look first for economy and convenience in a location, and second, for prestige. In most cases, a fancy address and lavishly decorated office will dazzle employees; but it will not do much to advance the new firm—it could even jeopardize your financial standing.

A second factor to consider in selecting the office is distance. Should you find an office close to home, close to your employees' homes, or close to clients? Each of these situations could have obvious advantages of its own, but it is quite unlikely you can find an office near all three parties. And who is to say that your clients and employees will not change or move at some future date? Aside from the highly unlikely prospect of moving your office every time a change in business occurs, there *is* a solution. The ideally situated office would be located near public transportation and major roads or highways. That makes it both easy to find and convenient to commute to for your clients, employees, and yourself, whether by car, train, bus, or plane.

HOW TO FIGURE COST AND SPACE REQUIREMENTS

The most crucial decisions involved in choosing an office have to do with renting the appropriate amount of space for your firm, based on income. This is especially true at the start of your private engineering practice.

when a major expense like rent can strongly affect the future of the firm. As was stated earlier, an appropriate estimate of moderate rental expense equals approximately 7 to 10 percent of the firm's gross sales.

Here is an example of how to figure cost-space requirements, while accounting for a growth projection for the firm:

Let's say that at the time you begin looking for an office, your firm has five persons and is taking in $150,000 a year in gross revenues. You have made a reasonable projection of 15 percent growth in sales per year and you intend to take a 3-year lease. With projected growth, the firm would be grossing about $230,000 by the time the lease expires.

One way to calculate rental cost requirements is first to take 7 to 10 percent of your present income and the same percentages of your projected income. In this case, the figures yield a range of $10,000 to $15,000 for current income and around $16,000 to $23,000 for projected income. A moderate cost then would fall roughly between the lower rental figure for projected income and the higher one for your present revenues, or approximately $15,000 to $16,000 per year. If you find it very difficult to obtain adequate office space at these prices, however, and your growth projections seem solidly ensured of success, you can consider a maximum rent equal to about 12 percent of your current income, or $18,000 per year. Conversely, for added safety, you might put the ideal rent figure at a low of about $14,000 per year.

In terms of employees, your income picture indicates that you might increase your staff by two or three persons during the 3 years the lease is in effect. There are many ways to arrange the office to accommodate the seven employees you expect to have. There are also various ways to figure out how much space you will need for them. One approach would be to shop around, consulting newspaper advertisements and rental agents to see what rates are currently popular in your area, then decide how much space you can reasonably expect to get for your money.

You will want to do this at some point in any event, but before you start shopping, you may benefit by using the common consulting guideline that predicts you will need 150 to 250 ft^2 of floor space per staff member (including storage space and room for conferences, visitors, etc.). Some firms have more than this and some have less, but it is roughly the average area allocated per person in most consulting firms. If you are expecting to build a staff of seven persons, then you would need 1050 to 1750 ft^2 of space, based on these estimates.

Now let's return to our cost requirements. In the example, we determined that a reasonable annual rent range would be between $14,000 and

$18,000 for an office which will eventually hold seven persons. Dividing the maximum cost by the smallest amount of space you will need and then dividing the lowest estimated cost by the largest amount of space needed gives a range per square foot of between $8 and $17.

Your chief sources of information about office space include rental and real estate agencies (many do not charge a fee to the office seeker), accounting and legal firms with which you do business, and the classified real estate section of local newspapers.

To give you a more detailed look at how a relatively small office can be used by a firm like the one in our example, Figure 1-2 shows the floor plan for a 625 ft^2 (58.1 m^2) office accommodating five persons (a proprietor, secretary, and three designers or drafters). The office shown would

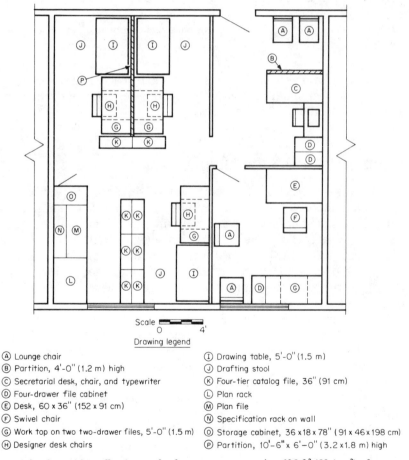

Scale
0   4'
Drawing legend

Ⓐ Lounge chair
Ⓑ Partition, 4'-0" (1.2 m) high
Ⓒ Secretarial desk, chair, and typewriter
Ⓓ Four-drawer file cabinet
Ⓔ Desk, 60 x 36" (152 x 91 cm)
Ⓕ Swivel chair
Ⓖ Work top on two two-drawer files, 5'-0" (1.5 m)
Ⓗ Designer desk chairs

Ⓘ Drawing table, 5'-0" (1.5 m)
Ⓙ Drafting stool
Ⓚ Four-tier catalog file, 36" (91 cm)
Ⓛ Plan rack
Ⓜ Plan file
Ⓝ Specification rack on wall
Ⓞ Storage cabinet, 36 x18 x 78" (91 x 46 x198 cm)
Ⓟ Partition, 10'-6"x 6'-0" (3.2 x1.8 m) high

Figure 1-2   Consulting office layout for five persons occupying 625 ft^2 (58.1 m^2) of space.

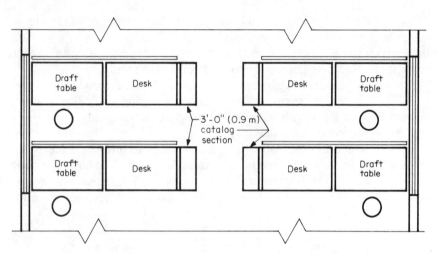

Figure 1-3 Floor plan section of design for office occupied by four drafters or designers (not to scale).

be considered a highly effective use of a small space. Figure 1-3 shows a basic layout that could be used to house four designers or drafters in an area of 350 to 400 ft^2 (27.9 to 37.2 m^2).

EQUIPMENT AND SUPPLIES

There are several factors to keep in mind while selecting office equipment. One important point is to see that you do not tie up an inordinate amount of capital in furniture and equipment which may not be vital for business. What sort of "solid" assets are truly vital to the consulting engineering firm? Obviously you will need basic office equipment such as a typewriter, desks, and telephone, plus some furniture for visitors. In addition, you will probably want to have storage cabinets and files for your correspondence, catalogs, and other literature.

But how much is enough? "Enough" is what is needed to satisfy your current basic business needs and the needs of your employees. Additional equipment can be purchased as the business expands. In the early stages of operation, a minimal amount of furniture will suffice. For instance, a secretary obviously needs a desk, chair, file space, storage cabinet, and typewriter. The reception area requires enough furniture for visiting clients and others to sit comfortably.

How expensive, durable, and attractive should the equipment be? Experience with many firms shows that a medium grade of long-lasting (5 to 10 years) furniture is quite sufficient. It should be in good shape and have a neat, serious appearance, not a far-out trend-setting stylishness (see Figures 1-4, 1-5, and 1-6, for example). And it need not be terribly expensive. For example, a typewriter should be electric, but it does not have to have an attached printer, video display terminal, and cost $10,000, as some word processing units do.

For some smaller supplies, such as pens, pencils, and rulers, some firms depend on the employees' personal items. Usually, though, it is best to expect the staff to bring very little to work other than itself. Except for calculators, drafting instruments, and pens which an employee might own and prefer to use over other equipment, the firm should try to outfit itself completely. Maintaining an adequate supply of essential office items will enable the firm to continue working smoothly even if an employee departs with her or his supplies or additional employees are hired.

Another factor to consider when selecting office equipment is the kind of impression you seek to create with visitors and staff. If a small firm stuffs its office with extremely expensive furniture, for instance, there is a good chance that some clients will wonder about the sobriety of the company's economic sense. Few clients want to deal with consultants who appear to spend extravagantly, so it is best to avoid making such an impression.

Some consultants feel that the best way to convey a classy, professional

Figure 1-4   Principal's office at Mueller Engineering Corporation. *(Photo International.)*

Figure 1-5 Partial view of design section at Mueller Engineering Corporation. *(Photo International.)*

Figure 1-6 Typical secretarial area at Mueller Engineering Corporation. *(Photo International.)*

image during the early stages of business (without overdoing it) is to use high-quality stationery and fancy engraved calling cards. Indeed, many law firms, architects, doctors, etc., do take this route—using expensive copy paper, 100 percent rag stationery, and so on. For a small consulting firm, this would be a sensible investment only if you plan to stay at the same address for several years because the main cost of printing the stationery comes from making new printing "plates" when you change items on the letterheads and cards. If you are fairly certain about staying at your location for a few years, an investment of several hundred dollars for impressive-looking stationery may be worth your while.

But generally, you do not need to go to great lengths to create an image for your firm. The quality of your work and your reputation will speak for themselves. There are a good many office accoutrements being advertised these days which may seem like they will build the firm's ego, but at the same time, they could damage the pocketbook. Perhaps the most popular items lately are word processing units, complex blueprinting machines, and superfast copiers. Obviously, these items could come in handy, if your firm really needs them. But it is rare for the newer consulting group to require state-of-the-art electronics in order to function efficiently.

If you do want to try these sorts of items, you may want to investigate leasing them from computer and office supply companies. Leasing is often a relatively minor expense, on a pay-as-you-go basis. Outright purchase, however, eats up your basic operating capital more rapidly. At the very minimum, leasing a computer, blueprinter, or copying machine will provide opportunities to experiment as you determine the firm's needs.

When a new firm does purchase supplies, it usually needs to establish credit arrangements with local office equipment dealers. These accounts can be worth quite a lot to your firm when clients seek documentation of your financial strength. Many clients will ask to see a financial statement and may want to know about businesses with which you have lines of credit. Since the number of credit accounts will be limited at the start of your firm, credit at office supply outlets may be a significant ingredient in building a positive financial statement to show to clients.

How much will it cost to outfit your office? Certainly, the days of opening an office for $500 or $1000 are over. Several inflation-fraught decades have passed since then, and they will most likely continue into the twenty-first century. We can make only some rough estimates here, and even these will be affected by interest rates and changing economic conditions in ways which we cannot predict now. In any case, the lowest figure a small firm could start with would be $5000. When you include all the ancillary items and general layout for an office like that in Figure

1-2, however, a more reasonable estimate jumps to double that, or $10,000.

This may sound prohibitively expensive, but it should not be so as long as you gauge financial resources wisely: Use leasing, loans, credit. Also, you can buy used furniture and equipment to cut costs while starting your business. As you hire new staff and expand your operations in the ensuing years, the older furniture, drafting tables, and other items can be provided for members with lower seniority, while employees with more time and experience with the firm can be rewarded with newer equipment.

INSURANCE COVERAGE

Now it is appropriate to examine briefly several kinds of insurance with which you should be acquainted when you start consulting. Further details on insurance and how it relates to PE practices will be presented in Section 5.

Professional Liability

Professional liability insurance (otherwise known as "errors-and-omissions" coverage) protects a firm from lawsuits which may be connected with errors made by the firm's consultants on a particular project or which could become a source of future claims brought by clients or others against the engineering firm.

How much of this type of insurance does a new firm need? There is no fixed answer, but one should be able to arrive at a reasonable maximum estimate for lawsuits which could potentially be lodged against a firm like your own by inquiring with PE societies and fellow engineers who have been in private practice for some time. Fortunately, lawsuits of this type are not widespread; but someone among your colleagues will probably have familiarity with liability issues and can most likely provide information on your area of engineering and any special conditions which may apply to firms practicing in your part of the country or state. As a guide, it is not unusual for medium-sized firms (those with 10 to 15 members) or even small firms in the architectural or structural fields to start with liability policies having total values of $500,000 or even $1 million.

Sounds outrageous? It's not. The reason for such seemingly high levels of coverage is, of course, that a lawsuit for errors or omissions can be most costly. Structural or architectural errors may involve sizeable losses to a client and can affect large parts of a building or other structure. If a client does attempt to sue, $500,000 and more is not an uncommon settlement for the client to seek.

In engineering disciplines other than those affecting public safety and large structures, however, coverage may be lower. Usually, the gross value would be in the range of $100,000 to $500,000. Many consultants feel that the risk of lawsuits for them is negligible, but one should be wary. It is difficult to predict what type of work one's firm may eventually engage in and what sorts of risks can arise. Therefore, it is best to be prepared, even though lawsuits seem unlikely to occur.

What sort of premium can you expect to pay for liability insurance? It is not feasible to quote actual costs for specific cases here, but an estimate for a $500,000 policy with a $5000-deductible clause would be about $5000 per year at the time of this writing. Like other types of insurance coverage, liability coverage has a deductible amount, usually about $5000. Premium costs can be reduced by asking for a higher deductible, up to $10,000 or even more.

Automobile Insurance

In most firms, the principals and employees will be driving their own cars to and from work. Travel costs incurred while on the job are usually paid for by the firm on a per-mile basis. The car owners generally have their own automobile insurance policies to cover most liabilities which could occur during pleasure or business use. But the company's owners and principals are in a different situation than most employees: They may claim most of their driving as business-related; therefore, they should have policies that specifically cover business uses.

In addition, the firm can buy a "floater," or excess coverage, policy, which will protect against claims that may exceed in value the amount of coverage provided by the policies held by individual car owners in the firm.

Property Insurance

Insurance coverage should be obtained for the following four areas if you rent or lease your office: fire, theft, valuable papers, and lost time arising from business interruption.

Fire, theft, and valuable papers insurance. The large quantities of paper used by engineering firms and the possibility of fires starting from various sources makes fire insurance a must. Theft coverage is similarly important. Fortunately, neither is highly likely to occur. The coverage costs are consequently rather low, compared to some other kinds of insurance.

Valuable papers are a different matter entirely. Losing blueprints or other plans can be an extremely serious and costly casualty. Drawings, reports, and other documents can be worth thousands of dollars and represent hundreds of hours of work. A simple accident like a window being left open to rain could destroy such materials, or a print could be damaged while passing through a blueprint machine. You will probably help protect yourself against such losses by duplicating documents while work is in progress, but it can still cost you hundreds of extra dollars to copy documents in an emergency. Because of the risks involved, many firms obtain coverage valued at rates as high as $10,000 or even $50,000 for valuable papers.

Business interruption insurance. Suppose there was a fire in the building you occupy. You might be relieved, initially, to find that the fire did not reach your office; but the threat to your business might not end there. Quite possibly, you will not be able to use the office for several days, while the building's owners try to investigate the cause of the fire and to work to restore services to the structure, such as power and ventilation. Unfortunately for you, your payroll will continue. And clients will expect their work on the date originally agreed upon.

Business interruption insurance is designed to help you handle such a situation. All firms should look into obtaining this protection against "loss time," especially those whose offices are located near establishments which have substantial fire risks, such as restaurants. Of course, fire is not the only potential cause of time loss. There are others including grave illness of a member of the firm and natural disasters.

Insurance against the Death of a Key Member or Partner

Besides causing grief and a sense of personal loss, the sudden departure or death of a partner or key member of the firm can affect business severely. A death, particularly, can exert great strain on the company's principals. At the same time it adds to the stress experienced by surviving relatives of the deceased.

The following case illustrates the kind of situation that can develop.

Two partners formed a structural engineering firm and took out a $50,000 insurance policy on each other about a year later. The policy provided that if one of the partners died while they were in business together, the partner's widow would get $50,000 plus a reasonable share of the remainder of her husband's earnings. The surviving partner would keep the business as his own. Two years later, the partners decided to raise the coverage to $150,000. Then, 5 years after they had started the firm, one of the partners did die. The task of determining the deceased partner's share of the business then went to accountants, who estimated that share—including cash, accounts receivable, and work in progress—at $300,000. The widow was amicable and understanding but was not about to settle for $150,000. After extended deliberation, the surviving business partner agreed to pay the balance of the $300,000 total over a 5-year period. He determined that he could, with great effort, manage to bring in $100,000 gross revenues per year to help himself comply with this schedule. Out of this $100,000, however, he would pay nearly $50,000 in taxes of all kinds, with $30,000 going to the widow. That left about $20,000 for support of his family, living expenses, and college tuition for two children.

In this case, the insurance coverage was just barely adequate. But it certainly helped. Grim as such examples may sound, there is hardly any better option for protecting the firm against the stresses that follow a death or sudden departure. The consequence could be worse if the firm did not have such protection.

Medical Insurance

Coverage such as Blue Cross, Blue Shield, and major medical policies are also crucial. They are not inexpensive, however. The premium may approach a total of $1000 per year for each member of the firm.

The major medical policy usually requires inclusion of a minimum amount of life insurance. Whether you pay part or all of the total premium yourself, having the coverage will be a factor in attracting and keeping personnel. A group purchase plan is the most economically convenient arrangement in most cases.

THE ACCOUNTING SYSTEM: TOOL FOR EFFECTIVE BUSINESS

Several considerations of equal importance come into play when you select and install an accounting system. The first is that you, as the head of a new or growing firm, will be directly responsible for the business.

Regardless of whether you will be employing someone to do the accounting for you, you need to develop a firsthand, working knowledge of both the accounting system and the flow of income and expenses. The best time to gain this knowledge is early in the firm's development, when there are fewer items to detail. Then you can assign accounting responsibilities to others and still be able to oversee your financial progress skillfully as the firm expands. Once you acquire familiarity with your accounting system, you will be able to look at the results of other people's accounting with a "practiced eye"; you should be able to recognize problem areas or discrepancies quickly.

A second consideration is that you should find an accounting system that is simple enough to understand, so it does not seem like a foreign language. If a client or government auditor finds your books a problem to interpret, the result will likely be displeasure for all concerned. The system therefore should be able to accommodate expert scrutiny, as well as the firm's growth, without needing extensive alterations.

A well-run consulting firm will have one or more people who are able to glance at a financial statement to determine the health of the firm, based on its income and expenses. Accordingly, an effective accounting system readily highlights major areas of income and expense. At your disposal is one invaluable aid: the professional accountant. She or he can help you set up a system suited to your particular operation.

Exhibit 1-3 shows a typical cash-basis income and expense statement (the example contains rounded figures to simplify computations). Alternatively, accounting could be done on an accrued basis instead of cash, in which case the values of work in process (i.e., work which you have performed but not yet billed) would be integrated into the direct cost area with its beginning and ending values. The accrued basis is an option that you and your accountant may want to explore, but we will use the cash-basis system for illustration here.

There are many ways of designing your records. The statement shown here is not intended to be seen as the best alternative. You may actually decide to list some items under wholly different headings from those shown in Exhibit 1-3.

Usually, the simplest method is to maintain a basic series of monthly ledger sheets, along with a series of additional sheets, each having an individual heading—for example, travel and office expenses. The most common kind of entry ledger is a three-column, lined paper which can be purchased at almost any stationery store. It can be set up as in Exhibit 1-4.

Update your ledger each month. If financial conditions are strenuous for some reason and the firm has very little money in the bank, however, more frequent updating is advisable.

EXHIBIT 1-3 *Income and expense (in thousands of dollars).*

Fee income			$165
Deduct cost of fee income			
Direct labor		$65	
Job supplies and expenses		5	
Special consultants		3	
Overhead expenses			
Clerical payroll	$10		
Administrative payroll	4		
Indirect payroll	6		
Car expense	5		
Depreciation	1		
Office expense	8		
Insurance	5		
Professional fees	2		
Dues, licenses, and publications	1		
Rent	15		
Payroll taxes	6		
Property taxes	1		
Telephone	3		
Travel expense	4	71	
Total			144
Gross profit			$ 21
Deduct other expenses—net			
Interest income		1	
Reimbursed expense		1	
Advertising		2	
Interest expense		3	
Promotional dues		1	4
Net profit before taxes			17
Federal taxes		2	
State taxes		1	3
Net profit			$ 14
Retained earnings			
Previous year			3
Total retained earnings			$ 17

EXHIBIT 1-4   A typical monthly ledger sheet.

ABC ENGINEERING COMPANY
July, 19XX

National Bank

(in thousands of dollars)

Date	Check no.	Item	Debit	Credit	Balance
7/1					50
7/2	5000	American Rents	2		
7/3		ACE Mfg. (A.C. layout)		4.5	

Your bank will send your account statement and canceled checks monthly, at which time you should compare your ledger and make any necessary corrections. Resolving discrepancies between the bank statement and the ledger as soon as possible is extremely important to you, your accountant, and state and federal tax agencies.

You will notice that Exhibit 1-3 shows two major categories of expenses: Overhead and Other Expenses. The reason for this delineation is that government agencies sometimes do not allow items like promotional costs to be classified as overhead. In an audit, you would most likely be asked about your overhead as a percentage of direct labor. In this example, direct labor is $65,000 and overhead amounts to $71,000, or about 110 percent. Such a value may be considered higher than normal, and the best evidence you can present to show that the estimate is accurate is the information contained in your records. If you were to include the "other expenses" with overhead, the overhead percentage would be even higher, and it is quite possible that an auditor would not permit you to include those items in that category.

Labor, however, should usually be divided into four types: direct, indirect, administrative, and clerical. Considered individually, they can be useful indicators of your firm's financial health. In essence, the categories are defined as follows:

• **Direct Labor:** Amount of time spent by you or your employees on a particular project (whether billable or not)

• **Indirect Labor:** Amount of time spent on any office task that is not billable to a client (includes time off from actual work for personal tasks, coffee breaks, and other time not devoted to work performance; it is sometimes referred to as "nonproductive effort" or NPE)

* **Administrative Labor:** Amount of time that you and any other partner(s) or principal(s) devote to managing and operating the firm (you may wish to include sales in this category, but many firms list sales separately)

* **Clerical Labor:** Amount of time used for secretarial work

Each employee's hours can be classified as either direct or indirect labor, depending on what tasks are performed at a given time and upon the circumstances under which they are performed. However, the firm's owners and principals may actually do some work in each of these categories—direct, indirect, and administrative. A more detailed discussion of these categories will be presented later. But let's take a brief look now at how an administrator's time might be classified.

Figure 1-7 shows a typical company time card, filled out by our consultant friend, John Jones, the principal owner of Jones Engineering. Mr. Jones worked 40 hours in the week shown on the card. His time was divided as follows: direct labor, 21 hours (12 hours on the Smith job and 9 on the Brown job); administrative labor, 16 hours (9 hours office management and 7 hours selling); and indirect labor, 3 hours (here, Jones calls it nonproductive effort—NPE—to distinguish this time from billable and administrative time).

Now, if you were to take a 52-week collection of time cards for both Mr. Jones and his employees, you could find the total amount of direct labor, indirect labor, and administrative labor for the entire year. Suppose, for example, that Mr. Jones pays himself $25,000 a year. At 40 hours per week, there would be 2080 hours in the entire year, including

Employee name _John Jones, Principal_

Week ending: Month _July_   Day _10_   Year _19- -_

Project name	Sunday	Monday	Tuesday	Wednesday	Thursday	Friday	Saturday	Total
Smith Job } Direct Labor		3	3	2	1	3		12
Brown Job		2	1	2	2	2		9
Office Management } Administrative Labor		2	1	2	3	1		9
Sales } Indirect Labor		1	1	1	2	2		7
Nonproductive effort (NPE)			2	1				3
Totals		8	8	8	8	8		40

Figure 1-7 Typical weekly time card for consulting engineering employee.

holidays, vacations, and NPE (yearly holidays and vacations are generally considered to be NPE). Mr. Jones has 300 hours total NPE time. He spends 1000 hours working directly on clients' projects and 740 hours on administrative duties. We can thus determine the percentages of total time devoted to each type of effort for the year by dividing the number of hours in each category by the 2080 work hours. Indirect labor constituted about 15 percent (300/2080) of the year's time, direct labor about 50 percent (1000/2080), and administrative effort about 35 (740/2080).

In discussing his income taxes with his accountant, Mr. Jones might now request that 15 percent of his salary, or $3750, be charged to indirect labor; 50 percent, or $12,500, be charged to direct labor; and 35 percent, or $8750, be charged to administrative labor. Next, the sum of Mr. Jones's time and that of his employees can be broken down into the proportions of the total work effort used in each type of labor.

These kinds of calculations can help you in several different ways. Besides providing easy-to-understand details for your financial statements to clients, it can tell you the percentage of total hours spent in each type of labor and its overall cost to the firm. The proportion of time used for indirect labor, particularly, provides a measure of the firm's efficiency. Using this method, you can actually see how much administrative effort you have expended to bring in a given amount of business in a specific time period.

PUBLICIZING YOUR SERVICES

After you finalize the basic operational plans of your firm, such as accounting, management, and financing, attention can be turned to publicizing your services. The first step involves announcing the opening of your business. The best and perhaps most common vehicle for the initial announcement is a professionally printed letter or postcard. Your local printer should have examples on hand for format and style.

One card should be sent to each potential client, to contractors, suppliers, banks, other engineering firms and fellow engineers, friends, accountants, lawyers, etc. You may develop a list of 200 addresses, but it is probably best to print 500 or more announcements. Many firms of modest size send out more than 500 cards. If you do put some extra effort into developing an exhaustive list, your chances for attracting clients early in the firm's life will be increased.

The cost of printing each card may be lowered by printing a large supply, but the overall expense will not be tiny after you include the cost of producing and mailing the announcements at first-class rates. This is one start-up cost which will not be repeated, though. It should be viewed as an investment in your future. It is also good to have some extra cards remaining after you do the main mailing since you may come upon names to add to your mailing list as time goes on. Nothing precludes your sending out additional announcements, even if your office has been open for several months.

Mail announcements should be followed up by supplementary letters, telephone calls, and selected visits to potential clients within 2 or 3 weeks of mailing. What to say and how to present yourself on such visits is a subject that Section 2 covers in detail. The importance of the follow-up cannot be overemphasized. Addressees can dispose of an announcement card arbitrarily, forgetting that they had ever received one. But a phone call or visit will confirm that you are indeed interested in developing professional contacts—that you mean business. Moreover, you will have a much better opportunity to describe your services and to "sell yourself" during the follow-up call.

There is at least one simple ploy you can depend on to help introduce yourself in the follow-up, just in case you do not have a sudden creative sales inspiration. Simply introduce yourself and tell your respondent that you were wondering if she or he received your letter. You may get some surprisingly blunt replies. But you will at least make sure that the addressee knows about your firm. This may not necessarily lead to gaining an instant client, of course, but it will spread your name effectively and help you secure goodwill to start your firm along the path to success.

Besides sending announcements by mail, many firms prepare a brief one- or two-page press release for local newspapers and professional organizations. These may include photos of the firm's owner(s). The newspapers may be quite happy to print your article, so it is well worth your while to write the release and send copies to as many publications as possible. A press release could read like this:

ABC Engineering announced that it will open an office providing professional engineering services in structural and mechanical engineering and related design at 170 Oxford Street, Suffolk, beginning March 1.

John Jones, ABC's owner and chief executive officer, said the firm will provide "a comprehensive approach to high-quality engineering services with a staff of experienced professionals."

Mr. Jones formerly worked as chief engineer for XYZ Corporation for 10 years and has a broad range of experience in all aspects of mechanical and structural engineering practice. He is licensed as a professional engineer in New York and New Jersey and is a graduate of New York University's School of Engineering.

CONSULTANT'S PROFIT KEY 2

Charleston C. K. Wang and Harold E. Parker, of the law firm of Groeber, Wang & Southard, have the following hints on serving as an expert witness in a variety of courtroom situations. Their hints were first published in *Chemical Engineering* magazine and follow:

You may never have been called to act as a witness in a trial or a hearing, but you never know when it might happen. You are good at your work, but will you be any good at expressing yourself as an expert witness? You should know what will be required and how you should present yourself.

The nationwide increase in product-liability lawsuits, agency rulemaking, legislative hearings, and other adversarial proceedings has created a growing demand for the chemical engineer to serve as expert witness. You may have heard dramatic anecdotes about brillant coups and humbling debacles experienced by engineers on the witness stand. To be an expert witness, especially for the first time, is an interesting, challenging, and sometimes frightening occasion. If done properly, the job can be personally satisfying and financially rewarding.

WHY HAVE AN EXPERT WITNESS?

An observer of an event is qualified to testify as a witness because of what he or she has seen. This person is called an eyewitness.

The expert witness, on the other hand, contributes something else: By virtue of his or her *specialized knowledge,* the expert witness has the ability to express *opinions* or make *inferences* beyond the knowledge or experience of the average person. The content of the expert witness's testimony need not be from direct observation of events. For this reason such a witness may be asked to comment on a hypothetical set of facts. Of course, an expert witness can also have first-hand knowledge of the facts of a case through direct observations; if so, he or she is allowed to describe what was observed and to give *technical inferences* from those observations.

Very often, the outcome of the controversy, especially if it is a scientifically complex one, will hinge upon the results of the battle between expert witnesses.

IN WHAT FORUMS DO WITNESSES WORK?

To be an effective expert witness, you must analyze the objectives of the forum in which you are testifying.

For instance, in a *civil trial,* the forum seeks first to determine the truth, and then to decide the legal rights and duties of the parties. In a *legislative hearing,* the forum seeks to determine the broad facts comprising a social controversy, in preparation for the drafting and passage of laws. In *administrative agency hearings,* the forum seeks to determine detailed facts that are beyond the expertise of the legislature, in preparation for promulgating detailed rules and regulations that have the force of law.

In all of the above proceedings, the expert witness testifies for a party who has a material interest in the outcome of the proceeding. You must understand the objectives of the party you are testifying for.

THE CIVIL TRIAL

Of the forums described, the civil trial probably gives the expert witness the greatest challenge. This is because a trial is a fully adversarial proceeding— in which opposing parties, through their attorneys, engage in logical combat, with the expectation that one party will prevail over the other in the end.

In addition, questions of fact, including scientific fact, are decided by a panel of jurors drawn from the local voter roster. This panel tends to consist of citizens who are not as well-versed in science and technology as the expert witness.

Because of the difficulties found in a civil trial, it will be used here to illustrate the techniques for being an effective witness.

The most important thing to remember from the start is that, in a civil trial, your job as an expert witness is to *help* the jury *understand* the technical aspects of the case, and to *persuade* the jury to accept your explanation of the technical facts. If you can master the civil trial, the other forums will be easily handled, as the issues are more diffused, and the outcome is less clear-cut. In addition, these tribunals usually have a greater grasp of the subject matter, especially if you are before a regulatory agency.

To be effective you should be credible, correct, clear, concise and candid. Let us now look at each of these five Cs in turn:

CREDIBILITY

Once admitted by the judge as an expert witness (something routinely done if you are a degreed engineer or have engineering work experience), you must set out to win the trust and confidence of the jury.

To this end, before you even step into the witness box, you will need to find out about the jury. Gauge its capacity for understanding scientific material. After your attorney has completed jury selection and made the opening statement, he or she can tell you a lot about the peculiarities of the jury. While your attorney is not the foremost technical expert on the case, he or she is skilled at evaluating and persuading people. Work with him or her to build an effective team that wins trials.

After you are seated in the witness box, your attorney will ask you a series of prepared questions about your professional qualifications, for the benefit of the jury. Answer by presenting your degrees, accomplishments and honors modestly, with minimum embellishment. Remind yourself that you are speaking to a group of average people, some of whom may harbor an anti-intellectual bias.

The way you answer gives the jury its first impression of you. This impression can very often be a lasting one. If you are accepted here, what you say subsequently will be accorded great weight. If you are not, the jury may unconsciously reject your testimony, in preference to that given by the opposing expert, even if you deliver a scientifically brilliant lecture.

Never forget that your job is to help the jury figure out the scientific facts on the jurors' own levels of understanding. To appear credible, do not over-play the "expert," but concentrate on being the "witness." A healthy dose of good humor and patience is guaranteed to enhance the receptivity of the jury to your opinions. However, do not go overboard by being conden-scending or self-effacing. Also, do not allow yourself to become overly cas-ual, but seek to retain a balance of formality. Jurors, after all, expect to see an expert. Consistently maintain your demeanor throughout, whether it be during the direct examination or the cross-examination.

CORRECTNESS

Needless to say, the expert witness must be correct in what he or she says. Do your technical homework. Be sure you are up to date. Prepare notes and commit the facts to memory. However, when you are actually testifying, you should be spontaneous and avoid notes.

Anticipate what the opposing expert witness can say. Look for weak-nesses in your theory, as well as the weaknesses of the opposing expert's theory. In modern lawsuits, surprises are discouraged through the use of pre-trial depositions. The attorneys take the testimony of the expert wit-nesses in advance of the trial. A transcript of the sessions will be prepared. You will have a chance to review your own testimony for correctness, as well as the testimony of your adversary. If you find an incorrect statement, make sure you are the one to be correct at the trial. More and more depo-sitions are taken on videotape in jurisdictions that allow them.

The factual correctness of your testimony will serve you well during cross-examination. If you have made the correct statements, the opposition will be hard-pressed to attack your testimony. It is at these times that you will thank yourself for having gone to the library to recheck your informa-tion that one last time.

CLARITY

In addition to having a pleasant demeanor and the correct facts, the expert witness must speak clearly. This means clarity in both the actual physical delivery of the testimony and the content of the testimony. Look directly at the jurors and speak audibly, slowly and with much thought. Have your voice project confidence and scholarship. Avoid technical jargon at all costs—something an expert may find difficult to do—otherwise the jury will lose all interest and understanding. They will end up irritated and confused.

While seeking clarity, do not become condescending. Use words you think the jury will understand and appreciate. As you can see, this may depend on your skill at sizing up the jury; often this is not easy, but it will come with experience.

Obtain a set of prepared questions from your attorney and practice answering them in front of your spouse, secretary or whomever you can hold captive in confidence. If they cannot understand what you say, it is likely that the jury won't either. Keep practicing until a number of different captive audiences understand you. As a safeguard, have your listeners explain back to you what you have tried to impart.

CONCISENESS

To the expert wishing to explain a point, conciseness may appear to be the antithesis of clarity. A balance must be struck. While it may take more words to ensure clarity, never wander from the point. Use the necessary number of words to explain a technical point and then stop. Give your attorney the chance to move on. Do not go back over a point, as it tends to give the impression that you are unsure. Never volunteer more facts than those asked of you. Never venture beyond the bounds of your qualifications, for if you do, the opposition will surely pounce on you and make you look the fool.

During cross-examination, do not get into a verbose exchange with the opposing attorney. Simply answer the question he or she has asked. Occasionally, the cross-examining attorney may become overly aggressive and unaccommodating. If this happens, your best strategy is to exercise patience coupled with firmness of reply. If the scientific data permit it, maintain your original position, although you may wish to concede just a little in order to get the hostile attorney to move on.

If the hostile attorney tries to lead you into a trap of logic, step back, pause to gain perspective, then explain why his or her line of questioning is scientifically faulty, and state what the correct question should be, giving the appropriate answer. Do not expect him or her to accept your answer as correct; it is foolhardy to try. The opposing attorney is not the one you must convince; direct your best statements succinctly and confidently to your jury.

It must be said that cross-examination is the most demanding and difficult part of being an expert witness, and the skill of fending off a hostile attorney perhaps can only be learned first-hand. Nevertheless, it may be comforting to know that your own attorney is poised, ever ready to object to unfair tactics.

CANDOR

We have come full circle to a matter of demeanor once again. To be candid means to be natural and honest. To be candid, you must be psychologically prepared; you must be relaxed and self-assured. You must not become defensive or bitter.

If the opposing expert witness or attorney succeeds in scoring a point against you, accept their victory gracefully. As an expert, you know that technical opinion often can take on a spectrum of shades and that there may not be a singular correct answer. Let the jury know that your point has validity of its own, despite the opposing statement. Try not to appear flustered, stubborn or recalcitrant; to do so is to concede your point.

Be an eternal optimist but, at the same time, show respect for the opposing witness and his or her attorney. Remember that while it is a battle of experts, you do not win by slaying your adversary or by making a fool of him or her, but you win when the jury chooses to accept your testimony over that of your opposition.

PRACTICE, PRACTICE, PRACTICE

While this maxim is constantly repeated by seasoned attorneys to novice attorneys, and by law professors to law students, its wisdom is equally applicable to the expert witness. Work with your attorney to prepare for the inevitable cross-examination. If you have time, attend an ongoing trial to observe the tactics and practices involved. Experience the heat of battle from a distance first so that you go in adequately hardened.

Ask your attorney for a prepared set of questions that he or she plans to use at trial. Practice answering them alone, practice on your captive audiences, and practice on electronic media. Use a cassette recorder at the minimum, and if you have video equipment, use that. The better law firms specializing in litigation will do one or a number of mock video sessions as a matter of routine. (In major cases, firms are known to assemble a mock jury for maximum realism.)

Your actual delivery at trial will be the encore performance.

SUMMARY

Section 1 provides a broad perspective of the major practical areas to be faced by the PE who wishes to become a consultant.

There are a great many considerations that are vital for success in the early stages of the private practice of engineering. Start-up should be preceded by a thorough evaluation of one's personal resources and goals. Substantial planning is needed for choosing the service the firm will render, organizing and managing the business, financing the firm's activities, meeting legal requirements, budgeting, and opening the office. Further plans are needed to make market projections; find prospective clients; set fees; and insure the firm against potential injury from interruption of business, theft, and professional error. Remember that:

• Private engineering practice requires individuals who enjoy working independently. The ideal consultant is willing to venture into a business world where extra effort and resourcefulness can bring more money and more personal fulfillment than most people find in working as salaried employees of large engineering firms.

- The licensed engineer usually has a distinct advantage over the nonregistered engineer; but there are some areas of consulting where the PE license is not required.

- Organizing the firm properly is as important as performing the engineering work itself.

- Start-up costs can vary, as can earnings, in the new firm. But most consultants can expect to spend at least $10,000 to open an office and get the business going.

- Every firm benefits from diligent record keeping. This can be a tedious exercise during the early phase of the business, but there is no better time to learn about your firm's accounting than when the company is still small. Once you set the accounting foundations, you can turn your attention to consulting, hire a skilled staff, and watch the business grow.

section

2

How to Obtain Clients for Your Consulting Practice

Good self-evaluation is as essential to building an engineering practice as it is to the decision to become a consultant. The fundamental tool you can use in building a power-packed list of happy clients is thorough knowledge of your firm's capabilities. This requires an appraisal of the professional performance and abilities of each member of the firm and of the collective team as well.

KNOW YOUR FIRM

To help you appraise your consulting engineering firm—either as you plan to run the firm, or as it exists today—we present an objective five-step appraisal procedure in this section of the Handbook. Certain parts of this appraisal procedure apply to new firms; other parts apply to existing firms. Answer each portion of the appraisal as it applies to you now. Use the other portions of the appraisal as a guide to what you might do in the future once your consulting firm is well established.

THE OBJECTIVE APPRAISAL

Start the objective appraisal of your firm—either new or established—by (1) having each member of the firm rate his or her consulting abilities, (2) having each member rate every other member of the firm on the same basis, and (3) having an informed discussion as a group about members' strengths and weaknesses and the team's overall capabilities. If criticisms are needed, have them submitted anonymously on paper.

The following list gives brief descriptions of major themes you might consider during your objective appraisal:

1. Exactly what types of work is the firm and each of its employees equipped to do? See Consultant's Profit Key 3 for a practical answer to this question.
2. Where is the market for your services (in terms of location, economics, engineering disciplines, etc.)?
3. What are the most efficient and viable ways to acquaint people with the firm's services? Who is best qualified to publicize the firm, and what can the team as a whole do to secure clients?
4. What sort of schedule and procedure should be designed to promote your consulting business?
5. What kinds of clients will make the best use of the team's full range of capabilities? How can you get a good client mix that will provide the firm with a broad base of support?

CONSULTANT'S PROFIT KEY 3

Roy Peterson, Creative Research Services, Inc., prepared this key for *Consulting Engineer* magazine under the title "Inventory Your Staff's Strengths." At the time he wrote it, he was Research and Planning Specialist at Barr Engineering Co.

One of the best investments a firm can make is a detailed inventory of its staff's qualifications—covering professional, other technical, and administrative employees. When a firm reaches from 30 to 50 employees or more, it should seriously consider the systematic inventorying of information that describes in useful detail the educational background, working skills, and project experience of the staff.

Without doubt, employees represent a firm's greatest asset. Nevertheless, it is fairly typical for data on employee qualifications to be incomplete and not up-to-date. The personnel files may contain some historical information on educational credentials and prior work experience, perhaps with more recent documentation of engineering registration or of an advanced degree.

Marketing files may include resumes of professional employees, but these, too, probably need updating.

Generally speaking, a small firm (e.g., 5 to 25 employees) seems able to function with a minimal amount of formally organized information on its employees. This reflects the close, daily communication in a small firm, which serves to convey new information and to reinforce prior knowledge of the educational and vocational activities of its staff members. But as a firm begins to outgrow its small status, it needs to document employees' qualifications in a systematic manner.

Some four years ago the president of Barr Engineering Co. decided that the time had come for a personnel qualifications inventory. The firm had grown severalfold from the small group of civil engineers who incorporated in 1966, and was reaching out for larger markets for its consulting services. The soundness of the decision has since been confirmed by further growth to an overall staff of 70 people, including more than 40 professional engineers and technical specialists who provide diversified services in the fields of civil engineering, resource planning, and environmental sciences.

When a literature search in 1978 indicated that no off-the-shelf forms and procedures were available for developing such an inventory, a special administrative project was initiated to develop forms and procedures appropriate to the firm. The author became involved in the project in mid-1979 and has since followed through with its final implementation.

GATHERING THE INFORMATION

The basic tool used in the Personnel Qualifications Inventory (PQI) is a detailed survey form. In its current version, the form has 278 specific items, including 69 relating to education, 53 descriptive of working skills, and 156 covering topical aspects of project experience.

The PQI operates on the honor system. A set of survey forms and instructions is distributed to the employee, who fills out the form entirely on his or her own. Only responses that seem grossly out of line are checked with the employee to verify their accuracy. The respondents are assured that all survey data will be used in an appropriate, responsible manner for maintaining the inventory of personnel qualifications.

The three major survey categories are broken down into the following subcategories (the number of specific survey items are shown in parentheses):

- Educational background (69): natural sciences (25), engineering disciplines (24), other disciplines (20).
- Working skills (53): field work (6), analytical methods (11), computer models and programming (8), general office (6), communication (8), business development (8), project supervision (6).
- Project experience (156): field data collection and investigation (7), data/information systems and computer applications (8), hydrological studies

(17), water supply evaluation and design (11), pollution control and waste disposal (19), resource planning and management (29), environmental monitoring and analysis (10), geotechnical analysis (13), municipal engineering (9), hydraulic and structural design and construction (33).

A numerical code is used by the survey respondents. A number from 0 to 5 is selected for each item to denote level of educational attainment, level of proficiency in working skills, or level of project experience.

The level of educational attainment in specific academic disciplines is based on the following:

- 1 = Some coursework (1 to 12 quarter credit hours), or equivalent.
- 2 = Minor specialty (more than 12 quarter credit hours), or equivalent.
- 3 = Major specialty: bachelor's degree, or equivalent.
- 4 = Major specialty: master's degree, or equivalent.
- 5 = Major specialty: doctor's degree, or equivalent.

The term "equivalent" takes into consideration educational contributions from noncollege coursework and self-teaching. The survey participant adds the letter "E" after the selected code number to indicate the use of the equivalent option. It was found from the initial survey results that the E option had been used with reasonable discretion; hence, the E responses are treated in the same way as the standard responses in evaluating and applying the PQI data.

Level of proficiency in working skills is based on:

- 1 = Some familiarity/limited experience.
- 2 = Useful know-how.
- 3 = Moderate proficiency.
- 4 = Well-established skills.
- 5 = Expert.

Similarly, numbers denote project experience in specified areas:

- 1 = Limited (less than one year).
- 2 = Significant (one to two years).
- 3 = Substantial (two to four years).
- 4 = Extensive (four to eight years).
- 5 = Very extensive (more than eight years).

The survey results were used to develop a series of matrices of the kind illustrated in Exhibits 2-1, 2-2, and 2-3. The basic results were analyzed further to derive general profiles of educational background, working skills, and project experience. Exhibit 2-5 illustrates a portion of the general profile that was developed for engineering disciplines within the educational background category. The table shows the distribution of code numbers selected

EXHIBIT 2-1   Educational background: engineering disciplines.

Staff member	Civil engineering	Structural	Water resources	Fluid mechanics	Soil mechanics	Sanitary	Hydromechanics	Geotechnical	Environmental	Engineering mechanics
John Jones	4	1				1	2			
Susy Carter	4	3	4E	4	3	3	4	4E	3	3
Robert Doe	2			1	1			1		
William Davis	3	2	1	1	1	1				
Sam Smith	3	1	1		1	3E			3E	
Wilson Brown	3	2		1	1	1				

EXHIBIT 2-2   Working skills: communication.

Staff member	Tech. writing	Tech. editing	Audio/visual	Prof. societies	Public relations	Expert witness	Court presentation	Public info. & educ.
John Jones	4	3				3	3	
Susy Carter	3	3	2	3	3	4	4	2
Robert Doe	4	3	3	1		1	1	
William Davis	2	1						
Sam Smith				3	3			5
Wilson Brown							2	1

EXHIBIT 2-3   Project experience: geotechnical analysis.

Staff member	Geotech. data	Seepage analysis	Slope stability	Struct. anal. tailing basin	Lab anal. soils & rocks	Struct. anal. earth dams	Soils & minerals excav.	Foundation analysis	Computer earth struc.
John Jones									
Susy Carter	1	4	2	4	2	3	1	1	1
Robert Doe	2				1				
William Davis		1					1		
Sam Smith					4				
Wilson Brown									

EXHIBIT 2-4   Summary of professional staff qualifications.

Staff member	Surface water hyd.	Ground water hyd.	Water supply sys.	Flood studies	Hydr./hydraul. models	Hydraulic engrg.	Dam inspec./repair
John Jones	●		○	○	○	○	○
Susy Carter	●	●	○	○	○	●	●
Robert Doe	○	●	○		●	○	○
William Davis			○			○	
Sam Smith	○			○	○	○	○
Wilson Brown							

● *Major qualification;* ○ *minor qualification.*

by professional staff respondents for specific disciplines such as civil engineering. A series of tables was developed to show distribution of numbers for all items.

EXHIBIT 2-5 *General profile of educational background: engineering.*

Discipline	Level of attainment code					
	0	1	2	3	4	5
Civil engineering	9	0	1	21	9	1
Structural engineering	15	6	12	6	2	0
Water resources engineering	14	13	5	5	4	0
Fluid mechanics	12	15	9	2	3	0
Soil mechanics	11	21	2	4	3	0
Sanitary engineering	13	16	9	3	0	0
Environmental engineering	20	10	8	3	0	0
Geotechnical engineering	27	7	1	1	5	0
Hydromechanics	30	5	1	0	4	1

A simple procedure was developed for combining the responses received for each survey item into an overall qualifications index. The number of responses received within each code category for a given item (as shown in Exhibit 2-5) was multiplied by the applicable code number (i.e., 0 to 5); the sum of these weighted responses for a given survey item is defined as the qualifications index. Since the code numbers increase to denote higher levels of education, skills, or experience, larger values of the index denote relatively higher overall qualifications. As illustrated in Exhibit 2-6, the values of the qualifications index were ranked within each subcategory; in the example shown, the values apply to specific areas of project experience involving resource planning and management. The values shown in parentheses in Exhibit 2-6 refer to the total number of "4" and "5" code number responses for the qualifications item; these parenthetical values provide a convenient indication of the firm's senior-level core capability within that specific area.

The survey results also provided the basis for developing a summary matrix of staff qualifications; a portion of this matrix is shown in Exhibit 2-4. The 32 specific qualifications included in the summary matrix represent a synthesis of the 278 items currently inventoried by the survey. The selection of specific major and minor qualifications was based upon the survey results, with consideration also given to other more recent information. The summary matrix can be used conveniently as an overview item in promo-

EXHIBIT 2-6   Project experience: resource planning and management.

Area of project experience	Qualifications index	
Watershed planning (general)	46	(6)
Floodplain management	39	(5)
Water resources planning (general)	37	(3)
Erosion control	37	(4)
Municipal drainage planning	32	(4)
Flood control planning	30	(4)
Multiple resource planning & management (general)	29	(3)
Environmental impact studies, assessments or statements	28	(1)
Lake/reservoir restoration & maintenance	26	(1)

tional brochures and proposal documents. In addition, the matrix lends itself to rapid adaptation for more specific uses by merely deleting the rows (staff members) and the columns (specific qualifications) that do not apply.

Development of a staff qualifications inventory should be tailored to fit the educational backgrounds of the staff, the working skills of importance to the operation, and the types of projects that the firm has been performing or expects to perform. This development process includes soliciting information and ideas from representative members of the staff; reviewing personnel and marketing files; and using college course catalogs and similar materials to help fill in details.

It should be possible to develop, without any difficulty, a list of 200 to 300 survey items, which would constitute the basis for building an inventory. A trial version of the personnel qualifications survey form (and instructions) should be distributed to selected senior staff members. Their survey responses and critical comments should be considered carefully, and necessary modifications made in the survey form and instructions.

Choice of a suitable period for systematic updating of inventory information depends upon a firm's evaluation of relative needs versus costs. In some instances, a two-year cycle of inventory updating may be adequate; in others, a one-year cycle may be more appropriate. More frequent updates for younger, less experienced employees are desirable because of the relatively high rate of growth taking place in their project experience and working skills.

Its tangible benefits for marketing and management applications make a staff qualifications inventory a good investment for any firm. In addition, the firm stands to benefit from the "Hawthorne Effect." Five decades ago, this famous industrial experiment showed that taking concerned interest in employees can contribute positively to staff morale and performance.

Recognize Limitations

Inherent in any good self-evaluation is a recognition of jobs one cannot do, as well as jobs one can do. For example, your firm might be quite capable of taking on a project in which you are commissioned by the local government to design parking facilities for railroad commuters. But if you have no one on your staff trained to work with electrical systems, your firm might not be able to design the lighting system for the parking fields. By recognizing your limitations ahead of time, however, you can help to maximize the chances for a positive outcome. Perhaps you will build strong contacts with electrical engineers who will help you in a joint venture when needed, or perhaps you will hire a new member who is skilled in the electrical area.

In asking yourself and others "Whom do we want as clients?" you may be tempted to answer: "Everyone!" Obviously, one needs to narrow down the choices a bit. Limitations, in this sense, can help you open doors.

A simple way to start building a list of potential clients and preferred areas of work is to consult the Yellow Pages of your telephone directory. Make a note of each type of business which might be able to use your company's services. Then review the list, possibly along with the other members of the firm present, and categorize each business as a *potential, undetermined,* or *unlikely* source of jobs. You can use other categories, of course, as you see fit.

Publications Give Leads

There are many different magazines and newsletters which carry information about potential clients for the consulting engineer. Experienced consultants subscribe to a variety of publications to keep abreast of the latest developments in their field. There are too many for us to be able to present a comprehensive list here, but the best known include: *Commerce Business Daily* (published by U.S. Department of Commerce); *Dodge Building Reports;* the *Canadian Journal of Commerce;* and *Constructor Magazine.* You should not underestimate the value of other sources, however, such as newsletters of state and municipal government agencies, trade magazines, technical journals, and the classified sections of local newspapers. These can sometimes give direct leads to clients needing the type of consulting that you can provide. At the very minimum, the publications and advertisements contained in them can help provide an overview of the markets available. In them you will find records of construction contracts, consulting projects, electronics and computer research jobs, real estate transactions, and all sorts of other useful information. If the volume of reading material becomes overwhelming, you can reduce it somewhat by hiring a local news clipping agency to scan various newspapers for articles on topics you request.

Potential Clients: The Wide-Angle View

To make sure that your appraisal of the general areas of potential consulting work is complete, here is a list of the types of clients available to most engineering firms:

1. Other consultants who may not have some of your capabilities
2. Architects of all kinds
3. Public works departments of all municipal governments
4. State agencies, such as transportation, environmental protection, urban development, and corrections departments
5. The Army Corps of Engineers
6. Naval Facilities Engineering Command
7. Air Force Engineering Command
8. Veterans' Administration
9. The following departments of the federal government: Defense, Interior, Energy, Transportation, Agriculture, Environmental Protection, Housing and Urban Development
10. All industrial plants in your vicinity
11. Contractors (either building contractors or individual subcontractors) who may need your firm's special expertise
12. Real estate developers who may use your type of services
13. Insurance companies and banks (They often require professional engineering assistance.)
14. Industrial suppliers of all types (They may need your help in presenting new systems using their products.)
15. Development programs for overseas countries (You can sometimes obtain leads in these areas through the United Nations' monthly "Pre-Investment News" and from regional heads of U.N. development programs. Also, the Department of Commerce has much useful information which can lead to profitable overseas work.)

Promote, Promote, Promote

Securing new clients is as vital to your firm's existence as doing the actual work they request—and doing it well. Obviously, if you do not promote your services, you will not be attracting new clients, although clients may sometimes seek out your firm based on information that they have found on their own.

Perhaps the worst mistake—the classic error—made by the rookie consultant is this: She or he promotes the firm, obtains a client, and then turns every bit of energy toward doing an exemplary job and finishing it for this client in a timely manner. But what happens to the firm's promotion in the meantime? It stops, lagging uselessly behind the current project until the consultant finishes that job and finds no other jobs planned. This often happens *unless* another member of the firm has been appointed to keep promotion going while the project is being completed (or—if there is only one consultant in the firm—the engineer spends part of each week in promotional efforts).

The "stop-and-start" marketing effort is a principal reason why many small consulting firms do not grow. Some promotion will occur, of course, as your reputation spreads by word-of-mouth from satisfied clients. But strong growth is unlikely unless you "Promote, promote, promote," on a day-by-day, week-by-week, month-by-month basis. If more work comes in than you can immediately handle, hire more people to get it done. That is the way to continuing success and expansion.

A crucial factor in good promotion is the selection of the member(s) of the firm who will direct the marketing efforts. People who are best suited for this task are those who have some expertise in sales and in dealing with the public and media and who are polite but unafraid of asserting themselves. They should also be able to work under pressure. Salespeople should be able to work consistently and methodically. A charming personality may not be as important in selling the firm's services as the attributes of resourcefulness and willingness to work hard.

It is reasonable to expect that your firm's sales expense will be from 5 to 10 percent of your total business fees. If you expect to sell $200,000 worth of services each year, you thus can expect to spend $10,000 to $20,000 to get those sales. This is a common way for firms to develop their sales expense budgets. The requirements for any sales effort include: hours of work by the marketing staff, purchase of materials and advertisements, secretarial time, travel to clients' offices, and record keeping.

Thorough records of all sales efforts are essential. Clearly, marketing does not come cheaply. Properly managed, it can pay off in ways that some engineers only dream about. So promote, promote, promote!

Defining Prospects for a Balanced Clientele

Definitions of the *big* client, *small* client, and *average* client vary depending on the size of the consulting firm itself. For example, a small firm consisting of one chief engineer and three or four other, part-time engineers, designers, and drafters might bring in $100,000 or $150,000 in yearly fees. For such a firm, a prospective client who might pay $40,000 in fees during the course of a year would be considered *big.* A large, 100-member consulting firm, however, which could perhaps generate $2 million in yearly sales, would consider the same client *small:* The fees it pays would amount to only 2 percent of gross sales.

It is important to define your firm's prospects in terms of big, small, and average-sized clients, chiefly because the number of each type of client that you obtain will affect business. How can this happen? One way is that a profusion of very small clients will inhibit efficiency and profitability because each client requires your firm to provide separate secretarial work, separate bookkeeping, separate billing, telephoning, and follow-up (among other things). Unfortunately, most consultants do have a disproportionate number of small clients, who may bring in projects worth only $1000 or so.

This is not to say that the small client does not deserve the same type of service as the big one, nor that the "little guy" is not appreciated. On the contrary, the small clients are in many cases the foundation for a new firm's growth and its staple source of revenues. But the engineering consultant who wants to build an efficient, profit-oriented business should aim his or her promotional efforts selectively, taking into account the wide panorama of prospects. Just as one should not depend solely on small clients for business, one should not depend on big clients alone. Obviously, it could be an enormous advantage to your firm if you manage to start up with one or two clients who will supply the firm's entire income for a year. But if anything should go wrong in the course of delivering the work to these big clients, your firm could regret the decision to limit its sales to them.

There is no general rule for the ideal range of clients. However, the key to sound business—in this regard as elsewhere—is to maintain a balance. A desirable mixture of clients might be to have one major account that supplies approximately 25 percent of your firm's business, plus several average-sized clients that provide about 50 percent of your firm's business. In this way, there will be no risky dependence on any one client or group of clients to keep the firm strong.

Further, there is the good possibility that average-sized accounts, handled with care, will eventually develop into large accounts. Your continuing growth may depend strongly on your performance with such clients. If they do not grow, it may be worthwhile to evaluate the situation to determine if the consulting firm can enhance its prospects for making the average client grow and, in turn, to benefit your firm. One way in which some consulting engineering firms tend to limit their growth—sometimes to the point of actually losing their average clients—is by spending an inordinate amount of time and effort on their larger clients. Unless you achieve a balance of service among your clients, you may induce a high rate of turnover among average-sized clients, with a consequent slowing of your firm's expansion.

Stand Up and Get Your Clients

One of your chief tasks in building your business is to create interest in your firm. Unfortunately, you cannot send out free samples of your work in little boxes like manufacturers of cereal and shampoo do. But you can stand up and be noticed in other ways. You have a variety of promotional tools at your disposal, such as telephone solicitations for interviews, advertisements, and mail campaigns. These are aids to the visibility of your services. Also, you need to build a reputation for providing high-quality engineering work.

Once your marketing efforts are underway, inquiries from serious, potential clients should start coming in. At least one person (preferably several) in the firm should be skilled in handling these inquiries professionally. That person should have a thorough knowledge of your firm's capabilities and of any areas which deserve special attention. The person who deals with potential clients directly can make the difference between gaining or losing an important account. Even if an inquiry comes from a source in which the consultants have no potential interest, the caller should be treated in such a way that she or he ends the inquiry with a greater knowledge of the firm's professional expertise than when the call was first placed.

MARKETING CONSULTING ENGINEERING SERVICES

Marketing is basically the process of selling. For the consulting engineer, it means presenting the firm's services, experience, and capabilities in a manner which will help the engineer secure work in dispensing advice and assistance to people who are willing to pay the engineer in exchange for this help.

In all fields, successful marketing has several key ingredients. Social psychology plays a large role, along with skill in communication and knowledge of economics. Engineers generally are more accustomed to handling facts and figures than they are used to the art of attention getting and persuasion. Yet the thrust of marketing probably has more to do with knowing "how to win friends and influence people" than with anything else. Many of the principles expounded decades ago in the book of that title by Dale Carnegie are still the standards for good marketing today.

Talking with Potential Clients

Anyone who intends to sell consulting services must communicate with buyers and potential buyers of those services at some time. The communication, when it occurs, must effectively promote the consulting firm. But few engineers have any formal training in marketing and communication. Some would rather not have anything to do with the "communication process" between client and consulting firm. Actually, a lot of the worry some engineers feel about communicating with clients is unwarranted. Acquaintance with a few simple ideas can help anyone sell a product or service. You need only put the ideas into action to be successful.

Every communication involves at least two people. The way in which some people fail to promote themselves or their company is that they talk *to* their potential clients, but do not *listen* well enough. The art of successful marketing involves keeping your ears open to everything the potential clients say, including things that he or she might not state directly. Good communication is best referred to as *talking with* the client, for only when you attend to the client's needs and concerns will you be displaying the skills of the experienced consultant.

The inexperienced salesperson, like the nervous teenager anxious to make conversation on a first date, spends a lot of time and energy telling the person sitting opposite her or him about herself or himself or the business. If the talker ever gets to the point of relaxing or of having nothing left to say, however, she or he frequently finds that the person to whom she or he has been speaking has a good deal to say. Certainly, your potential clients will want to know about your firm, but the fact is: The reason people seek consultants is because they have problems they want solved. Therefore, the best way for you to sell your services is—besides establishing the expertise of your firm—to find out what the client's problems are and what he or she would like done about them. The client may have some suggestions or may leave the solution totally to your discretion. You are the expert. As a salesperson, a major part of your job is to talk with the client, determine how to best satisfy the client's needs and desires, and show that your firm can do high-quality work with tact, efficiency, and economy.

There is a lesson to be learned from the marketing departments of big corporations, which often send along technical experts with their salespeople to meetings with prospective clients. The best marketing involves communicating information about the consulting firm *plus* data about the technical aspects of prospective work. The consulting engineer can often fulfill both roles at once. You can prepare for meetings with potential clients by having your salespeople confer with other experts in your firm to review the kinds of issues that clients might wish to discuss. Corporate salespeople, interviewers, and even politicians use such conferences and mock meetings to prepare for public dialogue. All that participants need do is to imagine the kinds of situations or questions that could arise in the actual meeting. Your salespeople should try to handle these developments as they would in the client's presence. Members of the firm might even have fun "playing the client." And they will certainly learn more about marketing in the process.

Make Sales "Sizzle"

The standard guideline of specialty selling is: "Don't sell the steak. Sell the sizzle." For the consulting engineer, the "meat" of the work is the job done for the client. The "sizzle" is the interest and excitement you can create when you show that you can do that job well, fast, and at a reasonable cost.

Your meetings with prospective clients will in some ways be similar to job interviews. Application forms will be filled out, experience and education will be discussed, and the needs of the client or employer expressed. But your firm will be selling a lot more than the person seeking work with an employer; you may be selling the services of several individuals (your staff) at once, along with their combined productivity, the project's starting and ending dates, budgets, and many other special requirements.

The prospective client's main concerns about most jobs will usually be:

1. Money
2. Work performance
3. Obtaining a satisfactory end result

Often the easiest issue to resolve is money. Fees are relatively straightforward and simple to establish, based on the standard rates or percentages developed by professional engineering societies. Concerns about work performance and end results are usually more difficult to overcome. They can evoke unlimited questions from the client about why your firm should be used.

To make sure your sales sizzle, you need some sort of sales demonstration. This can come in the form of references from previous clients or employers, exhibits of some of your firm's designs, specifications, and plans for completed projects, financial statements, records of employee performance, and even visits to the sites of projects on which your firm has had strong influence.

The principles of selling are actually rooted in basic human relationships. A good analogy to marketing the services of consultants is what happens in the sale of a car. For one thing, the prospective buyer will be much more likely to become interested in the car if she or he has some knowledge about the particular automobile and feels comfortable with the salesperson. Both the car and the dealer should be reputed to be reliable and efficient (it is doubtful that anyone would buy a gas-guzzler or, for that matter, that anyone would buy the services of consultants who spend clients' money lavishly). Further, the salesperson must be able to answer the customer's questions about the car and honestly and politely persuade her or him that the customer would be pleased with a purchase from this automobile dealer. Also, the customer usually wants to take a demonstration ride before buying and to know that the car will be delivered quickly and on the date agreed to by both parties. The salesperson who can demonstrate that the car fulfills the buyer's need for transportation, economy, personal prestige, and reliability will most likely make the sale. In consulting, the ingredients of good selling are based on the same principles.

SOME RULES FOR EFFECTIVE COMMUNICATION

In nearly all businesses, the chief responsibility for the development and direction of promotion strategies rests with top management. Specialists are assigned to carry out the selected strategies. The effort requires communication that is clear, concise, and telling—regardless of whether it is written or spoken. Steps the consulting firm should take to achieve effective communication are:

1. Review and understand the material to be promoted and the order of priority for promoting any of its elements
2. Gather data on the firm's work and analyze the organization's expertise in each project area and/or market area
3. Identify competing firms and their strengths and weaknesses
4. Identify potential clients and find out what they know and/or think about your firm
5. Learn exactly what potential clients want; evaluate the firm's ability to satisfy the prospect's needs

You can use the preceding steps to develop a program for promoting your firm and targeting sales efforts.

Here is an example of how a consulting firm might use these rules to achieve a specific goal.

Suppose that a new firm is interested in establishing itself as an expert organization in the large and growing field of energy conservation. Reviewing the capabilities of the firm's members, its general goals, and the market within a 100-mi radius of the firm's offices, the range of potential work is narrowed.

Obviously, "energy conservation" as a whole is too broad a term to provide a strong focus. But the firm may succeed in identifying several industrial plants in the local area which may benefit from installing systems to save electricity and oil. Most new consultants have experience in these fields and have a proven track record of high-quality work with former employers. Moreover, the firm can determine, through contacts with regional engineering societies and others, that there are few competing firms in the local area with the same kind of expertise. The consulting firm can now use all this information to its benefit and begin contacting potential clients.

Before doing this, though, the firm might also establish a goal for the amount of sales it would like to make, based on the preceding determinations. For instance, it might decide that a realistic aim to reach for would be to bring in $100,000 in gross fees, using the time of three members of the consulting firm for 1 year. In this way, both the market and the sales goals have been identified. The next step is to communicate effectively to receptive prospects. Your actual promotion efforts could include:

1. Developing position papers, press releases, and factual data for direct-mail distribution to prospective clients

2. Writing, printing, and distributing a small, separate brochure devoted entirely to energy conservation—complete with charts, graphs, and photography—all in an easy-to-read and easy-to-understand format

3. Running advertisements in papers and publications read by prospective clients

4. Providing expert lecturers on energy conservation topics to groups, officials, and societies to which prospective clients belong

5. Using follow-up mailings to clients and prospects, as well as telephone calls

6. Preparing and presenting slide-show exhibits at meetings with clients and prospects

Problems in Communicating

The foundation of good marketing communications is serious and thorough preparation. Where some salespeople fall short of their task is in not having enough information about the topic that is being discussed or in not expressing their ideas clearly. Further, some talk too much, with the result that they overwhelm, bore, or confuse the potential client.

Essentially, clients would rather be *sold* than be told. An atmosphere of mutual respect and give-and-take is vital to selling. Arguments do not promote sales. The potential client will become the happy client when he or she arrives at the decision to use your services under his or her own accord, with the salesperson's assistance. If the client has questions, it means that something needs to be clarified.

Conveying the Proper Tone and Clear Ideas

Examples of two very clear phrases that some people use in daily speech to mean the same thing are "Shut up" and "Be quiet." However, their tone and style are totally different. The former is harsh, informal, and could actually offend the person to whom it is addressed. The latter conveys the same message, but without negative connotations. Similarly, the words used in business communication can take on varying tones. Salespeople should always be attentive to the feelings which they want to convey to their audience and should select their words accordingly.

With diligence, almost anyone can learn to speak and write with a desirable tone. In business, the tone of most communications should be serious, though not so grave as to bore the listener or reader. A casual tone may sometimes be permissible. But you will get your best indications of how to proceed by listening attentively to the person to whom you address your statements.

A problem that affects many engineers in speaking and writing is that, by training, the engineer becomes accustomed to a mathematically rigid style that is overly dependent on jargon. Perhaps this appears most often when the engineer needs to depart from purely technical matters to explain an error, ask a favor, or reject a proposal of some kind. To avoid such problems, read your written material aloud as if you were addressing your audience and get a feeling for how it will sound in public. If necessary, delete complex and highly technical phrases and constructions.

Each speech, letter, or advertisement should make a clearly defined point. Use simple, clear, and direct statements. Avoid using terms like "systematized monitored capability," "behavior modulated schematics," and "architecturally coordinated environment." Phrases like these may sound as if they enhance one's reputation as a technical expert; but what use is your expertise if you cannot make yourself understood to the public?

Some younger engineers talk that way because they have become accustomed to hearing their superiors use convoluted sentences and technical jargon. The problem is that those superiors never learned how to communicate with maximum effectiveness. More than likely, they were able to get to the positions they occupy despite their use of convoluted phrases, not because of it.

One technical development that has given rise to a whole new language of gibberish is the computer explosion. Computer terminology is fine for one computer expert to use while talking about computers to another expert. However, if a consultant bombards prospective clients who are not computer experts with "computerese" and makes no attempt to define the terminology beforehand, the result may be misinterpretation and confusion of the consultant's words.

Plain English Is Best

Simple words and short sentences are the building blocks of good communication. Unfortunately, most of us do not always talk and think in that fashion. When writing or engaging in business discussions, ideas must be organized. General ideas can be broken down into topics and subtopics. Each topic should lead logically to the next.

Every sentence provides a transition, as do the beginnings and ends of paragraphs. In a business letter and in most advertisements, paragraphs should be only a few sentences long, each one containing a clear point. This will make the material more attractive to the reader's eyes so that it will be easier to understand.

In engineering there are times when you *will* need to present complex technical material. A good rule of thumb in this regard is to build your discussion gradually, starting with simpler words and ideas, leading up to the technical material in an orderly manner. Make sure to explain terms and concepts that may be unfamiliar to your audience. Clarity and courtesy work hand in hand, whether you are trying to make a sale or to turn down an unqualified job applicant. Ask yourself as you formulate your statements, "How would I feel if this were said to me?" Gauge all communications accordingly.

ADVERTISING AND PUBLIC RELATIONS

The basic goals of advertising and public relations are to plan, control, and coordinate your firm's communication with the public. Properly performed, it will optimize the outcome of all sales efforts. Unfortunately, advertising and public relations are not always highly regarded aspects of business in engineering circles. This is because a few overzealous people and companies have occasionally abused the freedom to use these tools by misleading people about their qualifications, fees, or other characteristics. That a handful of people have done so does not indicate that the system is faulty. The historic success of advertising and public relations in promoting both fact—and, in a few cases, fiction—attests to their effectiveness as promotional media. Clearly, the responsibility to use these powerful tools properly rests with the individuals who do the promoting.

Sincerity and perseverance are two of the most important factors in your promotion efforts. Ultimately, the truth about an advertiser—if it has been obscured initially—will emerge through the advertiser's dealings with the public. Perseverance often determines whether any benefits will accrue from a promotional program. Few, if any, engineers can afford to presume that their private consulting practice will be supported merely by potential clients' hearing about them from some disinterested sources. That is why care must be taken to inform the public accurately about your firm and its service. Strong promotion requires consistent effort and diligence.

As we have seen, there are many different ways in which you can promote your business to the public. The audience that you may want to reach is also varied. It can include: past, present, and future clients and employees; professional peers (competitive and noncompetitive ones); contractors; manufacturers and suppliers; public officials; and editors and reporters from the news media.

The Truth About Advertising

The majority of advertisements by consulting engineering firms consist of small, boxed entries in business-card style that appear in engineering or engineering-related trade magazines and newsletters. These ads usually indicate briefly (in the space of a few inches) that the engineering firm provides services in the fields of business in which most of the publication's readers engage. Such ads help pay for the publication of the magazines they appear in. Contrary to the vitalizing effect they can have on your business, this sort of advertising has a rather ominous-sounding name: It is often called "tombstone" advertising, presumably because of the shape of the ads. Figure 2-1 shows several such ads.

During the last few years, the U.S. Supreme Court has ruled that certain forms of media advertising, formerly restricted, can now be used by

Figure 2-1 Typical business-card ads placed by consulting engineers and consulting firms. *(Power).*

2-21

professionals in private practice. Generally, however, there are few consulting firms that feel their services would justify heavy mass-media advertising. Engineering firms, in particular, can utilize other promotional methods more effectively. Many such methods are not only less costly but are also easier to control than mass-media material.

For example, a consulting firm which specializes in mechanical and electrical services for architects may advertise its capabilities in magazines or newsletters read in industrial plants, for a wider audience. The result could be that the engineering firm starts receiving a whole range of new and valuable leads through responses to the ads. Alternatively, members of the firm could write a series of articles on specific engineering topics and submit them to magazines read by manufacturers. This can become another fertile ground for sales to potentially big clients. A big difference between articles and ads is in the way the inquiries are initially generated. One possible advantage of articles is that they can display your firm's expertise immediately to prospective clients, who might otherwise continue to look at ads.

For consultants who do advertise, several important considerations emerge:

1. **Ethics:**   What are the limitations, if any, placed on advertising by professional organizations to which you and/or your firm belong?

2. **Image:**   Be certain that the media vehicles you select (e.g., trade magazines and newspapers) have a public image that will not downgrade your firm. If possible, the publication should enhance your status.

3. **Content:**   As a general rule, an eye-catching ad (though not frivolous) with a few short sentences and factual, hard-hitting, but believable, ideas will get the message across best.

4. **Follow-up:**   All requests for information in response to your ad should be answered as promptly as possible. If you use direct mail (i.e., printed matter mailed to potential clients' offices) as your promotional device, it may be best to send out letters in groups of 25 to 50 each time and then follow up the mailings a week or two later with telephone inquiries.

Many consulting engineering professionals feel that media advertising tends to conflict with the traditional, reserved image of the professional organization. However, larger numbers of engineering professionals are using the media all the time. Some lawyers are even advertising divorce services on television.

It is the follow-up activities which often make advertising for the professional truly beneficial. Before placing an ad, you should always examine the effects that advertising in a particular medium can have on the company's image.

Basic Know-How for Public Relations

Public relations can be classified into two types: (1) active and (2) passive. In essence, everyone engages in passive public relations, which includes everyday behavior such as talking on the phone, writing letters, setting office policy, controlling the appearance of the office, and fielding questions or complaints. Even your personal appearance and ordinary conversations are part of passive public relations. Much has been written on these subjects, but the common theme of advice boils down to the golden rule: If you want to make a good impression, "Do unto others as you would have them do unto you."

Good active public relations utilizes the golden rule too. But it requires deliberate, positive planning and programmed communication. Active public relations does not occur by itself. *You* make it happen. Among the most important features of an active public relations program are (1) organized scheduling; (2) strong, sharp writing; (3) use of photography; and (4) skilled research of background data for public presentations.

A public relations program can be planned in the same basic manner that most aspects of professional practice are planned: by identifying major objectives and determining the best strategy for meeting these objectives. Writing news releases and announcements is a crucial ingredient of most active public relations programs. The first step in preparing written material is to research all data relevant to your topic. Then create an outline and highlight the important points. Public relations writers should plan to produce at least three drafts of each article—rough, second, and final—so that heads of the firm can review the material, delete extraneous data, and produce a polished, interesting piece.

The use of photography can help to illustrate the qualities of your work in a manner that words may not be able to convey. An effective ongoing photography program should include at least three or four good pictures of every project performed. Consideration should be given to developing photos as color slides because slides can easily be stored and retrieved for large-scale viewing. Both black-and-white prints and color prints can be made from the slides, if needed. Liberal use of photos, properly coordinated, will help support statements made in your letters and articles. Slide presentations to large audiences can also be very effective.

Awards programs can be a significant route to active public relations for your firm. The most progressive firms enter as many awards contests as they can. Clients should be informed if their projects are involved in a contest. Usually, a client will be especially delighted if an entry is a winner, as long as she or he has approved it first.

News Releases

The most frequently used public relations device for consulting engineering firms is the news release. A news release is essentially the creation of a newsworthy story about particular and interesting accomplishments of your firm. Set up your news release facilities by stocking a card file with the names, addresses, and phone numbers of publications to which you might aim your releases for public presentation. An important point to remember in press release writing is that "news" is whatever the publication's editor decides to print. Therefore, a good guideline is to write releases about the types of "events" that the news media carries. Such events include: obtaining a new (or particularly important) contract for an engineering project; winning an award; hiring new staff; opening a new office; appointments of firm members to the boards of business associations or civic organizations; and the publication of books written, edited, and/or researched by members of the consulting firm.

A well-written news release will have a much better chance of getting published than one that requires extensive rewriting by the publication's editors. The style should be—as in any other news story—concise and to the point. Its first paragraph generally answers the questions "who, what, when, where, why, and how" in a few short sentences. Since the release may be too long for space allocated by some publications, further paragraphs should carry material of descending importance so that they can be deleted if necessary. This type of organization will help editors to fit the release into the space available.

The simplest way to learn the journalistic writing style is to read carefully your daily newspaper. With concentration, you will soon be able to spot many stories that originated as news releases.

Your news release should be printed on plain paper (not your letterhead) for submission to publications. An alternative would be to buy paper prepared specifically for news releases. Such paper usually has the word "NEWS" printed at the top in large letters. Regardless of which type of paper you choose, the final release should be copied and mailed to all media organizations (including, if you like, local television and radio stations) you think might have reason to publish it. You can also deliver the release by hand, but the means of delivery may not have much effect on the outcome.

Here is an example of a typical press release:

NEWS

ABC Engineering Associates, 111 Oak Street, Anywhere, U.S. 00001
For more information, contact
Jean E. Doe, Director of Communications
Telephone: 678-9000

FOR IMMEDIATE RELEASE   June 1, 19XX

ABC Engineering Associates of Anywhere has been chosen to design the mechanical and electrical components of the 37-story institutional plaza office building planned for the revitalized downtown area. The choice was announced jointly by the financial group funding the project, National Bank, and the building developer, Therese Albizer.

Charles Bernstein, ABC's president, said, "The new office building will include a total of 398,000 square feet of office space with a separate parking garage for 1500 cars and a street-level shopping area. The project is one of the largest in this city in the past few years."

Bernstein noted that the building will have limited glass areas composed of insulated, tinted windows, sun screens, and solar collectors for hot water and partial heating supply.

Construction is planned to start this August and is expected to be completed within 2 years. The building will not be occupied until at least 19XX, but more than 25 percent of total floor space has already been rented, noted Albizer.

ABC is a mechanical, electrical, process consulting engineering firm which provides facility and utility engineering to many industrial and architectural companies in the metropolitan area. The firm has recently added a specialized Energy Conservation Group to its 38-member staff. ABC Engineering has been in business in Anywhere since 1959.

—END—

Publicity Through Magazine Writing

As we noted earlier, another effective public relations tool for consulting engineering firms is the technical magazine article. Most engineers who write magazine articles generally do so for magazines published only for their peers. While this type of writing is a valuable contribution to engineering, its public relations value for your services is limited. A reprint of such an article may impress a potential client somewhat. But oftentimes the topic and its mode of treatment do not relate directly to the basic needs of the client.

For optimum results, the article should appear in a magazine serving the interests of potential clients. For example, a new system of high bay lighting is of interest to illuminating engineers. It is also of interest, however, to those who direct facility engineering and who read magazines like *Chemical Engineering.* Such an article could be prepared by your firm or by a freelance writer. An advantage of using a freelance writer is that he or she can describe your services and praise your firm as an outside observer. Thus the freelance article may give more weight to the article's characterization of your firm.

Generally, you have three points to guide you in developing magazine articles. (1) List the potential clients whom you wish to attract. Be specific. (2) Research and describe some topics those clients would like to read about. (3) Identify the publications that carry this type of material which your potential clients read.

There is no substitute for studying those publications and learning what sort of articles they publish. You may be surprised at the number and variety of magazines that your clients subscribe to.

To map out the market for your writing, several different directories can be consulted, such as *Ayer's Directory of Publications* and the yearly *Writer's Market,* available in most libraries. The subject index in these books can help direct you to publications oriented to your clients' needs. Magazines that have the largest circulations are sometimes the most desirable markets. But do not limit your possibilities—consider smaller magazines also.

Editors of technical magazines constantly need articles. While many articles are written by in-house staff, the editor is usually well aware that there are many good article ideas for which she or he does not have time or staff to pursue.

It can sometimes be most helpful for you to simply pick up the phone, call the editor of a magazine with which you have become familiar (publication directories often list names of editors), identify yourself, and state in clear, simple language that you have an idea for an article that may be of interest to the magazine's readers. (One word of caution: If the article is about your firm only, it would be best to have a freelance writer make the call, so you do not appear to be looking for a free advertisement.) If your article interests the editor, she or he will quickly direct the conversation to a description of what you should do next, whether she or he wants you to send in the article or rewrite it with a new focus or to do something else.

The editor may say that most or all the articles are written by the magazine's regular staff. Do not let this discourage you. Merely ask if you can act as an information source. If you function as a "source information" provider, as it is sometimes called, your article may eventually be presented with pictures and perhaps only a few alterations. The PR result can still be quite profitable.

Inquiries about interest in your article can be mailed to the editor. The goal is the same as it is when you inquire by phone, but a letter must "sell" the article to the editor. In a brief paragraph or two, the letter should describe what is unique about the article and how it may benefit the magazine's readers. This cannot be done unless you have developed a solid idea of what its readers are accustomed to seeing in the publication. Once an editor makes a commitment to examine and/or print the piece, you should make sure that it is written in the tone and style of articles currently appearing in the magazine.

Use Your Articles for Additional Public Relations

Reprints of your published articles can often be obtained from the publisher, or you can have a printer produce them. (Be certain to get the publisher's permission before distributing reprints.) To use your articles in a full-fledged PR program, include a copy of the cover of the issue containing your article. Reprints should be sent to clients, potential clients, plus other members of your public, along with a brief note. You can also produce slides of the article to use in large presentations.

The time and cost of producing magazine articles is substantial but not overwhelming, especially when compared to the volume of new business that it can help you bring in. A professional writer would probably take 10 to 15 hours to write a medium-sized article (less than 15 double-spaced manuscript pages). On that basis, a budget of $300 or $400 might logically provide for the overall effort, including payment to the professional writer.

Other Writing Markets—Newspapers

Besides writing articles for engineering publications and specialized trade magazines, some consultants expand their PR programs by including the writing of feature stories. Articles by consulting engineers occasionally appear in the "home building" sections of Sunday newspapers (some papers have such sections on weekdays too). The types of stories newspapers present usually have a more general orientation than the technical articles presented in trade magazines. Stories aimed at newspapers should deal with practical topics such as energy conservation; construction and enhancement of building foundations, roofs, and roof supports; lighting systems; and a variety of other items that would appeal to the layperson and engineer alike.

As with most freelance writing, one of your first steps in preparing a feature news story is to query editors about their papers' needs. Then, proceed according to the editors' suggestions. Feature articles in general-interest publications can be valuable to your marketing efforts since they may help to display the expertise of your firm at the same time that they attract attention from a wide range of readers. Of course, similar articles in specialized publications read exclusively by your potential clients can sometimes reach your desired audience more directly.

Letters to the Editor

If engineers wrote letters to editors about every technically inaccurate article that appeared in newspapers, the papers would probably have little room left for news. But such letters are written infrequently, and they appear in print even less often than they are written. Still, a letter to the editor of a newspaper, magazine, or newsletter can be a good way to get yourself published (and, thus, to advertise your firm).

Letters to the editor should be brief (less than 500 words), dispassionate, and not overly strident. Editors prefer praise to criticism, like anybody else. Therefore, you need not confine your comments to complaints. But it is a good practice to try to correct false impressions that an article may have generated. But if your letter is written in defense of one or more of your clients, it is wise to send a copy of the letter to the client(s) concerned. Even if the letter does not get published, it can provide a positive outcome and in some cases will help you to learn to write better.

In writing letters that concern the defense of an action taken by you or other engineers, it is wise to consult a lawyer first. Any defense of the engineering profession itself is best made by spokespersons appointed by professional engineering societies and lobbying organizations.

Notes to Clients

Many businesses send cards and notes to clients for holidays and other events. These help remind clients that their business associates want to maintain cordial, ongoing relationships. For consultants, it is a good practice to follow, even though it can at times be somewhat burdensome. Generally, clients will be happy to know that you have been thoughtful enough to take the energy to send them a holiday greeting or a congratulatory note when they earn acclaim for particular accomplishments. To keep track of your note sending and important dates, keep a list of the special times of the year when it is appropriate to send notes to clients.

Company Newsletters

As your consulting firm grows, you may find it worthwhile to start a company newsletter. A good in-house publication will keep all members of the firm, as well as clients and other business contacts, informed of important developments affecting your firm and the people involved with it. Many large corporations consider the newsletter to be one of the most important elements in their public relations program.

Newsletters need not be slick or elaborate, although those that are professionally produced tend to give an added touch of class to the companies publishing them. Your newsletter should be published regularly. A newsletter can be a highly effective public relations device when it covers specific projects of your firm in a journalistic manner, with articles illustrating ways in which particular aspects of your design or other services have helped clients save money and time.

The number of pages in professional newsletters varies, depending on the amount of information presented and the amount of time and money the firm can devote to the publication. Some newsletters are as short as 1 page. Others run to 24 pages, or more. Most consulting firms do not change the number of pages in their newsletters from issue to issue.

You can rarely, if ever, present articles on all of your clients and projects in every issue of your newsletter. It is best to focus on particular clients, events, or designs separately in each issue so you present the information in a balanced and easy-to-read format.

Books

Engineers who write books and get these books published generally recognize that the accomplishment carries both reward and risk. Writing a book requiries much time, money, and energy, and it also requires finding a publisher, and sometimes, a literary agent. Some writers do publish their books themselves—usually at a cost of more than $20,000 per title. For most consultants who write books, the whole effort becomes a second job, or at least, a lengthy spare-time project.

A book can give its author much added prestige, and—if the book is useful to prospective clients—it can often generate new business. Many consultants who have published books find it profitable to give complementary copies to potential and current clients, along with a brochure describing the consulting firm.

Oftentimes, however, engineering books are too technical for the general client audience to understand. An author risks appearing presumptuous if he or she gives copies of a book to clients who may not know how to interpret the technical contents. Therefore, in many cases, the book becomes a "soft-sell" item. It should be shown to clients selectively in a casual manner. One should not appear to be trying to sell the book to the client. What you want to sell, of course, is your firm's services. For that, the best publication to show the potential clients could very well be your company brochure.

PREPARING WINNING BROCHURES

One idea on which most professional consulting engineers agree is that a descriptive brochure is a major sales tool in private practice. The brochure is a booklet or pamphlet that describes, promotes, and subtly sells the firm. Anyone who doubts this should remember that nearly all the products that engineers specify come from brochures and catalogs.

Brochures that are well organized, detailed, and professional-looking are the ones used most frequently by clients and the general public. Engineers and other consumers frequently buy goods through brochures and catalogs, though little or no additional sales efforts may be made by representatives of the companies whose products are described. Except for one major difference, the virtues of brochures apply equally well to consulting firms as they do to sellers of supplies and equipment. Most consulting engineers will benefit from developing and distributing a brochure for their firm.

You must remember, however, that an engineer is selling a relatively intangible product—consulting services. This is a product that depends, for sales, on clients' inferences and opinions about the information in the brochure. The supply catalog, in contrast, describes physical objects whose appearance and specifications alone might cause a client to buy. Thus, the job of the engineer's brochure is to describe the performance record of her or his consulting firm in a way that demonstrates the firm's ability to meet clients' requirements.

An engineer's performance on a client's project depends almost totally on the work of skilled people, not on mechanically produced objects. But clients may have a hard time deciding which of several consulting firms has the most skilled and reliable consultants. Clients often hear nearly identical claims from different engineering companies. Thus, for the engineer, the brochure may *tell* clients about the firm; but further sales efforts are often needed to actually sell the services. Consequently, the brochure should be seen as one tool in the overall sales and public relations program.

Still, the usefulness of a good brochure should not be underrated. At the very least the brochure can get clients interested enough in your firm to ask for more information and for a meeting to discuss potential projects. In effect, the brochure sells the meeting. Then comes the real opportunity to sell your services.

Finding Brochure Experts

Initial steps in developing a brochure involve three basic questions:

1. What should be the focus of the brochure?
2. What kind of company image should it create?
3. What types of clients will it be directed toward?

To handle the general task of brochure preparation, the consulting firm should appoint at least one individual as brochure director. Services of experts from outside the firm can also be used.

All ideas for the brochure should be collected, screened, and organized by the project director. The basic guideline used by the director is to seek to answer the preceding three questions definitively.

Experts who can assist in developing winning brochures generally come from one of two categories: (1) specialist companies and (2) freelancers. Specialist firms include public relations consultants, advertising agencies, graphic design firms, photographers, printers, and typographers.

Some public relations consultants work only on press releases and media gatherings, while others offer complete services for all aspects of PR, from writing and producing brochures to holding press conferences. When seeking help in brochure preparation, you may want to inquire at various PR firms about the range of their services and request cost estimates for brochures.

Advertising agencies are usually oriented more toward graphics and ads, while graphics firms provide layouts, primarily for ad agencies. Photographic services may be needed for your brochure if there is no photography expert on your staff. As was mentioned earlier, all engineering projects should have at least three representative photographs taken of them. Some of these photos can be used directly in a brochure, or they may suggest additional picture taking of particular projects to emphasize certain aspects of the consulting firm's capabilities. Spending money for a professional photographer is usually well worthwhile.

For many consulting firms, a printer offers the most economical source of outside brochure services, including typesetting. A knowledgeable, well-organized brochure director can work with a wide range of printer's services to produce a high-quality, attractive brochure.

Further, freelancers may be available for many aspects of brochure preparation, such as graphic design and type specification. But the task in which freelancers frequently do the most good is in writing. A good writer who can remold an engineer's stilted, cumbersome writing into an interesting and accurate description can be very helpful.

The Brochure's Appearance

There are many different ways to assemble a brochure. The final choice is a matter of personal preference and cost, particularly for the binding. Here are some of the methods for binding a brochure:

1. **Saddle Stitching:**   Refers to the placement of staples in the center of a page or of several leaves folded in half (this is the type of binding used in most magazines). *Note:* If a brochure requires frequent additions or other changes in information, saddle stitching may be impractical because changing one page necessitates reprinting or altering the other three pages attached to it.

2. **Comb Binding:**   Consists of a plastic "spine" that holds documents together with teeth inserted through slits along the edges of pages. This type of book can be assembled on an in-house machine, permitting frequent alteration. Some designers feel comb binding is unattractive, however.

3. **Screw Binding:**   Requires a machine to punch holes along the edges of pages. Holes are then sealed with an edge binding which locks the sheets together. The process makes for convenient alteration of documents, but like the comb binding, it requires the purchase of punching and sealing machines.

4. **Plastic Slide Covers:**   Consists of a tapered plastic edge holder or "slide," which is flexible and inexpensive. Users often prefer to staple entire document and plastic cover before putting on the slide in order to secure pages.

5. **Perfect Binding:**   Uses several sets of glued sections held together by a glued-on cover and flat spine. This method is expensive and usually used only for very large brochures and large printings.

6. **Hardcover Binding:**   Puts document together like a hardbound book, providing great durability but little flexibility. Expensive.

Most professional brochures have flexible covers, which tend to be less expensive and easier to use. Clear plastic-slide-bound covers printed with the company logo or other designs are becoming more and more popular, although many engineers prefer other forms of edge binding. One particularly useful modification of the plain brochure is the addition of a pocket on the inside back cover. This pocket enables the firm and its clients to put loose pages into the brochure securely and allows additions to the basic printed contents. Choose your brochure format based on actual tests with full-sized materials.

Brochure Contents

Unlike annual reports, professional brochures generally do not contain a table of contents or letters from the firm's president. A brief introduction and history of the firm is sufficient but not entirely necessary.

Personnel descriptions should usually be limited to the company's principals and officers, accompanied by photos of the individuals. Operating personnel are usually not included in the main body of the brochure, for two reasons: (1) They are often not as certain to stay with the firm as are the company heads. (2) If they leave, extensive alteration of the brochure could be required. Further, the special expertise of individual employees may not be relevant to every client's needs. That is why many firms choose a binding that permits them to alter brochures conveniently. It is also why a back cover pocket is handy. Separate descriptions of appropriate key personnel can be produced for insertion when needed.

The next item that appears in most brochures is a description of the firm's services. Services can be summarized and/or accompanied by photos of outstanding projects. If desired, services can be listed and described in depth with a more thorough accompanying narrative. Or a combination can be used: a list of services with photos and narrative description. For instance, the firm can devote whole sections of the brochure to detailed descriptions of particular services.

Selection of your brochure's format depends on what kind of image you want to convey and on what sorts of clients you are appealing to. Usually, concise listings and short descriptions, along with an appropriate photo or two, will get your basic message across.

Most firms also list specific projects in which they have been involved. Like the service listing, the project section can be emphasized with lots of detail, or it can be presented as a short summary. For example, when the federal government seeks proposals for work from engineers, it often asks for a listing of 10 recent projects similar to those proposed. In a brochure, a reasonable estimate would be to list 10 to 50 recent projects, giving the type of work performed, plus location, name of client, date of completion, and perhaps a paragraph or two of description.

For a firm that does have a large list of impressive clients, a separate client listing can be valuable. If your firm has worked on several projects for only a few very large clients, the listing of the projects under the clients' names can also be helpful.

The role of photos in professional brochures can hardly be overemphasized. When a client says, during or before a meeting, "Let's see what you've done," the client means exactly that. Photos, rather than words, sometimes make the strongest impression.

Putting It All Together

The material assembled for an effective brochure is often twice as long, before editing, as the final product. The first step in editing is to photocopy all the proposed material. Then select appropriate portions, arrange their sequence, size the photos, and plan the page layout. While an outside professional can refine the written copy, the basic information should come from in-house personnel. For most professional brochures, the sequence of sections is: introduction, description of the firm's principals, services available, projects, and clients.

Most brochures emphasize projects and clients. They illustrate clearly and concisely how the consulting firm's services, staff, and general history relate to high-quality performance. For a general engineering firm with 25 or more members, however, this could easily result in a finished product that is too long to maintain reader interest. Therefore, as was noted earlier, many firms opt for using a binding that enables them to add or omit pages conveniently—in accordance with the type of project and/or client being solicited at a given time.

Further information on preparing brochures is available from many professional public relations agencies and large printing firms. One of the more complete sources of data for engineers can be obtained from the American Consulting Engineers Council, 1155 15th Street NW, Washington D.C. 20005. Ask for their publication titled "The Brochure on Brochures."

The Last Word

The brochure is indeed a real sales tool for engineering consultants. But the salesperson has the last word in making a sale. Consulting services cannot be sold by mail like garden seed or file cabinets. In a face-to-face meeting with a potential client, the brochure can effectively present the consulting firm in "white tie and tails," so to speak—even though the firm's sales representative may be wearing a drab gray jacket and black pants. (We will examine sales meetings in depth later in this section.)

Since the brochure often makes the first impression, every word, claim, and photo within it must be aimed at characterizing the firm and its capacities accurately and professionally. Statements made by a salesperson can be amended rather easily, if necessary; but written words will help or haunt a company for as long as they circulate.

An example of the contents of a brochure used by one consulting engineering firm appears immediately following this section.

Brochures for New Consultants

How does a new engineering consultant prepare a brochure without ever having obtained a client or completed a project? The answer to this is:

1. Use the same general outline for your brochure as given earlier, i.e.:
   a. Introduction
   b. Description of firm's principals
   c. Services available
   d. Projects
   e. Clients
2. In parts *d* and *e,* Projects and Clients, use—with suitable permission—data on Projects and Clients you handled while on the payroll of another firm.

   Thus, under Projects you might say:

   > While employed by XYZ Engineering Corporation, Mr. Doe was Chief Project Engineer for the $950-million Sky High Nuclear Power Plant. He also served as Project Engineer for a number of other large design assignments, including those for the Safe Chemical Company, West Oil Company, and Daily Newspaper Corporation.

   Under Clients you might say:

   > Clients served by Mr. Doe while on the staff of XYZ Engineering Corporation include Sky High Nuclear, Safe Chemical, West Oil, Daily Newspaper, and others.

Once your firm has a record of finished projects, you will insert these in place of those you worked on while employed by someone else. You will also insert your own client names once you have enough of them. There may be a period during your first 2 years in business in which you will have to list both your projects as an employee and those as an independent consultant. The reason for this is that either list alone might not have sufficient selling power with prospective clients.

HANDLING MEDIA EVENTS

Consulting engineers sometimes feel they would like to gain public attention through the mass media on radio, television, and in the press. For some kinds of professionals—movie stars, stock market analysts, and politicians—techniques of dealing with the media are a basic aspect of daily public relations. But mass-media appearances for the engineer are rare. When an engineer does appear on television or radio, it is usually in the context of providing brief background information on some unique event of community interest.

Such events could include celebrations commemorating the opening of new buildings or pollution-control facilities, for example. But they can also include catastrophes such as structural failures of a bridge or power plant or other problems affecting public safety.

During the reporting of such events by the media, the risk of the engineer being misunderstood or misquoted is substantial. Therefore, even if you're an expert in the field related to the occurrence (but not directly involved in the event or its causes), the best course of action is to avoid comment to the media. But, if you are directly involved—as the chief expert hired to resolve the problem, for instance, or as a member of a firm connected with engineering aspects of the relevant structures—you might have to talk to the media. Keep in mind, however, that adverse effects can result if the matter is not handled with great care and restraint.

Here are guidelines to help you, as a consultant, avoid potential problems in dealing with the media:

1. Unless you are totally prepared to provide reliable and detailed information about the event, do not comment on it; avoid any interview situation. Simply tell reporters and other media inquirers that you are not prepared to discuss the event.

2. If you *are* prepared with adequate information, first confer with your attorney about granting interviews on the topic. Get the attorney's approval before agreeing to submit to interviews; then ask that all meetings be held at your office. This will give you greater control over the interview.

3. Never agree to speak "off the record." A reporter's use of this phrase does not guarantee that your remarks will be kept confidential.
4. Record the interview on tape so that you have a word-for-word record of your statements.
5. Be fully prepared for any questions or other challenges which could arise. Be sure to know all facts thoroughly and attribute any information obtained from other sources to the appropriate parties. Always consider how your words will be interpreted on film, radio, or in print, *before you speak.*

A recent public-utility accident is infamous for the uproar and confusion that accompanied its reporting by the media. Nearly everyone involved lacked preparation for discussing the event publicly—government officials, utility company spokespersons, and reporters. The result, in at least a few cases, was near hysteria among part of the public. Consultants and others who have had contact with the media have since learned that good preparation and restraint provide the best shield against problems in dealing with news events.

SPEAKING TO GROUPS

Still another facet of public relations is the opportunity to promote your firm through public speaking. Too many people reject this opportunity because they fear they will not make a favorable impression in front of an audience. However, public speeches are actually quite similar to the type of passive public relations in which we engage daily. The chief difference is that a speech requires turning one's passive PR skills into active ones, often with a larger audience. Many engineers discover that speaking to a local group on topics that they know sells makes a favorable impression. It can be a highly rewarding feeling to gain personal recognition through public appearances.

If you start your speaking efforts by making short speeches to civic or fraternal groups to which you belong, you will soon master the ability to get up before an audience and talk with confidence. Before long, you should find it enjoyable to speak to other groups with whom you may not initially be as well acquainted. Many local libraries, commercial firms, and professional organizations are happy to sponsor speeches on topics of interest to their members.

A public presentation can be prepared in much the same way as a magazine or newspaper article. The writer identifies topics he or she feels people would like to learn more about, contacts representatives of the groups which might be interested, then prepares an outline of the speech. Next, a rough draft is written. He or she can practice the speech in front of family, friend, or selected members of the firm. The speaker should be familiar enough with the talk to recall its major points. This will enable him or her to avoid having to read from notes while speaking.

Good speeches contain examples of actual case histories to illustrate particular points and get the listeners involved. Brevity is also important. Usually, 20 minutes or so will be enough time for you to get your points across, while interesting your audience and leaving them intrigued and wide awake.

Further, you or the group sponsoring the speech can hand out news releases, brochures, and other printed information during the presentation. News releases should be accompanied by photographs of the speaker (5×7 or 7×10 in) and any other newsworthy aspects of the topic discussed, plus copies of the speech itself. Even if the media does not cover the event, your speech and documents can lead to added business for your firm.

PROFESSIONAL MEMBERSHIP

As we noted in Section 1, membership in professional organizations can be valuable to the consulting engineer. Engineers are rather notorious for joining such groups and subsequently avoiding meetings and other kinds of active participation. Much personal and professional recognition can be garnered by the engineer who does take an active role in a professional organization. One's position in such a group can carry prestige that will impress colleagues as well as clients. The contacts the engineer establishes in the course of performing services for the group can lead to interviews with potential clients who might otherwise be hard to approach. An active professional group member also derives satisfaction from the contributions that her or his participation makes to the profession and to society in general.

SOME CONCLUSIONS ABOUT PUBLIC
RELATIONS PROGRAMS

Consulting engineers know that their relations with their public—fellow engineers, clients, friends, the media, and others—hinge on professional behavior. But too many consultants think that the essence of professionalism in public relations is the ability to wine and dine their potential clients and business acquaintances. This idea may have held some weight in the past; but nowadays wining and dining and gift giving are often viewed as common but superficial ways of conducting business.

Common sense and social custom do dictate that one should offer to buy lunch or dinner for potential clients or give small presents to business associates—at appropriate times such as luncheon meetings or holidays, respectively. However, an elaborate dinner or lavish gift may do more harm than good. Wise consultants know that it can be a turnoff to a potential client if one appears to be trying too hard to impress. The general guideline, therefore, is that one should not rely heavily on surface amenities; when in doubt about the appropriateness of such an action, it is best to exercise restraint so as not to risk looking overly lavish.

A well-coordinated, professional public relations program uses an integrated approach. Starting with the appointment of a skilled PR director to spearhead the consulting firm's overall public relations and advertising efforts, the firm utilizes everything from advertisements, articles, direct mail, brochures, and letter writing to meetings with clients, media contacts, public appearances, and follow-up phone calls (among other tools of the trade).

Often the public relations director—a person with the ability to write and speak well and to organize and handle information creatively and sincerely—can be appointed from within your own staff. If your firm seems full of great engineers but not full of writers, then the search for a public relations head might extend to friends and business contacts or to agencies that specialize in public relations work.

Once you have chosen your PR director (who might even, of course, be yourself), then you can begin identifying your markets and draw up a long-range plan (1 to 5 years) of operations based on the principles we have examined, and your future goals for the market.

EMPLOYEE RELATIONS

Relations among the members of the consulting firm are another important concern, closely related to PR in general. Every consulting firm is only as good as the people it employs. Moreover, employees work best when they are happy with their situation in the firm. Wages and benefits are not the only factor affecting employee satisfaction. Also important are management's concern for the employees' personal welfare, their ideas, and their advancement in their careers.

Good employee relations start with a clear, logical, and sensitive statement of company policy. Ideally, all policies should be written and printed in the form of a handbook for staff members. Such a handbook should introduce new employees to the general ways in which the firm operates. It should also provide a clear explanation of all the firm's rules and regulations, without sounding too rigid, like a military manual.

Another crucial ingredient of employee relations is the staff conference. Meetings can take many different forms, depending on the specific purpose and situation in which they are held. For example, one could hold an open discussion for employees to talk about current projects or any topics that happen to come to mind; or particular company policies, new job prospects, or promotions could be discussed. In addition, suggestions can be solicited for improving production, among other topics. Regardless of the content, staff meetings should be held fairly regularly.

Important news affecting the firm can and should be disseminated to employees quickly, through meetings or bulletins. In nearly all types of consulting firms, the regular meeting can be a gratifying source of personal recognition, education, and even relaxation for the staff.

CLIENT RELATIONS

The clients of consulting engineers can be said to encompass three basic categories: (1) clients that operate in the private sector of commerce and business; (2) clients that work in the public sector (e.g., local government agencies); and (3) fellow professionals (such as other engineers and architects). Let's examine some characteristics of each and how these factors affect consultant-client relations.

The Private Sector

Many engineering firms are capable of satisfying the needs of clients in the private sector. Generally, the firm that a client selects is the firm that is known best by the client's selection committee. Obviously, the consulting firm that can get in touch with the client and can demonstrate superior problem-solving skills and reliability in meeting challenges will become a prime candidate for the project being considered. Frequently, the same firm will be the first to receive additional projects and referrals to other clients.

The main task of client relations for the private sector is to explore, explain, and document the consulting firm's abilities. Client relations should be maintained on a personal, one-to-one foundation, in which the representatives of both the client and consulting firm feel comfortable speaking to each other on a first-name basis.

Clients in the private sector can usually be placed in two classes. Large clients often have their own technical and engineering staff, while smaller clients do not. The difference carries advantages and disadvantages. For instance, the engineering staff in a larger firm is often more knowledgeable and more understanding about the consultant's work than the head of a smaller firm that lacks an engineering group. At the same time, the staff of the larger firm is likely to be more opinionated. The smaller client, who has no engineering staff, may permit the consultant to exercise greater authority over design; but the same client may present more problems in settling fees and accepting engineering concepts with which he or she has not had much experience.

Regardless of which type of private-sector client you seek, your best sales will probably come through a third party—i.e., referrals from previous, satisfied clients. Such referrals are cultivated through continual communication with all good clients. Since having more than one job with a given client will testify that you have good relationships with them and that your firm does high-quality work, your sales team should emphasize repeated sales to one or more clients.

One technique that seems to bring larger numbers of repeat sales is the use of the same consulting personnel in both the start-up phases of client relations as in project completion.

The people best qualified to oversee client relations for the firm will be the senior managers or principals. Clients will have greater confidence in your services if they know that your firm is interested enough to communicate with them from the highest levels within the firm. This will help assure clients that the whole firm is committed to each project—as opposed to responsibilities being parceled out to lower-ranking members in a fragmented manner. Whatever the potential client's problem is, the presence of senior consulting personnel in client relations can often turn the problem into rewards for both the consulting firm and the client.

The Public Sector

Since the process of selecting a consulting firm for each job depends on the size of the public agency (whether it is local, state, or federal), the complexity of client relations varies too. As in the private sector, the consulting firm's public relations representatives need to acquire and maintain political contacts at as many levels as possible.

There are two main advantages in establishing such contacts and personalizing them. First, it is a relatively simple way of spreading the word about your firm in the large public-sector market. Second, the staff members of many public agencies are usually helpful in defining agency needs, procedures, and project requirements.

Much has been written about government competition with private firms, but the issue is not as critical as the writings would make it appear. Good public relations by the consulting firm and the use of personal contacts with the right agencies and people can often lead to some rewarding project assignments.

An important aspect of public relations with government is the need for clear explanation of the consulting firm's capabilities and limitations. Strict record keeping by public agencies often mandates that engineering firms working on government jobs detail their services through all phases of a project.

Additional points to keep in mind for good client relations in the public sector include:

1. Public-sector clients are often more difficult to satisfy mainly because they feel a great deal of public pressure to complete projects quickly and at minimal cost.

2. The firm must maintain constant contact with the public agencies' staffs in order to demonstrate its continuing interest in the clients.

3. The consulting firm's public relations staff should generate fresh, creative ideas continuously and should show the public officials that the firm knows how to enhance their public standing.

4. Despite an old cliché that says only bad work will be noticed, high-quality engineering will be remembered by each client and will likely bring additional assignments.

Fellow Professionals

When working and corresponding with architects, engineers, and other colleagues, one should always strive to understand clearly how much energy, time, and effort these fellow professionals have devoted toward getting their own clients. In many cases, one must recognize that particulars of design, scheduling, cost, etc., have been agreed upon by the professional and the client before outside consultants are hired.

What, basically, would the fellow professional want from your firm? Certainly, he or she wants appropriate engineering skills (which may be lacking in his or her firm for specific jobs), plus competence and the motivation to do excellent work on schedule and within the cost budget requirements.

To secure work with fellow professionals, the consulting engineer must continually seek to solidify his or her reputation as a consistently professional producer. Constant involvement of sales staff and management personnel is mandatory. The architect or engineer who hires other consultants will not easily tolerate talk that is not followed by concrete action.

Perhaps the most important consideration of client relations in this regard is fee—i.e., negotiation and payment. Regardless of the fee structure used (hourly, percent of cost, or lump sum), strong, personal, one-to-one client relations must be maintained. This will be helpful not only in avoiding fiscal complications but also in keeping track of the limits of service.

The second most important consideration, many feel, is the scope and quality of the services rendered. Good client relations in regard to services rendered are based largely on the consultant's ability to listen to the client. The essential ingredient therefore is familiarity with the client's needs. Incidentally, listening can help you to build more profitable relationships than many other efforts combined. Some of your best project leads could easily come from fellow professionals.

MARKETING YOUR PROBLEM-SOLVING SKILLS

The chief role of the consulting engineer is as a problem solver. If clients never had problems, there would be no need for professional engineering consultants. Even when the consultant's client is a fellow engineer, the consultant is unique: She or he is the person who can help remedy the client's difficulties. In marketing your problem-solving abilities, a client will become eager to "buy" if persuaded that your firm can solve the problem at hand efficiently, effectively, and economically.

Most beginning consultants have one or two sources of work lined up before they open their private practice. From that foundation, the number of projects may expand somewhat, based on the quality of the performance on initial jobs. This evolution, however limited, is the beginning of company growth. But more effort is needed for the firm to attract increasing amounts of new, lucrative work.

The first step, discussed earlier in this section, is to create a list of 500 to 1000 names of possible work sources. These should be categorized by type of potential work and by locations of the potential work and by locations of the potential clients' offices. (Location is important for two reasons: How far from your office can you afford to travel for jobs? And how can you provide services that will be superior to those of firms located closer to your potential clients?)

Usually, it is not feasible to travel to each office on your list and ask if your services might be needed. The first logical step in actual promotion is to write a letter to each potential client and follow it up with a personal phone call.

Let's examine the probable outcome of a typical mail campaign. Traditionally, responses to mail requests usually come at a rate of from 1 to 3 percent of the number of letters sent out. Thus, if you write 100 letters, you can expect between 1 and 3 replies. Furthermore, the normal proportion of replies that will become sales is around 30 percent. Consequently, your 100 letters might result in no new clients. This should not be a cause of discouragement, though. The consultants who profit from letter writing are those who are willing to persevere.

Estimates show that 1000 letters might generate 10 to 30 replies and could produce between 3 and 9 clients. But that is rather optimistic. Your letters, however well written, might arrive at inopportune times, be mislaid, and possibly not get read. That is why a sensible approach is to write 15 to 25 letters at a time; then, after a 2-week waiting period, call the clients to whom the letters are sent. This procedure is easier to control and gives you more time for thorough follow-up, while providing a greater balance between mail promotion and other daily tasks.

Introductory Letters and Interviews

In preparing the introductory letter, it is important to recognize that potential clients are not waiting breathlessly for promotional letters to arrive. Clients targeted by consultants typically receive between 5 and 25 unsolicited letters every day. Such letters fill wastebaskets. Addressees particularly do not relish the idea of wading through long letters of several pages. Clients tend to throw out most unsolicited advertising (as you may well know from your own habits).

These facts of life lead to some clear guidelines. The promotional letter competes for attention. Unless yours stands out, it is likely to get thrown out. Therefore, your letter must be brief, concise, eye catching, and personally directed. It should be typed with no visible corrections and signed by hand, not by machine. (Nearly anyone who receives both hand-signed and machine-copied letters—usually from politicians at election time— can attest that the difference is obvious. Since you are not writing to a neophyte, it is safe to assume the recipient of your letter can spot the differences.)

Ideally, your promotional letter will be no more than three short paragraphs, with each paragraph about four lines. If you can put the appropriate words into one succinct paragraph and send it to the president of the company you wish to contact, that is fine. Even if you plan to describe your company and its services, it must be done in 100 words or fewer—preferably fewer.

In preparing the letter, ask yourself, "What is the specific purpose of this letter?" For consultants, the goal is to get work, to tell people how good the firm is, and to obtain an interview with the potential client.

A series of letters could do all these things and much more. Barring a miracle, however, none of the letters will automatically get you a commission by return mail.

The point is that for the practicing consultant, selling is not necessarily very difficult, if only the engineer can get the opportunity to present his or her case. That should be the aim of the introductory letter.

The opening paragraph of your letter should present cogent reasons why the potential client might want to meet with members of your firm. Contrary to what some people think, there is no secret formula or set of words that guarantee instant success. The main emphasis should be on positive ideas rather than on neutral or negative ones. Consider the following examples of possible opening sentences:

1. When you require professional structural engineering services, we can . . .

2. We are practicing professional structural engineers who can . . .

3. If you need a professional structural engineer, we can . . .

The first opener makes no bones about its purpose. It says that you *can* help and that you are available; it carries a positive statement. The second example starts off by blowing the firm's own horn and continues to boast. That is not likely to get a positive response. The third opener pleads, "I hope you have a job for me." That is quite negative.

At this point we could continue looking at examples of promotional letters. But readers who tried to follow the style of the examples would wind up generating lifeless, "canned" letters. Promotional letters should have all the vitality and personality the writer can give them. Therefore, it is best for each consultant to explore her or his own ideas, based on the general guidelines we have discussed.

Telephone Follow-Up

Why should you telephone for follow-up? First, because it enables you to find out if your letter was received. And if you are politely persistent, you might learn the effect that the letter had on its readers. Based on these findings, you can decide to ask for an interview, send an additional letter (if the first was not received), or change some of the wording in future letters.

Selling by telephone is a major facet of American business. Even though the salesperson can leave his or her firm's phone number when the person being sought is not available, very few calls are actually returned. Salespeople often become hardened to the many kinds of brush-offs that can occur on the phone. Consultants, too, run into all the typical problems—such as reluctant recipients of calls and the formidably protective secretary. But most callers ultimately can reach their party, if they are unobtrusively persistent.

Still, some engineers feel they are at a loss for words on the phone, especially when trying to sell themselves.

There are no easy solutions. But simple honesty is hard to brush off. To the secretary who wants to know why you wish to talk to Mr. Jones, you might say, "I'd like to discuss the letter I wrote him." For Mr. Jones himself, if all else fails, you can start by asking, "Did you receive my letter?" At the very least, the sound of Mr. Jones scrambling through his pile of correspondence should encourage you to continue. If you next state simply that you would like to meet him, with no obligations on his part or on his company's part, the odds are fair that your wish will be granted. This is generally true for about 10 of 25 follow-up calls.

Salesperson Training

Sometimes a prospective client will request that certain types of personnel from the consulting firm not be present at a meeting. The client may even make it clear that she or he does not want the salespeople to attend. That may mean that the person who wrote the letter, who made the phone call, and who sold the interview is asked to stay away. Regardless of which member(s) of your firm finally attends the meeting, the initial job is to sell the representative as a sharp, knowledgeable member of the company.

For your firm to be fully prepared to meet the requirements of all sales situations, all personnel should be trained to deal personally with potential clients. Remember that every contact anyone in the firm has with people outside the company can affect its overall image.

One of the best tools in training each member of the firm to become an effective salesperson is to analyze and discuss the reasons for the loss of any recent potential projects to competitors. This can be referred to as the lost-jobs analysis. Many consultants find that the reasons given at first are simply excuses, such as "The competitor has too much clout," or "We don't have any connections." Discovering the real reasons may require a considerable amount of probing and judging—one becomes sort of a "sales detective." Oftentimes, salespeople, secretaries, and even suppliers can—knowingly or unknowingly—provide helpful clues. Exhibit 2-7 shows desirable characteristics for salespeople.

Most failures are based on what the potential client perceives as a lack of competence in the firm. Even in cases where political affiliations account for much of the ability to acquire a particular project, proven competence emerges as the final determinant for who gets the work. In political situations, noncontributors to the party in power rarely get the job; but the biggest contributor is not always the one to land the contract. The contract winner is usually the firm that shows itself to be the most reliable and professional.

The logical assumption of the lost-jobs analysis is that something said or done by a member of the consulting firm is a factor in losing a job. A related factor might be the failure to say or do something.

The goal of the lost-jobs analysis is not to name the person(s) responsible but to find the cause of errors and to correct them before such errors

EXHIBIT 2-7   Sales applicant rating form.

Desirable characteristics	Comment	Numerical rating
1. Is the applicant a self-starter?	Prodding salespeople into action is self-defeating.	20
2. Is the applicant a good analyst of both people and situations?	Applicant must instinctively feel what people want and learn to recognize both favorable and hopeless selling situations.	20
3. Is the applicant persistent and consistent?	Most consulting selling is "cold calling."	15
4. Is the applicant confident, and does she or he have a neat appearance?	Too many potentially good salespeople are overly loud in appearance and destroy potential clients' confidence in them.	10
5. Does the applicant have a thorough knowledge of engineering?	The salesperson must be willing to learn engineering or already know it. The applicant should work on commission if he or she must learn engineering.	20
6. Does the applicant know what the firm can and cannot do?	The applicant must recognize the firm's skills and limitations.	15

Note: In general, the closer an applicant scores to 100 on this form the more productive an employee he or she will be.

happen again. The golden rule of judging mistakes is that the only person who makes *no* errors is the person who does not do much of anything. Virtually everyone in every firm makes a mistake at some time or other; generally, if one continues to work, one will continue to make occasional mistakes. With good preparation through the lost-jobs analysis, mistakes can be minimized effectively in many cases. Further, a good sales presentation will negate the effects of many errors. A regular program of meetings to discuss positive aspects of the firm's sales program and ways to improve it will further enhance promotion efforts.

The Value of Sales Reports

The amount of detail and the depth of sales records is less important than the benefits accrued by keeping the records themselves. Highly elaborate records do not serve much of a purpose; but simple, straightforward reports are a valuable exercise in improving sales.

Each principal of the firm should have copies of weekly sales reports. The purpose of the copies is not to check up on salespeople, although that may likely happen. The real purpose is to enable the firm to pinpoint areas that can be improved and areas in which particular individuals can lend further assistance—things as simple as the identification of colleagues and business associates who can act as intermediaries and contacts with potential clients.

There are two basic types of sales records every firm should maintain. These are: (1) weekly reports of general sales efforts and (2) records of specific efforts directed toward an individual client.

A typical weekly report indicates briefly the actions taken every day, plus the results of those actions. Figure 2-2 is a typical weekly report, containing raw working data. Such a record could also include informa-

Weekly Sales Report		
Date	Action Taken	Result
10/1	Sent letters to 21 prospects.	Will start calls 10/15.
10/1	Talked to Brown Co. (J. Harris)	Asked for data on company and about new plant addition. Will send letter and brochure.
10/2	Prepared letter and brochure for Brown Co. (J. Harris).	Will call for appt. 10/16.
10/3	Phone calls to ...	Have appt. w/Jones 10/16.

Figure 2-2 Example of typical weekly sales activity report.

tion on company conferences and other items such as job analyses. One copy of this report should be bound in a permanent file and kept for at least 10 years.

The second type of sales report is a record of all efforts devoted to a specific client. Here, too, the firm may have active and inactive files; but regardless of the level of activity with a particular client, the file should be kept for several years.

Since the records of contacts may get lengthy, a hardcover three-ring loose-leaf binder (8½ × 11 in) can be used, with alphabetical tabs, so information can be added and deleted with ease. Additional copies can be made and kept in permanent files, while the original copy can be kept in the sales office. This will help the sales personnel to update the permanent files so they reflect the current status of each client. Pages can have a simple format, as in Figure 2-3.

Brown Company (Name, address, telephone no.)		
Date	Action Taken	Results
10/1	Talked w/ J. Harris, Plant Engineer.	Asked for more data.
10/2	Sent letter to Harris.	Will call 10/6.
10/6	Called Harris.	Out. Will call 10/7.
10/7	Called Harris.	Interview appt. for 10/22.

Figure 2-3 Sales report for a typical company prospect.

Too many firms expend extreme amounts of energy keeping highly elaborate sales records. Since the object is to secure added projects and clients, the firm should not place too great an emphasis on complex reports. Efficient, concise records like those shown in Figures 2-2 and 2-3 will provide an overview of promotion efforts, which will help the firm and its sales team to determine how many calls, letters, meetings, etc. are warranted in a day's work.

One might find, for example, that it takes 600 letters and follow-up calls to secure five new clients (a reasonable estimate). This might be done in less than 6 months. And remember—without the continuing promotion efforts, sales may shrink.

Practice Sales Sessions

One way to train an efficient sales team is to set up mock sales situations. Each mock-up can deal with specific expected sales problems and ways to resolve them. When mock-ups are first attempted, your participants may feel somewhat awkward. But it is certainly better to feel awkward during practice than in real sales situations.

The goal of any mock session is to establish a sense of familiarity, facility, and confidence in everyone who may someday be involved in either face-to-face sales meetings with clients or in telephone conversations with clients and other people from outside the firm. One session might be devoted to using telephones, while another could cover in-person meetings.

In telephone sessions, participants can practice securing interviews. The salesperson tries to get an interview with the prospective client, while the person who plays the client role offers reasons for rejecting the interview. Many valuable techniques can be learned from a mock situation, especially if the dialogue is recorded and later listened to by the sales team. Mistakes often become glaringly obvious on the recordings.

The other type of practice session, the mock interview, can also be quite helpful. Members of the firm will get a chance to play the roles of both client and salesperson. Oftentimes, such sessions point up a lack of sufficient information on the part of members appointed to tell the client about the firm and its capabilities in specific areas. Another problem that often shows up in these sessions is inappropriate handling of answers by salespeople. In many cases, the salesperson's replies to questions are incomplete, unclear, or may even sound arrogant.

Client-Project Analysis

When an interview is secured with a potential client, the consulting firm should analyze the particular situation involved and determine items that require specific coverage during the sales presentation. Some areas one might consider are:

1. What is the client's problem?
2. What other goals does the client have? Goals can be complex. For example, they can include such peripheral concerns as ego satisfaction and public attention.
3. In what specific ways can the client benefit from the proposed work?
4. What will be your tools for communicating with the client (e.g., a prepared talk, brochure, plans, or renderings)?
5. In what order will you approach each aspect of the interview?
6. What areas of potential resistance and agreement could arise during the presentation?

7. What does your firm think the client actually needs?

8. What is the limit of your financial capability to carry design costs before payment? What is the client's limit?

9. How might you respond to the client's desired dates for project completion? Is the completion date definitely feasible in the current situation?

10. What areas of resistance might arise when you are engaged in closing the sale?

Rehearse the Presentation

Once your analysis is completed, the next task is to select personnel, materials, and data that will be needed to make the sales presentation. After this is done, the firm can organize and sequence the presentation's contents. The leader of the sales team can be designated (if he or she has not already been chosen), and speaking roles and necessary support positions can be assigned.

Next, the firm can rehearse the interview. The rehearsal should focus on background information and on finalizing priorities. Among the topics that can be analyzed in the rehearsal are:

1. Any additional questions or areas of uncertainty that members of the sales team have about the project and client

2. Any necessary modifications that should be made in the personnel, materials, or other aspects of the presentation

3. Planning for all possible questions that the client could reasonably ask

4. Planning for unexpected absence of a team member

5. Planning for an unexpected loss or failure of presentation materials and equipment

6. Planning for recourse in case the firm begins to develop any doubts about the likelihood the project will actually become available (e.g., contingency plans for the possibility that a pending environmental impact permit may not be issued on a given project at the expected time)

7. Determining if any outside consultants should be included in the sales presentation or in the work itself.

The last two items are both serious concerns. Their implications for the present and future status of a potential project are strong. For example, facts may develop which indicate that further pursuit of the project would be a waste of time and money for the consulting firm. All possible precautions should be taken from the beginning of each promotional effort to prevent such situations from occurring. But the consulting firm cannot control all factors affecting every project. So it is best to examine the project's ongoing status before making the actual sales presentation. Conditions for both the consulting firm and the client could become seriously complicated if the consulting firm takes on a project and later discovers that the job will not be feasible.

In contrast, if the firm determines that some modifications must be made in the presentation—such as calling in an outside consultant—then other actions must be taken. If a consultant from outside the firm is needed, the terms of her or his relation to the project should be immediately worked out in writing. When such a step is taken, the direction of those parts of the project that are related to the outside consultant's expertise usually go to that consultant and her or his firm, automatically. Further, if the project is such that few firms could do the entire job themselves, then some type of joint venture can be sought. Because the added elements of a joint venture can break the flow of the sales presentation, extra time and effort should be devoted to developing and practicing a smooth, integrated sales talk. It is important that the sales group appear as a single, coherent entity to the potential client—even if the project will involve more than one consulting firm.

The Winning Performance

After the practice sales presentation, the promotion team should be ready to present the firm and its services, experience, and capabilities to the potential client. At this point it can help to follow a few simple guidelines to turn the actual interview into a winning performance for the team.

One of the requisites for establishing a thoroughly effective line of communication with the interviewer(s) representing the client organization is to cultivate a relaxed atmosphere of mutual cooperation. Salespeople should take care to appear sharp and knowledgeable but not overly aggressive. A well-planned, structured presentation can then proceed, allowing both flexibility and an ordered approach to topics of varying priority.

Whenever possible, the consulting firm's team should seek an individual interview—as opposed to the multiple consultant interview. The multiple interview usually involves several consulting firms, each giving their presentation to the client on the same day, one after the other. Such mass-production interview sessions tend to leave the client with little more than a blurred impression about the whole group of firms rather than a distinct recollection of the characteristics of individual presentors. Oftentimes, the team that wins the sale is the one that enters last. Multiple interview sessions cannot always be avoided, however, since many clients consider them to be more economical.

When you must engage in the multiple interview session, the chief guideline to follow is the same one that should guide every sales presentation. A consulting firm's best hope is to present its ideas in ways that will emphasize the group's uniqueness and its special capacity to fully satisfy the client. How does your firm differ from competitors? What makes it stand out from the crowd in its ability to complete particular projects economically and effectively? These are the sorts of questions that the sales team should focus upon.

The sales presentation should pinpoint the specific project(s) under consideration and demonstrate clearly the firm's ability to meet all relevant challenges. Salespeople should not resort to gimmicks, jokes, or other quirks which may divert attention from crucial aspects of the session. Every appointed speaker on the team should be allowed an appropriate amount of "floor time" for his or her portion of the overall presentation. Practicing the presentation beforehand will help the team gain control over the order and content of the presentation.

One of the best sales pitches a consulting firm can have is a well-reasoned demonstration of how the firm is uniquely qualified to do the job at hand, based on past projects. However, the temptations to claim greater capacity or expertise than the firm actually has can be substantial. Interviewers' questions sometimes seem to lead the respondents in that direction by stressing the most impressive characteristics of particular firms. It is extremely important that the team does not at any time make claims that the consulting firm will not be able to meet.

There are several other important *do*s and *don't*s for the sales presentation. One is that salespeople should maintain a pleasant demeanor and agreeable facial expressions. Frowns and scowls, intended or unintended, can produce negative impressions about the consulting firm. Also, the team should avoid arguing with the client's representatives. The team, after all, is not present to tell the client what is needed but to sell what the client wants. Further, the team should treat the potential job as a project that the firm would like to undertake but not as if the survival of the consulting firm depended upon it. Neither should the job be treated as a low-priority item on a massive backlog of current commitments.

When additional consultants are called in from outside the firm to assist in meeting the client's needs, the sales team should take precautions to dispel any notion that the use of outside consultants indicates some sort of weakness in the firm. The ideal response to such notions is to show clients that the added assistance effectively makes a good firm superior to competitors (even superior to those who would employ only in-house talent). Outside consultants broaden the firm's vision and creativity in ways that might not occur with a staff from one firm only.

The same point can be made for joint ventures, even though some clients hesitate to engage in joint-venture projects usually out of (1) uncertainty about who has the ultimate responsibility in the venture and (2) about what would happen if one of the firms involved failed to uphold its end of the deal. As was noted earlier, joint venturers should prepare a definitive description of their respective roles and obligations in advance. Such preparation may help ease the qualms of potential clients who are unsure about joint ventures.

Regardless of managerial arrangements, all questions should be answered clearly and conclusively. If no specific time for the end of the interview has been established, the presentation team will need to decide when all items have been covered properly and all questions answered. If some of the materials used during the meeting are to be left with the client, they should be distributed, with the explanation that the materials can be used as the client further considers the consulting firm. Otherwise, all materials should be collected, and the client should be informed that the materials are not for public distribution.

At that point, there is no reason to continue the meeting. There are many ways to conclude the interview, but the simplest is still to thank the interviewers for their time, smile, and say good-bye. Certainly nothing will be gained by appearing hesitant or awkward in departing.

After the interview, the sales staff should allow a few days to pass (or wait until the date agreed upon with the client for response) and then contact the potential client by phone. The standard approach at that point is to ask if any additional information is needed. It is best not to prod the client for a decision about the project, although this may be your primary concern. The conversation should be kept brief and unobtrusive. If the client has made a decision, you will most likely be informed about it, whether or not you ask the question directly.

Winning, Losing, and Learning

Clients frequently have an enormous amount of difficulty in selecting the winning consulting firm for a given project. For instance, the client may have initially selected a pool of 30 or more well-qualified firms, then narrowed this list to include 5 firms for the interview. Once the 5 firms have been interviewed, the client's troubles really multiply because now all 5 may seem like they could definitely do the job satisfactorily.

The final selection is often determined by relatively minor impressions and recollections about the interview. At that stage, personal quirks, mannerisms, and remarks of the consulting firm's salespeople can play a large role in determining which firm will get the job. Generally, selections are based primarily on a serious, carefully reasoned evaluation of each firm's presentation. But the final results often make subjective characteristics such as personal appeal strong swayers of opinion. That is why a calm, confident, and pleasant manner can be one of the strongest assets that members of the sales team possess. This is equally true in cases where the salespeople and interviewer are already acquainted with one another (as often occurs when the firm has offered consulting services to the same client in the past). Figures 2-4 through 2-7 show typical rating forms used to evaluate consulting firms and their proposals.

Once the interview is completed, it is not unusual for members of the sales team to feel elated—or, in some cases, depressed—in regard to the team's performance. It will not help win a job if the salespeople start moaning and groaning about "what could have been done," however. But it can be helpful to analyze the presentation to ask, "Did we cover all that we needed to cover?" and "How, if at all, could the presentation have been improved?"

All members of the firm should be prepared to handle winning or losing the project with tact and with an eye to the future. If the client calls to say that your firm has been selected, you should not react ecstatically but should indicate that the firm is thankful for the opportunity to serve the client.

Conditions are still somewhat tenuous immediately after the client announces the selection. That is why one should not convey the impression that the firm desperately needs the job to survive since doing so might damage the firm's image in the eyes of the client.

If the project is lost, the firm should attempt to find out—as quickly and as unobtrusively as possible—why it was not selected. An interviewer or other client contact may provide the answer or at least a clue. Your sales team (or you, as an individual consultant) can learn as much from sales losses as from contracts acquired. If the firm can find out why it lost a particular job, avoiding that mistake in the future should be simpler and should substantially raise your chances of gaining new projects.

When a project is obtained, the consulting firm should follow up with some action that will help to solidify the relationship with its new client. This action can be the signing of the contract or submission of preliminary proposals to get the work underway.

But if your firm was not selected for the job, it does not mean that you have lost the client forever. It means that only one project has been

		Outstanding	Very Good			Good		Adequate		Marginal		Inadequate		
	Points	10	9	8	7	6	5	4	3	2	1	0	Multiplier	Score
(GENERAL)	General Competence												x 0.75	
	Overseas Experience												x 3.00	
	Home Office Support												x 1.25	
(STUDIES)	Environmental Assessment												x 1.50	
	(As Required) Planning												x 8.75	
	(As Required)												x 1.00	
	Management Staff Organization												x 0.50	
(DESIGN)	(As Required) Design												x 1.00	
	(As Required) Design												x 1.25	
	(As Required) Design												x 2.50	
	(As Required) Design												x 1.25	
	Foundation Investigation												x 0.50	
	Structural Design												x 0.50	
	Electrical Design												x 0.50	
	Construction Supervision												x 0.75	
													ITEM A TOTAL	

(a)

The proposer's understanding of the problem and the effectiveness of the proposed approach and work plan is evaluated with reference to the criteria below:

	OUTSTANDING	VERY GOOD			GOOD		ADEQUATE		MARGINAL		INADEQUATE		
Points	10	9	8	7	6	5	4	3	2	1	0	Multiplier	Score
Understanding Problem— Quality of Overview												X 6	
Suitability of Methodology Thoroughness of Work Plan												X 9	
Coverage of All Tasks												X 5	
Division of Effort by Task and Discipline												X 4	
Realistic Manpower Estimate (Excluding final design)												X 6	
											ITEM B TOTAL		

(b)

Figure 2-4 (a) Experience and expertise rating form for evaluating a consultant for a specific type of project. (b) Approach and work plan rating form by which a client can rate a consulting firm seeking work. *(From Dr. Louis Berger, Louis Berger International, Inc. & FIDIC.)*

PERSONNEL

(450 Points)

Payments:	Qualifying Criteria				Total Points	Task Importance Multiplier	Maximum Score Possible	Score
	Education 0-20	Pertinent Experience 0-50	Developing Nations Experience 0-30					
MANAGEMENT								
1. Administrative						0.15	15	
2. Technical						0.60	60	
PLANNING								
3. Environmentalist						0.35	35	
4. (As Required)						0.15	15	
5. (As Required)						1.85	185	
6. Management Staff Advisor						0.25	25	
(As Required)						(0.25)	(25)	
(As Required)						(0.25)	(25)	
DESIGN								
7. Sanitary/Mechanical Engineer						0.25	25	
8. Civil Engineer						0.15	15	
9. Foundation						0.15	15	
10. Architectural/ Structural						0.15	15	
11. Electrical						0.15	15	
12. Cost/Specification						0.15	15	
SUPPORT								
13. Home Office Technical						0.15	15	
NOTE: Reviewer to attach Computations							ITEM C TOTAL	

Figure 2-5 Qualifying criteria for a consulting firm's personnel. *(From Dr. Louis Berger, Louis Berger International, Inc. & FIDIC.)*

missed. So this may be the time to regroup your sales force and plan future strategies that will have greater chance of success.

Perhaps the greatest underlying value of the sales interview is in learning: (1) how to present the firm in the most favorable way, (2) how to handle questions and objections, and (3) the best ways to control the

	CONSULTANT			CLIENT ESTIMATE
	Expatriate	Local	Total	Total
STUDIES In (Country)				
Abroad	___	___	___	___
Sub-Total				
FINAL DESIGN In (Country)				
Abroad	___	___	___	___
Sub-Total				
TOTAL PROJECT In (Country)				
Abroad	___	___	___	___
Sub-Total				

Figure 2-6 Breakdown of a consultant's key staff worker-month estimate. *(From Dr. Louis Berger, Louis Berger International, Inc. & FIDIC.)*

course of a sales presentation. These are skills that cannot be taught in any book but can be learned and practiced only in real-life situations.

After the Contract Is Signed

For most consulting firms, the days immediately following the signing of a contract become a time of serious questioning. Often—unknown to the consultants themselves—the new client may have similar questions or worries.

The engineering firm may wonder about such potential trouble sources as short deadlines, which of its members should be assigned to the job, and whether the client will pay bills on time. At the same time, the client may be having second thoughts about whether she or he chose the right firm, whether the firm will work quickly and effectively, plus a multitude of other concerns.

The consulting firm's actions during the first few days and weeks of the new project will make a lasting impression on the client and will affect the public image of the firm's competence and professionalism. If the consultants treat this period with care, the client will be pleased with the

	Nationality
Principal Firm: _____	_____
Joint Ventures or Associates. _____	_____
_____	_____
_____	_____
_____	_____
_____	_____

SUMMARY OF RATING

I. BASIC QUALIFICATION   Qualified _____ Disqualified _____

II. RATING	Maximum Points	Minimum Acceptable	Score
A. Experience and Expertise of Proposer	250	125	—
B. Approach and Work Plan	300	150	—
C. Personnel	450	225	—
TOTAL	1000	500	☐

Figure 2-7   Rating chart for consultants summarizing the various scores. *(From Dr. Louis Berger, Louis Berger International, Inc. & FIDIC..)*

foundation of the relationship and may convey his or her satisfaction to other potential clients. Otherwise, the engineering firm risks damage to its reputation, plus a loss of future jobs with the client and possibly even the loss of other clients who have contact with the first client.

Perhaps the simplest and most appropriate initial step for the firm to undertake is to select the staff members who will handle the job and arrange a short agenda of opening information (some of which must be procured from the client's staff). The basic goal of this initial effort is to show the client that your firm is taking quick, effective action on the project.

The agenda should consist of solid data that will be used as the work proceeds. To present the agenda to the client, set up a meeting with the client or client's representatives at either firm's offices.

The specific purposes of this meeting are: (1) to further assure the client of your firm's capabilities; (2) to meet the client's personnel and get a preliminary idea of the experts and the personalities with whom you will be cooperating; (3) to secure some useful additional data and to learn how well prepared the client is to start work on the project. The keystone of this early effort is *involvement*—that is, to get every person who will be working on the project accustomed to the idea that they will be cooperating to solve the client's problems.

One impression that the consulting firm should be cautious not to create, however, is an appearance of being overanxious. Some clients will not yet have chosen all the members of their staff for the project nor how their roles will be delineated. Therefore, the consulting firm should present the agenda with restraint and flexibility but also with confidence.

Your first meeting or series of meetings should also strive to define clearly any areas of the contract that appear vague or subject to misinterpretation. The period immediately following contract signing—when conditions are still relatively fluid—is the most appropriate time to clear up any residual questions. As the project begins, attitudes and ideas will start to solidify and become more difficult to change at a later date.

Questions about methods of payment, billing, project submissions, and related matters should be raised and answered during the early meetings. Reopening a contract, when necessary, may not be a simple matter; but many consulting firms have done so during the preliminary discussions. Usually they have great success. Such situations are best handled with the involvement of the same salespeople who helped close the deal, as well as the appropriate principals of the firm.

The sales team's continuing follow-through is vital to the support of a mutually satisfying client-consultant relationship during the duration of every engineering project. Thorough involvement will help prevent a tremendous number of potential problems as the work is completed.

BUILDING AN EFFECTIVE SALES TEAM

Now we will delve somewhat further into the techniques of selling professional engineering services. We will start by discussing some of the common guidelines used to build dynamic sales departments.

Throughout this section, we have referred to the "sales team" and "salespeople" as the members of the consulting firm who represent the company and its services to clients—through personal interviews, telephone conversations, letters, etc. The sales team may consist of (1) personnel whose only job is to sell, (2) members who devote part of their time to sales and part to other duties, or (3) a mixture of personnel. Owners and principals of a firm can also act as salespeople. Selling may be the primary function of some company officers, particularly in smaller firms. And in the one-person firm the seller is usually both the officer and the owner.

Any consulting firm will gain the maximum benefits from marketing if *all* members of the company receive training in sales techniques. For many firms, however, a time eventually comes when the business is too large for sales to be handled by only one or two principals and a handful of engineers and drafters. At that time, the firm may need to hire additional people whose main job will be to sell.

Two cautions are warranted as the sales department is built. (1) Principals or owners of the firm who have led marketing efforts up to this point should continue to communicate—as members of the sales team—with clients, in order to effect a smooth transition to the expanded team. Doing so will help to prevent alienation of existing clients and train the new salespeople in the firm's operations. (2) The same principals or owners must retain control over their other duties as they direct the sales transition. Your goal is to integrate your market with your new sales department, while maintaining a coordinated approach to delivering professional engineering services in the context of an expanding private practice.

All new salespersons should become fully acquainted with important details of past dealings that the firm has had with clients. They should also be introduced to plans for future marketing and promotion efforts. Ideally, the new employee(s) will be experienced in the sale of consulting engineering or other technical services.

Whom you select for the sales position will be vitally important. Not only will the sales department help to determine the overall direction of the business, but its members will have to deal with the principals, staff, and clients of the firm on a daily basis. Thus, personality traits, appearance, age, and manners become as important as experience and skill.

There are few standard criteria for selecting salespeople. The candidate for the job *must* have a proven ability to sell. Further, her or his personality should mesh smoothly with the personalities of the rest of the consulting firm's staff.

Perhaps the most common problem for new salespeople is a lack of definition of what kinds of actions they can and cannot initiate and approve or disapprove with clients and staff. It is important that salespeople be informed clearly of the limits of their responsibilities and authority. This should be done in writing and verbally when a new person starts work.

The consulting firm can solicit candidates for its sales department through advertisements in trade journals, newspapers, magazines, and contacts with employment agencies and business associates. Agencies can usually provide people who are well suited to your hiring criteria. But some firms prefer to advertise, using large display ads to emphasize the quality of the position(s) and of the job criteria.

In preparing to interview candidates, it is helpful to prepare a rating format, based on the firm's specific requirements. For example, for a firm doing electrical design, interviewers might ask the candidate if he or she has experience selling consultant services and whether the candidate would enjoy working in the electrical engineering field. For these and other attributes, the candidates are graded on a numerical scale or on levels such as "excellent," "good," "satisfactory," or "unsatisfactory."

Job candidates should always be fully informed in writing of the duties of their position. Performance standards should be defined explicitly, especially at the time of hiring. The consulting firm should clearly state in writing what will be expected of the salesperson in each phase of the work.

How many phone calls, letters, and visits should the salesperson make per week? Where will she or he develop and concentrate promotion? If questions like these are not answered as the sales department is built, a good salesperson's efforts may be thorough but misguided. The ultimate responsibility for building an effective sales team rests with the chief officers of the firm.

A key to achieving marketing excellence is a careful determination of salary and other compensation for the sales personnel. Salespeople in engineering firms can rarely function on the basis of commissions alone, as many other types of salespeople do. The sale of engineering services is a long-term effort, replete with dry spells and ups and downs. This could lead to wild irregularities in the receipt of commissions and to the probable resignation of salespeople paid only on a commission basis. The best solution may be to provide the salesperson with a regular salary, to cover normal living expenses without unduly burdening the firm's bank account. At the same time the firm could offer commissions on sales as an incentive to excellent work. Many firms find that this arrangement combines the best ingredients of both kinds of compensation.

Alternatively, the firm could provide bonuses to salespeople whenever the fees for services sold exceed a certain level. It is important to award bonuses on the basis of the fees, rather than on the basis of profits from the services sold, because the responsibility for profits rests primarily with the firm's top managers while the chief domain of the sales department is to promote sales. A competent sales group will seek only profitable work, of course, but the administrative branch of the company usually makes the final decisions on the work it will undertake and how it will profit thereby.

The Role of Company Officials

Can the principals of a consulting firm afford to disconnect themselves from the sales effort once a sales department has been set up? It is highly doubtful they can and still run a lucrative business. Regardless of the size of the firm's marketing branch, the role of the principals in selling remains critically important. Not only will the officers' administrative responsibilities keep them involved in overseeing sales, but the need for providing personalized consulting services requires contact between company principals and clients.

Engineering clients generally expect the members of consulting firms to show keen professional interest in each project undertaken. In cases where a client grows dissatisfied with a particular firm, the cause is often attributed to a lack of attention by the principals and staff in regard to the client's concerns.

But what if the principal is simply not a good salesperson? Certainly, it is well known in the profession that engineers are not, by training, accustomed to selling. However, there is no factual basis for presuming that any engineer—with time, effort, and practice—cannot become a salesperson. Salespeople are made, not born (though a little talent helps).

Some engineers fail to recognize their sales skills merely because the skills have never been identified as such. Any successful engineer must have succeeded in selling his or her capabilities, personality, and services to employers in the past, whether or not it was perceived as "selling" at the time. The logistics of selling consultant services are essentially the same.

Still, the bounds of responsibility and authority for sales efforts should be made thoroughly clear in the consulting firm. The sales department should not have the authority to make any sort of arrangement or commitment that will be contrary to the best interests of the company. Conversely, the administrators of the firm should allow the sales department to exercise a certain amount of freedom and independence in securing new clients. The salespeople certainly deserve to be able to explain their view of the merits of particular potential clients and projects—even when some members of the firm may disagree. However, such discussions should be confined to the offices of the consulting firm, rather than surfacing during meetings with clients. If a project must be rejected by the firm, the reasons for the decision should be described to the client only *after* the decision has been finalized.

Performance Standards and Records

Just as the establishment of the new sales department causes the consulting firm to redefine staff roles and responsibilities, it also necessitates new record-keeping methods and standards. As personnel changes occur, firms having more than one principal, or partner, may find individuals who had previously taken little active interest in selling suddenly expecting to influence decisions in these areas.

The person best equipped to decide standards for salespeople and the methods and content of record keeping is the principal or manager who had directed selling efforts before the new salespeople were hired. That person will be better acquainted with the requirements of the job and with the indicators of marketing performance.

Based on her or his past experience, the former sales director can work in concert with the new salespeople to design records that will measure performance fairly and accurately. These records, in turn, will help the company gauge progress and make projections of work load and future income. Sales and marketing reports should not be so extensive as to interfere with the actual act of selling, of course. Today many firms have a hard time getting their salespeople to maintain thorough records. Therefore, emphasis should be placed on the value of producing simple, brief reports. Examples of three forms for such reports are shown in Figure 2-8.

Many successful firms find that daily logs of marketing activities provide the most useful sales records. Similar to the weekly sales reports described earlier, the daily reports allow the firm to maintain an effective grasp of the progress of marketing efforts, on a day-by-day basis. The principals of the company (in particular, the former sales director) can monitor the reports to oversee the performance of the sales team and to develop new goals for subsequent months and years.

Basic categories of sales records include those that provide an abbreviated listing of each salesperson's office activities for the day (e.g., telephone calls to prospective clients, brochure mailings, in-house meetings); those that give a list and description of field work and in-person meetings with clients; and those that provide reports of follow-up calls and plans for the future. Since most selling is done in person, the second type of report provides an important measure of sales efforts. The customary way of setting up such reports is to title the pages to be used, according to purpose: "Daily Sales Activities—Office Effort," "Daily Sales Activities—Outside Effort," and "Follow-up Sales File," for example. Each can be arranged with three columns providing the date of each action, a summary notation of the action taken, and a short report of results and/or follow-up required. For each action, a few words suffice to produce a telling record. Rarely does it pay to maintain additional, more complex records.

What if a review of a salesperson's records indicates that performance has not been up to company standards? In such a case, the records will help the firm to find areas of difficulty. Company officials can examine the records to see the direction and extent of sales efforts and, possibly,

Office Effort

Name _____   Date _____

Date	Action Taken	Follow-up Required

Outside Effort

Name _____   Date _____

Date	Action Taken	Follow-up Required

Follow-up Sales File

Name _____   Date _____

Date	Action Required	Result of Action

Figure 2-8 Daily sales activities report forms.

to find the reasons for poor performance. The problems can be brought to the employee's attention, while further reviews are made each week.

What if the poor performance continues? In the beginning of a new salesperson's employment, considerable tolerance should be exercised. It takes between 3 and 6 months for salespeople in small and medium-sized consulting firms to become accustomed to the operations of the company and to start bringing in sales. To facilitate the monitoring of a given employee's performance, the supervisor can maintain his or her own record of the person's progress. If no improvement is shown in the first few months following the adjustment period, the employee can be warned that a serious problem exists and should be rectified. Unless the situation improves within the next 3-month period, it may be necessary to terminate the employment.

In record keeping, emphasis should be placed on consistency and regularity. It is perhaps one of the more comical characteristics of marketing personnel that the best salespeople often keep sparse or sloppy records, while the mediocre ones may maintain voluminous listings. When this occurs, it can be quite disconcerting to supervisors since the thorough appearance of the records can obscure the facts about performance.

In considering the importance of sales records, it is worthwhile to remember:

1. Daily records provide a standard for measuring the performance of salespeople.
2. Record keeping should not take up an inordinate amount of time; reports should be brief but telling.
3. The firm's sales efforts should be reviewed on a weekly basis.
4. All aspects of the company's marketing programs should be reviewed annually. During this review, use data from the sales records to analyze the promotional work, correct deficiencies, and plan for the future.

EIGHT MAJOR MARKETS

In the beginning of any engineering practice, the amount of experience the firm can show is quite limited. Usually, the principal(s) has a background of working on various jobs with other firms. Consequently, one of the most convenient and effective ways to begin building a profitable reputation for the firm is to seek work from established consultants who may need the special expertise that you have. For the beginning consultant who wishes to work with other professional firms, several markets are available.

1. Representing Large Firms Locally

Large engineering companies often seek the help of small firms with a few members. Usually the large firm asks new ones to represent it on a local level for major projects, particularly in cases where the large firm does much of the work in offices at a substantial distance from the actual project site. Competition for this type of work is much less severe than in other areas of practice simply because most of the established firms want to avoid helping competitors—regardless of size—and do not advertise the availability of the work.

There are some limitations in working for larger firms. Thus, if you agree to take part in the project for a particular firm, you are required to carry out the wishes of the company doing the design, even if the ideas do not coincide with those of your firm (that is, provided that the plans are ethical, legal, and based on sound engineering principles). Improvements, changes of design, and errors are the responsibility of the firm hiring the local consultant. On-site engineers are usually allowed to make suggestions and alterations only after obtaining appropriate authorization from the larger firm. The situation is a serious one since unauthorized changes carry the risk of lawsuits against the parties making those changes.

A major advantage of representing a large firm locally is that much experience—in how to do particular kinds of jobs, becoming familiar with the components of various projects, and learning the characteristics of relationships among client, engineer, contractor, and others—can be acquired, literally at the expense of the larger firm. The new engineering firm has an excellent opportunity for first-hand observation of many of the practical norms that exist in professional consulting engineering.

Generally, the reason a larger firm will seek assistance from smaller ones is that it may not be able to handle all the work planned or its members may not have the necessary education, experience, or capacity to direct work in the field. Further, a large firm may choose to be represented by a smaller firm because the larger one does not wish to risk being involved in potential conflicts at the project site. If a conflict does arise, the smaller firm may be able to resolve that conflict to the benefit of the larger firm without requiring undue attention from the major firm. As a precaution, the smaller firm should investigate all job situations thoroughly before getting involved—to be certain that there are no problems of law, ethics, or work quality. Participation in projects that do have such problems should be avoided.

Most state engineering societies and the National Society of Professional Engineers have manuals and other data on typical fees for this sort of interprofessional practice. The rates the manuals give are usually quite reasonable and provide a fair idea of what can be expected by the smaller firm. Negotiation over compensation may still occur, but the manuals help to minimize disagreement.

New firms usually solicit interprofessional work by letters to the larger firms, followed with phone calls and meetings at the larger firm's office. The initiative belongs to the new firm, which must ask to be considered for each job it is interested in. Larger consulting companies rarely advertise for the help of smaller firms, for various reasons such as the need to avoid attracting attention from competitors.

A large number of opportunities for interprofessional work are overlooked by new consultants sometimes because the consultants feel they
should solicit their own projects from the start. Those who insist on doing
so limit their ability to gain recognition and important experience
quickly. Interprofessional work is a strong sign of a new consultant's initiative and independence. Each project will add to the beginning firm's
expertise. After taking on such projects, the consultant can tell prospective clients about the interprofessional jobs. The result could be that your
firm will gain a more respected reputation in one year than many firms
manage to attain after several years in other kinds of consulting.

2. Services to Contractors

Contractors in the process of installing a large design often need specialized expertise in preparing alternative proposals and in designing work
situations based on the actual plans. Unfortunately, some engineers feel
it is somehow demeaning to work for contractors. This notion is totally
unfounded. If the engineer works professionally for legitimate compensation, the work performed should not be considered belittling—regardless of the type of client. Contractors of all types are professionals, like
the consultants themselves. The common fallacy that interprofessional
work needs to be confined to the engineering field still creates problems
for some engineers, however.

Contractors usually require consulting engineering for only small parts
of a project. In such situations, professional stamps should be used with
caution in order to minimize the risk of interprofessional conflict. For
example, a legally binding relationship may exist between the design engineer and the main client on a given project. The contractor, however, has
a relationship to the main client that is separate from that of the design
engineer. Therefore, the new engineer should be very careful to avoid
being used as an alternative design source for the contractor since doing
so would be to the detriment of the design engineer—a possible breach
of ethics.

Such situations can sometimes result in lawsuits between the contractor and the original design engineer, usually regarding the proper design.
The unfortunate person in the middle would be the new consultant, who
may be used by the contractor as a sort of football to be thrown at both
the main client and design engineer.

In each project you consider, the legal and ethical aspects of the situation should be investigated carefully to make sure that the engineering
designs will be used as improvements or alternative solutions—rather-
than as causes of competition between professionals.

If you as a consultant stamp any drawing, you become legally liable for its accuracy. Therefore, you should be certain that drawings represent your judgment (not the contractor's). The engineer who stamps any drawing made by a contractor may become legally implicated in a faulty design and will have little or no basis for defense if a lawsuit arises over the design.

As long as these precautions are kept in mind, work with contractors can be an excellent way to build a good reputation throughout your area. Contractors, particularly, will become familiar with your firm. This is an especially valuable asset: Not only do contractors perform the actual work at project sites, they also can influence clients' selections of consultants.

The market is very large for this type of consulting engineering. There may be many such sources of work in your area. Not only will it show you the separate viewpoints of contractors, design engineers, and others but it will also be valuable in helping you to do your own designs. Moreover, working with contractors often tends to reduce idealism in consultants. The result is a more practical, workable outlook on engineering projects.

As a new consultant, you should recognize that your authority is strongly limited in such situations. An engineer must be careful to control the use of her or his expertise so that involvement in a particular project does not encompass areas beyond those agreed upon by all parties who are initially engaged in the work.

3. Design-and-Build Teams and Project Management

The term "design-and-build teams" usually refers to organized groups of engineers, architects, and contractors who provide total design and construction services for particular projects. The premise of the group approach is that the team can simultaneously control costs and decrease the amount of time a project requires—from the beginning of the design to construction and completion of all work.

People who choose the design-and-build approach say it has dual benefits for the client: lower cost and faster delivery. But the general structure of such teams tends to favor the constructors since building usually incurs the largest expenses for the clients. Sometimes this tendency leads to decision making that pays more attention to total dollar values than to the client's best interests. While the design-and-build organization does assume overall responsibility for virtually all aspects of a project, it usually does not oversee the manufacture of equipment and the components of construction. Hence, more time is required for competitive bidding on these external items.

Consulting engineers often prefer another supervising approach—the project management method. This method performs the same basic functions as the design-and-build team arrangement, though in a different fashion. Some consultants feel that project management allows a more coordinated, more balanced approach to meeting all of a client's needs.

In a project management approach, the independent consultant provides complete management services and tailors them to the particular project and client requirements. The approach is based on the premise that the design phase is the heart of each construction project since it is in that phase that the client's needs are defined in detail. Design is also the area in which the greatest opportunities for savings exist.

The basic components of project management are:

1. Overall project control
2. Conceptual design
3. Detailed design
4. Preliminary studies
5. Value analysis
6. Equipment and material purchasing
7. Bid document preparation
8. Construction coordination and control
9. Contract selection and negotiation
10. Start-up services

The project manager usually directs both the chief of design and the chief of construction, emphasizing the need for clear definition of client requirements and utmost attention to meeting these needs. The project manager may also direct other services that may not normally be included in design or construction work.

In essence, a project manager supervises two submanagers—one for design and one for construction. The basic idea behind this arrangement is that the best construction value will be obtained through a coordinated approach that gives freedom of choice to the designers and builders with a range of options to the client. At the same time, costs are kept under rigid control. The consultant acting as project manager usually has no financial interest in equipment manufacturers or in field construction companies. Project managers generally do not earn commissions on the purchase of such equipment or services.

The differences between project management and design-and-build teams may appear small. Clearly, many consulting firms can participate in the team approach. But not all are capable of rendering the broad range of services required for project management. The project manager looks at the total job and may thus become involved in planning, budgeting, scheduling, selecting contractors, controlling quality and cost, giving purchasing assistance—all in addition to general administration. As a result, people in design and construction can retain their independence. At the same time, the client can be kept informed of all aspects of the work through a central source. The design-and-build team may not be able to offer these advantages.

4. Industry's Special Needs

The number and variety of consulting projects available in the industrial field are unlimited. This is the largest source of work for consultants. Consulting services are often needed in the design of products, processes, buildings, facilities and utilities, services, overall maintenance, and plant layout.

Consultants who consider work in the industrial field should carefully examine the range of services they can offer to the industrial firms and should decide which firms might be interested.

Daily newspapers, phone books, and other publications often list companies that supply engineering personnel for industry. Such companies operate more like employment agencies for part-time workers than like consulting firms, however. Certainly a consulting firm could offer its staff members on a similar basis, supplying technical personnel as the industrial companies require. But there is a risk in such arbitrary "loaning" of personnel. Thus, the consulting firm tends to give up control over its employees. When the consultant disagrees with the content or design, he or she may want to refuse to stamp the drawings. If the consultant stamps them anyway, the employee and the consulting firm may become liable for the results of the drawings—not a desirable position.

Consultants in industrial work usually seek lump-sum compensation. If the project's time requirements are very difficult or impossible to predict, however, an hourly fee may be preferred. Rates are generally the same as in other types of work. Occasionally they are a bit higher. One should take care not to charge fees merely on the basis of direct labor cost and a small overhead—as the employment agencies sometimes do—because overhead and administrative expenses are usually difficult to project in industrial work.

Industrial projects can be sought through the same means the consulting firm uses in other fields—letters, interviews, sales presentations, etc. Sources for such work are nearly everywhere. All industrial firms within

your area can be considered potential clients. The only problem that
some consultants have in the industrial field is that they do not spend
enough time and energy actively soliciting the work.

5. Joint Ventures

When a consulting firm lacks sufficient capacity to compete against larger
firms, a joint venture may provide the needed power. Beneficial organizational arrangements can be made for particular projects, for a series of
jobs, or for general promotion.

Joint ventures have a few special legal ramifications, including the
need to define the responsibilities of each member of the arrangement
and to obtain a separate errors-and-omissions insurance policy for the
joint venture itself. The members should agree upon the division of fees
for the arrangement, based on the efforts each is expected to contribute
and/or on the actual value of the work that will be performed. Proper
preparation will document the agreements and will allocate a specific sum
for sales efforts, a specific sum for administrative work, and specific sums
for the value of each member's contribution to design. The assistance of
an attorney can prove highly valuable during these steps.

Perhaps the biggest potential drawback to the joint venture is the attitude that motivates many engineers to enter into consulting: a drive for
independence. Certainly one of the main reasons that small engineering
firms thrive is that they provide freedom to their members for design
decisions. Independence need not be forfeited for a joint venture, however, as long as the firm's members are willing to work together to elect
the director(s) of the project to be undertaken.

Occasionally a client may seek a joint venture when she or he feels
acquainted with the expertise of some small firms but requires more
extensive services than any one of them can provide alone. In such cases,
selling efforts should be directed at developing the type of joint venture
the client desires and using the staff members whom she or he selects.
Such arrangements can provide lucrative work and can lay the foundations of growth for the firms involved.

6. Architectural Work

Small engineering firms do more than half of their work for architects and
architectural firms. Nonetheless, engineers and architects tend to remain
rather distant from one another, perhaps as a result of different professional viewpoints. Many engineers wonder why they should bother to

seek architectural projects. The answer is not complex. The design of most buildings requires the services of a firm that can plan the buildings' arrangements to meet clients' needs. Of course, the architect is usually the person who is best suited to the task. Architects thus often establish the primary contact with a client and assume the responsibility of overseeing the design and construction efforts.

Engineers can sometimes get jobs that require architectural work directly from a client, however. There is nothing wrong with this, but professional risks may develop if the engineering firm becomes tempted to have the architectural work done by people who are not licensed professional architects. If licensed architects are not used, it is possible that the consulting firm could be accused of engaging in deceptive, dangerous, and/or illegal practices.

To avoid such risks and to take full advantage of the opportunities that exist in architectural work, consulting engineers should at all times adhere to the guidelines described in national and state professional engineering society manuals of interprofessional practice. Every engineer should have at least one copy of these manuals. The data contained in them—about interprofessional business arrangements, laws, ethics, and related topics—can be invaluable to the engineer wishing to establish rewarding relationships with architects and other professionals.

What kinds of consulting services do architects need? All types; architects use consulting engineers for many projects, in every area of engineering—including structural, mechanical, electrical, civil, and other disciplines. Architects are continually on the lookout for engineering firms that can handle a project with thrift, speed, and excellence. Perhaps more than anything else, architects want consulting engineering that adds to the quality, usefulness, and reliability of their designs.

The best way to solicit architectural work is through personal interviews with the architects themselves. Such interviews are relatively easy to obtain and to conduct since most architects are interested in learning about new engineering consultants who have the characteristics and capabilities they seek.

Fees can be paid on a lump-sum basis, on an hourly basis, or as a percentage of construction cost. The fees are determined mainly by the interprofessional requirements and by the expected performance of the engineer. While some engineers think of architects as difficult to rely upon for payments, problems do not occur as often as such myths would seem to indicate.

Perhaps the most important point for new consultants to remember in soliciting architectural work is that one should not lower fees from their normal level merely because one's firm is not yet well known. In some cases, it may be worthwhile to reduce one's fees to gain entry to a particular project or field; but this should be done only when it is absolutely necessary and when certain conditions have been agreed to by the architects involved. For example, it can be helpful to develop and write an agreement in which any engineering efforts beyond those called for in the contract will be specifically paid for during the first few commissions accepted from a particular architectural firm. The consulting engineer should explain clearly that such arrangements are a temporary, precautionary measure, to be used until the firms become better acquainted with each other's work and with the requirements of specific projects.

Like engineers, architects can exercise a fiercely independent attitude toward design and project direction. The need to learn how to best cooperate with fellow professionals is something that virtually every engineer and architect faces at various times in his or her career. Unfortunately, the first one or two projects performed with outside professionals can often be problematic. An architectural firm's job coordinators may seem to burden the engineer with extra work and hassle. The antidote to such feelings is the knowledge gained from experience, mixed with a generous helping of preparation for each new project. The engineer should take on a particular job only when he or she feels certain of meeting all project requirements. If you can show a potential client that you have this certainty, it will become a major ingredient in your sales approach.

The engineer and architect both need to be cautious in using each other's name as a reference and as a selling aid with clients. There are at least a few members of each profession who will solicit work while listing people from outside their firms as staff members. When the outsiders are not informed about the use of their name or about the job opportunity itself, the conduct of the soliciting firm can be seen as deceptive and unprofessional.

There is nothing unethical about using architects as references—as long as they have been informed and have given you permission to use their names. If a client were to query an architect about your firm when the architect has no knowledge about being used as a reference, he or she may not comment favorably about your firm. Or the architect may, due to sheer surprise, leave the impression that relations with your firm are negligible.

The engineer who works with architects is part of a larger design team. He or she has responsibilities to that team, in addition to other consultant responsibilities. If one member of the team makes an error in dealing with others or makes an error on the project itself, every member could be held responsible. Thus it is not uncommon nowadays for lawyers to bring third-party liability suits against each member of a design team that has been implicated in some kind of legal problem. For example, if the roof of a building began leaking, a mechanical engineer may become involved in a lawsuit that charges that the design of the building's mechanical system may have affected the structural capability of the roof, possibly causing the leak. Consultants should be aware of these sorts of risks and guard against them when possible.

Another source of concern for some consultants is that politics may determine arrangements with architects in some government projects. If a design team is formed on the basis of political maneuvering, the result may be a forced association of people who cannot work together freely. Worse, the situation may incur a breach of professional ethics or illegal activities.

The engineer's role in political life will be discussed at length later in this Handbook. For now, let us say only that one should be highly cautious about such involvement. Providing that the architect's and/or the engineer's political activities are legal and ethical, the members of the team should develop an agreement in the same way they would with any other client. The team should make a cooperative effort toward satisfying the requirements of the government agency involved—whether it be a town, city, state, or federal body.

7. Commission Agents—Where Do They Stand?

Several years ago an engineering magazine carried an anonymously written article titled: "I Gave Up Ethics to Eat." The article has since become famous in the engineering field for its description of how its author/consultant turned from near starvation to affluence after hiring an undercover commission agent who had political connections. The move enabled the consultant's firm to grow from a one-person business to a 100-member organization. It also got him involved in bribery and other illegalities. As much as some people would like to deny the seriousness of such stories, the truth is that commission agents have had their hands in government contracts since the time contracts were first awarded. They will probably continue to exist for as long as the government continues granting contracts.

Commission agents include not only the "5 percenters" (those who earn fees equal to 5 percent of engineering fees by specializing in client influence) but also people who hold regular positions as manufacturers' agents, suppliers' representatives, and contractors. Further, a number of engineering firms have employed politicians, lawyers, lobbyists, business brokers, and foreign contacts as commission agents. Some do it secretly and some do it openly. Regardless of how it is done, however, the use of commission agents implies the unscrupulous use of someone's position to derive personal profits. This is a breach of ethics and possibly a breach of law as well. There may be no way to prevent commission agents from continuing—unless all engineering organizations police the field diligently.

Some commission agents exert their influence without even appearing to act as agents. For example, the same engineer who deplores the use of commission agents might accept "free" engineering from a manufacturer's employee in exchange for pledges to use the manufacturer's products when a particular construction contract is awarded. This sometimes happens when the engineering firm needs outside help on a project. Not only is the practice unethical but it may also place a dangerously disproportionate amount of the consulting firm's quality control in the hands of the manufacturer's representative—a risky choice.

Another way a commission agent might influence sales is by acting as a client's representative to solicit favors, cash, gifts, free designs, or other amenities from a consulting firm in attempts to influence client decisions. There is also the situation where the consulting firm might pay a finder's fee to a person who helps the firm obtain a job. The finder's fee is usually not viewed as a commission—in theory. But oftentimes the finder will not actually find a job for the firm. He will merely find you an opportunity to approach the potential client. Nonetheless, the finder expects to get paid.

Despite the ubiquity of commission agents, various state and federal laws have been enacted in recent years to reduce the value of payola, kickbacks, and political influence in both public and private sectors. Few engineers doubt that some government workers overemphasize the value of free dinners, gifts of tickets to sports events, and the like; but the influence of actual graft has diminished greatly since the days when it was practiced freely near the turn of the century. Today officials are very wary of accepting gifts from engineers and others who might engage in publicly funded work. Problems occur mainly when an inexperienced engineer presumes naively that it is normal to buy influence.

On the contrary, consultants do not need to use commission agents to succeed, nor do they need to resort to gift giving or payola. Equally important is the awareness that attempts to bribe and curry favor with government officials may lead to legal actions against the company initiating such efforts. Regardless of whether the firm's principals are charged with committing a crime, found innocent in court, or given a fine or reprimand, unethical attempts to win influence can ruin a firm's reputation for many years.

8. Government Contracts

Consultants can find a wide range of opportunities in government contracts—with villages, towns, cities, states, and the federal government—

in all engineering disciplines. Among the many kinds of agencies that employ engineers are: public-works departments, engineering departments, building agencies, school administrative offices, fire and police departments, offices of mayors and municipal managers, and corrections agencies.

Federal agencies that use consulting engineers include the navy (under the auspices of the Naval Facilities Engineering Command) and the army (under the Corps of Engineers and the Department of Defense), plus virtually every other department.

Government contracts should be sought via personal interviews whenever possible—especially on state and local levels. The engineer and/or company representatives must develop a working relationship with the government agencies so the agencies have detailed information about the consulting company on file. Letters, phone calls, and personal follow-up are mandatory. Also, political activity on the part of the engineering firm or its members may have an indirect—and sometimes direct—influence on the government's selection.

It is a generally accepted fact of political and business life that government agencies tend to select consulting firms that are not only highly qualified but that also display loyalty to agency officials' political parties. Elected and appointed officials try to avoid making overtly biased decisions but are sometimes less likely to favor organizations that have shown support for parties that oppose their own.

Another well-known aspect of application for government contracts—particularly with the federal government and some state agencies—is the need for the engineer to fill out specialized questionnaires regarding prospective work. Typical forms will be discussed later in this section, but one observation is appropriate at this point. Many engineers complain about the paperwork and red tape associated with government contracts. What they fail to realize in many instances is that the government workers may consider the paperwork as much of a burden as the engineer considers it. Still the forms must be filled out and the regulations complied with. Anyone can balk at paperwork; but only those who do the work have the chance to gain new contracts.

Special Problems in Public-Sector Jobs

Government contracts carry a few other influences that are not found in private-sector work. For instance, while the chief administrators of each state agency are usually appointed by incumbent party officials, the majority of staff members of the agency generally belong to a civil-service union or organization. Thus the incumbent officials and their appointees may not share the perspectives of the majority of the agency's staff members (even though the state employees may adhere to all decisions made by the agency's administrators regarding contracts, allocation of money, etc.)

The mixture of perspectives can thus complicate the consultant's sales efforts. Therefore, the marketing pitch should first be aimed at the administrators and then tailored to each state employee who will be reviewing applicants for work.

Situations get even more complex when the engineer seeks work with an agency in a state outside of his or her own. This complexity arises because state agencies may feel pressured toward giving all their projects to engineers who reside in their own state. There are some states in which nonresident engineers cannot secure a contract at all. But jobs can be obtained in most places if the engineer's capabilities match the project requirements.

Many states have specialized application forms that consulting firms must complete and update annually to be considered for future projects. Some of the forms resemble federal forms No. 254 and No. 255, which we will examine shortly.

Sources of Information

Information about government contracts and engineering projects can be obtained through the U.S. Department of Commerce's *Commerce Business Daily,* the *Dodge Building Reports,* newspapers, newsletters of state agencies, and personal contacts with government officials and agency staff members. Further, The World Bank (The International Bank for Reconstruction and Development) and some smaller banks and real estate agencies carry information about government use of consulting engineers. Other sources could include people who have ongoing relations with government projects: contractors, suppliers, architects, plus other engineers and professional organizations. Exhibit 2-8 shows a typical newspaper ad soliciting consulting engineering services.

At the local level, it may be worth your while to try to obtain copies of proposed or approved budgets for nearby towns and villages. The budgets often list the kinds of engineering-related work that the municipalities are considering or planning. Sometimes the budgets are produced before engineering firms are chosen and even before consultants have heard about the projects.

In many cases, copies of new annual budgets can be obtained from municipal offices. Or they may appear in local newspapers. You can get a head start on soliciting various kinds of work if you learn when the annual budgets are usually developed, examine them immediately after

EXHIBIT 2-8

ARCHITECT ENGINEER SERVICES

Provide architectural and engineering services to the Federal Aviation Administration (FAA), Eastern Region, located in Jamaica, New York, on a task order basis for a multi-year period. Scope of services required, in order of importance are as follows: Preparation of contract documents for a major modernization of a large facility, in a 24-hour a day live environment; design of electrical and mechanical systems and equipment; site surveys and investigation of existing conditions; civil and structural engineering design; architectural design and services; electronic engineering design. (Electronic design is not required for task #1, but may be required for future tasks and has minimum weight in this selection process). Task order #1, which will begin with the award of a contract to an A/E firm, is Phase I of the modernization of the New York Air Route Traffic Control Center (ARTCC) located in Islip, New York. Construction cost for task order #1 is estimated to be between $1.5 to $3.0 million.

A/E selection criteria will include: (1) specialized experience of the firm in the type of work required in order of importance; (2) qualifications of professional staff for performance of described services; (3) whether the required work will be done in-house or joint venture; (4) capability of the firm to accomplish the work in the required time; (5) location of the firm in the general geographic area of the FAA Eastern Region Headquarters in Jamaica, New York; (6) past experience, if any, of the firm with respect to the firm's performance of Government contracts; (7) volume of work previously awarded by the Federal Government to the firm. This proposal is not set aside for small business; small business status, however, will be a factor. Future design effort may include the following:

1. Phase II Modernization at New York ARTCC—$1–$5 million.
2. Site Adaptation of Standard FAA design—$0.5–$1.0 million.
3. Various design projects for airports in the following states:
   New York, New Jersey, Pennsylvania, Delaware, Virginia, Maryland, and West Virginia.—$1–$2 million.

Architect-engineering firms which meet the requirements described in this announcement are invited to submit completed SF-254 and SF-255, U.S. Government Architect-Engineer Qualifications. Response to this announcement must be received at the address indicated below, no later than COB 8/16 in order to be considered. This is not a request for proposals.

Department of Transportation
Federal Aviation Administration
Eastern Region, Fed. Bldg.
John F. Kennedy International Airport
Jamaica, New York 11430
ATTN: AEA-55

you obtain them, and then contact the government offices for further information about the engineering needs of particular projects. Even if you do not obtain work for the upcoming jobs, you might begin to establish contacts that will help you get other contracts in the future. Follow-up calls and/or letters are therefore essential.

In soliciting work with the federal government, it can be helpful to obtain letters of recommendation from members of Congress and legislators. Not many consulting firms are fortunate enough, however, to have direct contact with state and local officials; nor do those who have such contacts necessarily obtain every government job they seek. But public officials do wield influence with one another, and a letter of recommendation from a legislator can probably enhance your firm's sales efforts.

THE SELECTION PROCESS

Most government agencies have special selection committees for choosing the engineering firms to be used on specific projects. Since the committees at higher levels of government tend to be larger and tend to have more civil-service employees than do committees in small municipalities, observers generally agree that the committees in state and federal agencies are less influenced by politics.

Also, some engineers feel that political clout usually does not come into play in obtaining government work except for projects involving fees in excess of approximately $100,000. It is a good idea to try to hold brief, unobtrusive but strong interviews with officials of a few government agencies once a year. Doing so may help your firm stand out from the hundreds of other engineering companies that are also applying for projects with the government.

Federal agencies, in particular, may get 50 or even 100 applicants for their smaller projects. How do they select the right engineering firm for the job? Anyone involved in the task will probably tell you it is not an easy process. First, the selection committee and its clerical staff must collect all the questionnaires completed by the competing firms, catalog the data, and begin to sort out those that seem to have the general qualifications for the project from those that do not. Of course, prior to the selection process the project itself must be defined in detail. Firms that do not address the specifics of the project under consideration or that do not show their ability to meet the requirements through their responses on the forms are usually quickly eliminated from consideration. Thus the list of potential firms may be reduced to include very few of the original applicants.

Second, the prospective companies are contacted and the scope of the project is discussed with them in further detail. The final selection is usually based on additional discussions and on a vote by committee members.

Some consultants wonder if all the "red tape" they must go through to get government contracts—the questionnaires, the selection committees, interviews, follow-ups, and regulations—does not make government work a waste of time. While such facets of the public market may seem unduly burdensome, there can be little doubt that engineers can profit through government work. Those who perform services in the civil, environmental, aeronautical, and related fields make their living predominantly through government projects. Engineers who do high-quality work often develop a steady source of lucrative practice through government agencies. The red tape is thus seen as a necessary part of continued success.

Negotiating Fees

After the government agency's selection committee chooses the consulting firm for a particular engagement, the firm's representatives and agency members negotiate the fee and the terms of the contract agreement. Negotiations can be handled in personal meetings, by phone, and sometimes by letter. Regardless of how they are carried out, they offer the opportunity for all parties involved to finalize work plans, scheduling, organizational arrangements, and financial decisions.

The negotiation phase is a time when clients may exert their strongest efforts to keep the cost of engineering services low and subject to ongoing control. The situation can work to the engineer's advantage if handled properly. Perhaps the most helpful advice for the engineer in fee negotiations is "Don't paint yourself into a corner." In other words, one should not allow oneself to get locked into a particular position in which the firm cannot suggest alternatives or exhibit flexibility with respect to fees.

Flexibility can be a great sales tool. While it is true that current federal laws encourage U.S. government agencies to pay lump-sum fees in most cases, the government negotiators are likely to feel they are getting a bargain if the engineer agrees to lower his or her sum from the initial proposals. For example, if your goal is a $9200 commission, it might be advisable to set the fee initially at $9800 and reduce it to $9200—if appropriate—by the end of the negotiations.

Much discussion in negotiations revolves around the consulting firm's overhead. In some cases, the client may even ask to audit the firm's accounting structure to develop background knowledge for its calculations on the planned project. Engineering experts generally estimate overhead at around 100 percent of direct labor costs, while placing administrative requirements at 25 percent of the total hourly fee. Profit is also estimated at about 25 percent of the total fee. So the engineer's hourly rate usually works out to approximately $2\frac{1}{2}$ times direct labor.

Many federal and state agencies request consulting firms to submit a statement of projected overhead based on the government's own guidelines. Some engineers consider this additional step to be excessive, but it is nowadays regarded as a normal part of fee negotiation. Under government guidelines, overhead is sometimes calculated to be a bit lower—or 80 percent of direct labor costs.

The goal of the negotiation process is to develop mutually satisfactory plans for compensation to the engineer and for the specific services she or he will provide. In the unfortunate and somewhat unlikely event that the government representatives refuse to agree to the fees the engineering firm seeks, the firm can call for extended negotiations or reject the contract altogether.

Questionnaires

Of all the questionnaires that consulting engineers complete during the course of their careers, federal forms 254 (Figure 2-9) and 255 (Figure 2-10) are among the most important. The two questionnaires act as standard guides to consultants for defining the capabilities of their firms and for describing the expertise of the company's personnel, its background, and its experience relative to specific federal projects.

Both forms must be completed by engineers, architects, and related professionals seeking engagement in federal work. The questionnaires are available through the U.S. General Services Administration (GSA), Washington, D.C. 20405. Copies can also be obtained from regional branch offices of the GSA (most major cities and state capitals have federal offices that carry copies of the forms).

As was mentioned earlier, government projects attract a large number of applicants. Therefore, the goal of the engineer completing the questionnaires should be to emphasize his or her firm's capabilities in ways that make the firm stand out from the multitude. One reason the list of applicants is often very large is that many federal and state agencies have been encouraged in recent years to hire smaller firms when possible. This policy is pursued so that large companies do not have an unfair advantage in obtaining government projects. Certainly the opportunities for small firms make it worthwhile to file forms with the federal government.

Architect-Engineer
and Related Services
Questionnaire

Form Approved
OMB No. 3090-0028

Purpose:

The policy of the Federal Government in acquiring architectural, engineering, and related professional services is to encourage firms lawfully engaged in the practice of those professions to submit annually a statement of qualifications and performance data. Standard Form 254, "Architect-Engineer and Related Services Questionnaire," is provided for that purpose. Interested A-E firms (including new, small, and/or minority firms) should complete and file SF 254's with each Federal agency and with appropriate regional or district offices for which the A-E is qualified to perform services. The agency head for each proposed project shall evaluate these qualification resumes, together with any other performance data on file or requested by the agency, in relation to the proposed project. The SF 254 may be used as a basis for selecting firms for discussions, or for screening firms preliminary to inviting submission of additional information.

Definitions:

"Architect-engineer and related services" are those professional services associated with research, development, design and construction, alteration, or repair of real property, as well as incidental services that members of these professions and those in their employ may logically or justifiably perform, including studies, investigations, surveys, evaluations, consultations, planning, programming, conceptual designs, plans and specifications, cost estimates, inspections, shop drawing reviews, sample recommendations, preparation of operating and maintenance manuals, and other related services.

"Parent Company" is that firm, company, corporation, association, or conglomerate which is the major stockholder or highest tier owner of the firm completing this questionnaire; i.e. Firm A is owned by Firm B which is, in turn, a subsidiary of Corporation C. The "parent company" of Firm A is Corporation C.

"Principals" are those individuals in a firm who possess legal responsibility for its management. They may be owners, partners, corporate officers, associates, administrators, etc.

"Discipline", as used in this questionnaire, refers to the primary technological capability of individuals in the responding firm. Possession of an academic degree, professional registration, certification, or extensive experience in a particular field of practice normally reflects an individual's primary technical discipline.

"Joint Venture" is a collaborative undertaking by two or more firms or individuals for which the participants are both jointly and individually responsible.

"Consultant", as used in this questionnaire, is a highly specialized individual or firm having significant input and responsibility for certain aspects of a project and possessing unusual or unique capabilities for assuring success of the finished work.

"Prime" refers to that firm which may be coordinating the concerted and complementary inputs of several firms, individuals or related services to produce a completed study or facility. The "prime" would normally be

regarded as having full responsibility and liability for quality of performance by itself as well as by subcontractor professionals under its jurisdiction.

"Branch Office" is a satellite, or subsidiary extension, of a headquarters office of a company, regardless of any differences in name or legal structure of such a branch due to local or state laws. "Branch offices" are normally subject to the management decisions, bookkeeping, and policies of the main office.

Instructions for Filing (Numbers below correspond to numbers contained in form):

1. Type accurate and complete name of submitting firm, its address, and zip code.

   1a. Indicate whether form is being submitted in behalf of a parent firm or a branch office. (Branch office submissions should list only personnel in, and experience of, that office.)

2. Provide date the firm was established under the name shown in question 1.

3. Show date on which form is prepared. All information submitted shall be current and accurate as of this date.

4. Enter type of ownership, or legal structure, of firm (sole proprietor, partnership, corporation, joint venture, etc.)

   Check appropriate boxes indicating if firm is (a) a small business concern; (b) a small business concern owned and operated by socially and economically disadvantaged individuals; and (c) Women-owned; (See 48 CFR 19.101 and 52.219-9).

5. Branches of subsidiaries of large or parent companies, or conglomerates, should insert name and address of highest-tier owner.

   5a. If present firm is the successor to, or outgrowth of, one or more predecessor firms, show name(s) of former entity(ies) and the year(s) of their original establishment.

6. List not more than two principals from submitting firm who may be contacted by the agency receiving this form. (Different principals may be listed on forms going to another agency.) Listed principals must be empowered to speak for the firm on policy and contractual matters.

7. Beginning with the submitting office, list name, location, total number of personnel and telephone numbers for all associated or branch offices, (including any headquarters or foreign offices) which provide A-E and related services.

   7a. Show total personnel in all offices. (Should be sum of all personnel, all branches.)

8. Show total number of employees, by discipline, in submitting office. (If form is being submitted by main or headquarters office, firm should list total employees, by discipline, in all offices.) While some personnel may be qualified in several disciplines, each person should be counted only once in accord with his or her primary function. Include clerical personnel as "administrative." Write in any additional disciplines—sociologists, biologists, etc.—and number of people in each, in blank spaces.

9. Using chart (below) insert appropriate index number to indicate range of professional services fees received by submitting firm each calendar year for last five years, most recent year first. Fee summaries should be broken down to

Figure 2-9 Architect-Engineer and Related Services Questionnaire. *(U.S. General Services Administration.)*

Architect-Engineer and Related Services Questionnaire

reflect the fees received each year for (a) work performed directly for the Federal Government (not including grant and loan projects) or as a sub to other professionals performing work directly for the Federal Government; (b) all other domestic work, U.S. and possessions, including Federally-assisted projects, and (c) all other foreign work.

Ranges of Professional Services Fees

INDEX		INDEX	
1.	Less than $100,000	5.	$1 million to $2 million
2.	$100,000 to $250,000	6.	$2 million to $5 million
3.	$250,000 to $500,000	7.	$5 million to $10 million
4.	$500,000 to $1 million	8.	$10 million or greater

10. Select and enter, in numerical sequence, **not more than thirty** (30) "Experience Profile Code" numbers from the listing (next page) which most accurately reflect submitting firm's demonstrated technical capabilities and project experience. **Carefully review list.** (It is recognized some profile codes may be part of other services or projects contained on list; firms are encouraged to select profile codes which best indicate type and scope of services provided on past projects.) For each code number, show total number of projects and gross fees (in thousands) received for profile projects performed by firm during past few years. If firm has one or more capabilities not included on list insert same in blank spaces at end of list and show numbers in question 10 on the form. In such cases, the filled-in listing **must** accompany the complete SF 254 when submitted to the Federal agencies.

11. Using the "Experience Profile Code" numbers in the same sequence as entered in item 10, give details of at least one recent (within last five years) representative project for each code number, up to a **maximum** of thirty (30) separate projects, or portions of projects, for which firm was responsible. (Project examples may be used more than once to illustrate different services rendered on the same job. Example: a dining hall may be part of an auditorium or educational facility.) Firms which select less than thirty "profile codes" may list two or more project examples (to illustrate specialization) for each code number so long as total of all project examples does not exceed thirty. (30). After each code number in question 11, show: (a) whether firm was "P." the prime professional, or "C," a consultant, or "JV," part of a joint venture on that particular project (New firms, in existence less than five (5) years may use the symbol "IE" to indicate "Individual Experience" as opposed to firm experience); (b) provide name and location of the specific project which typifies firm's (or individual's) performance under that code category; (c) give name and address of the owner of that project (if government agency indicate responsible office); (d) show the estimated construction cost (or other applicable cost) for that portion of the project for which the firm was primarily responsible. (Where no construction was involved, show approximate cost of firm's work); and (e) state year work on that particular project was, or will be, completed.

12. The completed SF 254 should be signed by a principal of the firm, preferably the chief executive officer.

13. Additional data, brochures, photos, etc. should not accompany this form unless specifically requested.

NEW FIRMS (not reorganized or recently-amalgamated firms) are eligible and encouraged to seek work from the Federal Government in connection with performance of projects for which they are qualified. Such firms are encouraged to complete and submit Standard Form 254 to appropriate agencies. Questions on the form dealing with personnel or experience may be answered by citing experience and capabilities of individuals in the firm, based on performance and responsibility while in the employ of others. In so doing, notation of this fact should be made on the form. In question 9, write in "N/A" to indicate "not applicable" for those years prior to firm's organization.

Figure 2-9 *(Cont.)*

Experience Profile Code Numbers for use with questions 10 and 11

001 Acoustics; Noise Abatement
002 Aerial Photogrammetry
003 Agricultural Development; Grain Storage; Farm Mechanization
004 Air Pollution Control
005 Airports; Navaids; Airport Lighting; Aircraft Fueling
006 Airports; Terminals & Hangars; Freight Handling
007 Arctic Facilities
008 Auditoriums & Theatres
009 Automation; Controls; Instrumentation
010 Barracks; Dormitories
011 Bridges
012 Cemeteries (Planning & Relocation)
013 Chemical Processing & Storage
014 Churches; Chapels
015 Codes; Standards; Ordinances
016 Cold Storage; Refrigeration; Fast Freeze
017 Commercial Buildings (low rise); Shopping Centers
018 Communications Systems; TV; Microwave
019 Computer Facilities; Computer Service
020 Conservation and Resource Management
021 Construction Management
022 Corrosion Control; Cathodic Protection; Electrolysis
023 Cost Estimating
024 Dams (Concrete; Arch)
025 Dams (Earth; Rock); Dikes; Levees
026 Desalination (Process & Facilities)
027 Dining Halls; Clubs; Restaurants
028 Ecological & Archeological Investigations
029 Educational Facilities; Classrooms
030 Electronics
031 Elevators; Escalators; People-Movers
032 Energy Conservation; New Energy Sources
033 Environmental Impact Studies, Assessments or Statements
034 Fallout Shelters; Blast-Resistant Design
035 Field Houses; Gyms; Stadiums
036 Fire Protection
037 Fisheries; Fish Ladders
038 Forestry & Forest Products
039 Garages; Vehicle Maintenance Facilities; Parking Decks
040 Gas Systems (Propane; Natural; Etc.)
041 Graphic Design

042 Harbors; Jetties; Piers; Ship Terminal Facilities
043 Heating; Ventilating; Air Conditioning
044 Health Systems Planning
045 Highrise; Air-Rights-Type Buildings
046 Highways; Streets; Airfield Paving; Parking Lots
047 Historical Preservation
048 Hospital & Medical Facilities
049 Hotels; Models
050 Housing (Residential, Multi-Family; Apartments; Condominiums)
051 Hydraulics & Pneumatics
052 Industrial Buildings; Manufacturing Plants
053 Industrial Processes; Quality Control
054 Industrial Waste Treatment
055 Interior Design; Space Planning
056 Irrigation; Drainage
057 Judicial and Courtroom Facilities
058 Laboratories; Medical Research Facilities
059 Landscape Architecture
060 Libraries; Museums; Galleries
061 Lighting (Interiors; Display; Theatre, Etc.)
062 Lighting (Exteriors; Streets; Memorials; Athletic Fields, Etc.)
063 Materials Handling Systems; Conveyors; Sorters
064 Metallurgy
065 Microclimatology; Tropical Engineering
066 Military Design Standards
067 Mining & Mineralogy
068 Missile Facilities (Silos; Fuels; Transport)
069 Modular Systems Design; Pre-Fabricated Structures or Components
070 Naval Architecture; Off-Shore Platforms
071 Nuclear Facilities; Nuclear Shielding
072 Office Buildings; Industrial Parks
073 Oceanographic Engineering
074 Ordnance; Munitions; Special Weapons
075 Petroleum Exploration; Refining
076 Petroleum and Fuel (Storage and Distribution)
077 Pipelines (Cross-Country—Liquid & Gas)
078 Planning (Community; Regional, Areawide and State)
079 Planning (Site, Installation, and Project)
080 Plumbing & Piping Design
081 Pneumatic Structures; Air-Support Buildings
082 Postal Facilities
083 Power Generation, Transmission, Distribution
084 Prisons & Correctional Facilities
085 Product, Machine & Equipment Design

086 Radar; Sonar; Radio & Radar Telescopes
087 Railroad; Rapid Transit
088 Recreation Facilities (Parks, Marinas, Etc.)
089 Rehabilitation (Buildings; Structures; Facilities)
090 Resource Recovery; Recycling
091 Radio Frequency Systems & Shieldings
092 Rivers; Canals; Waterways; Flood Control
093 Safety Engineering; Accident Studies; OSHA Studies
094 Security Systems; Intruder & Smoke Detection
095 Seismic Designs & Studies
096 Sewage Collection, Treatment and Disposal
097 Soils & Geologic Studies; Foundations
098 Solar Energy Utilization
099 Solid Wastes; Incineration; Land Fill
100 Special Environments; Clean Rooms, Etc.
101 Structural Design; Special Structures
102 Surveying; Platting; Mapping; Flood Plain Studies
103 Swimming Pools
104 Storm Water Handling & Facilities
105 Telephone Systems (Rural; Mobile; Intercom; Etc.)
106 Testing & Inspection Services
107 Traffic & Transportation Engineering
108 Towers (Self-Supporting & Guyed Systems)
109 Tunnels & Subways
110 Urban Renewals; Community Development
111 Utilities (Gas & Steam)
112 Value Analysis; Life-Cycle Costing
113 Warehouses & Depots
114 Water Resources; Hydrology; Ground Water
115 Water Supply, Treatment and Distribution
116 Wind Tunnels; Research/Testing Facilities
117 Zoning; Land Use Studies
201 _____
202 _____
203 _____
204 _____
205 _____

Figure 2-9 (Cont.)

STANDARD FORM (SF)	1. Firm Name / Business Address:	2. Year Present Firm Established:	3. Date Prepared:

254

Architect-Engineer and Related Services Questionnaire

1a. Submittal is for ☐ Parent Company ☐ Branch or Subsidiary Office

4. Specify type of ownership *and* check below, if applicable.

A. Small Business
B. Small Disadvantaged Business
C. Woman-owned Business

5. Name of Parent Company, if any:

5a. Former Parent Company Name(s), if any, and Year(s) Established:

6. Names of not more than Two Principals to Contact: Title / Telephone

1)
2)

7. Present Offices: City / State / Telephone / No. Personnel Each Office

7a. Total Personnel

8. Personnel by Discipline: (*List each person only once, by primary function.*)

Administrative
Architects
Chemical Engineers
Civil Engineers
Construction Inspectors
Draftsmen
Ecologists
Economists

Electrical Engineers
Estimators
Geologists
Hydrologists
Interior Designers
Landscape Architects
Mechanical Engineers
Mining Engineers

Oceanographers
Planners: Urban/Regional
Sanitary Engineers
Soils Engineers
Specification Writers
Structural Engineers
Surveyors
Transportation Engineers

9. Summary of Professional Services Fees Received: (Insert index number)

Last 5 Years (most recent year first)

19___ 19___ 19___ 19___ 19___

Direct Federal contract work, including overseas
All other domestic work
All other foreign work*
*Firms interested in foreign work, but without such experience, check here: ☐

Ranges of Professional Services Fees

INDEX
1. Less than $100,000
2. $100,000 to $250,000
3. $250,000 to $500,000
4. $500,000 to $1 million
5. $1 million to $2 million
6. $2 million to $5 million
7. $5 million to $10 million
8. $10 million or greater

Figure 2-9 (*Cont.*)

10. Profile of Firm's Project Experience, Last 5 Years

Profile Code	Number of Projects	Total Gross Fees (in thousands)	Profile Code	Number of Projects	Total Gross Fees (in thousands)	Profile Code	Number of Projects	Total Gross Fees (in thousands)
1)			11)			21)		
2)			12)			22)		
3)			13)			23)		
4)			14)			24)		
5)			15)			25)		
6)			16)			26)		
7)			17)			27)		
8)			18)			28)		
9)			19)			29)		
10)			20)			30)		

11. Project Examples, Last 5 Years

	Profile "P", "C", "JV", or "IE" Code	Project Name and Location	Owner Name and Address	Cost of Work (in thousands)	Completion Date (Actual or Estimated)
1					
2					
3					
4					
5					
6					
7					

Figure 2-9 *(Cont.)*

8											
9											
10											
11											
12											
13											
14											
15											
16											
17											
18											
19											

Figure 2-9 *(Cont.)*

20													
21													
22													
23													
24													
25													
26													
27													
28													
29													
30													

12. The foregoing is a statement of facts

Signature: _____ Typed Name and Title: _____ Date:

Figure 2-9 *(Cont.)*

Form Approved
OMB No. 3090-0029

STANDARD FORM (SF)

255

Architect-Engineer and Related Services Questionnaire for Specific Project

Purpose:

This form is a supplement to the ''Architect-Engineer and Related Services Questionnaire'' (SF 254). Its purpose is to provide additional information regarding the qualifications of interested firms to undertake a specific Federal A-E project. Firms, or branch offices of firms submitting this form should enclose (or already have) on file with the appropriate office of the agency) a current (within the past year) and accurate copy of the SF 254 for that office.

The procurement official responsible for each proposed project may request submission of the SF 255 ''Architect-Engineer and Related Services Questionnaire for Specific Project'' in accord with applicable civilian and military procurement regulations and shall evaluate such submissions, as well as related information contained on the Standard Form 254, and any other performance data on file with the agency, and shall select firms for subsequent discussions leading to contract award in conformance with Public Law 92-582.

This form should only be filed by an architect-engineer or related services firm when requested to do so by the agency or by a public announcement. Responses should be as complete and accurate as possible, contain data relative to the specific project for which you wish to be considered, and should be provided, by the required due date, to the office specified in the request or public announcement.

This form will be used only for the specified project. Do not refer to this submittal in response to other requests or public announcements.

Definitions:

''**Architect-engineer and related services**'' are those professional services associated with research, development, design and construction, alteration, or repair of real property, as well as incidental services that members of these professions and those in their employ may logically or justifiably perform, including studies, investigations, surveys, evaluations, consultations, planning, programming, conceptual designs, plans and specifications, cost estimates, inspections, shop drawing reviews, sample recommendations, preparation of operating and maintenance manuals, and other related services.

''**Principals**'' are those individuals in a firm who possess legal responsibility for its management. They may be owners. partners. corporate officers. associates, administrators, etc.

''**Discipline**'', as used in this questionnaire, refers to the primary technological capability of individuals in the responding firm. Possession of an academic degree. professional registration, certification. or extensive experience in a particular field of practice normally reflects an individual's primary technical discipline.

''**Joint Venture**'', is a collaborative undertaking of two or more firms or individuals for which the participants are both jointly and individually responsible.

''**Key Persons, Specialists, and Individual Consultants**'', as used in this questionnaire, refer to individuals who will have **major** project responsibility or will provide **unusual or unique** capabilities for the project under consideration.

Instructions for Filing (Numbers below correspond to numbers contained in form):

1. Give name and location of the project for which this form is being submitted

2. Provide appropriate data from the Commerce Business Daily (CBD) identifying the particular project for which this form is being filed.

   2a. Give the date of the Commerce Business Daily in which the project announcement appeared, or indicate ''not applicable'' (N/A) if the source of the announcement is other than the CBD.

   2b. Indicate Agency identification or contract number as provided in the CBD announcement.

3. Show name and address of the individual or firm (or joint venture) which is submitting this form for this project

   3a. List the name. title. and telephone number of that principal who will serve as the point of contact. Such an individual must be empowered to speak for the firm on policy and contractual matters and should be familiar with the programs and procedures of the agency to which this form is directed.

   3b. Give the address of the specific office which will have responsibility for performing the announced work.

4. Insert the number of personnel by discipline presently employed (on date of this form) at office specified in block 3b. While some personnel may be qualified in several disciplines, each person should be counted only once in accord with his or her primary function. Include clerical personnel as ''administrative.'' Write in any additional disciplines—sociologists. biologists. etc.—and number of people in each. in blank spaces.

5. Answer only if this form is being submitted by a joint venture of two or more collaborating firms. Show the names and addresses of all individuals or organizations expected to be included as part of the joint venture and describe their particular areas of anticipated responsibility. (i.e. technical disciplines. administration. financial. sociological. environmental. etc.)

   5a. Indicate by checking the appropriate box. whether this particular joint venture has worked together on other projects.

Figure 2-10 Architect-Engineer and Related Services Questionnaire for Specific Project. *(U.S. General Services Administration.)*

Architect-Engineer
and Related Services
Questionnaire for
Specific Project

Standard Form 255
General Services Administration,
Washington, D. C. 20405
Fed. Proc. Reg.(41 CFR) 1-16 . 803
Armed Svc. Proc. Reg. 18-403

Each firm participating in the joint venture should have a Standard Form 254 on file with the contracting office receiving this form. Firms which do not have such forms on file should provide same immediately along with a notation at the top of page 1 of the form regarding their association with this joint venture submittal.

6. If respondent is not a joint venture, but intends to use outside (as opposed to in-house or permanently and formally affiliated) consultants or firms, as well as their particular areas of technical/professional expertise, as it relates to this project. Existence of previous working relationships should be noted. If more than eight outside consultants or associates are anticipated, attach an additional sheet containing requested information.

7. Regardless of whether respondent is a joint venture or an independent firm, provide brief resumes of key personnel expected to participate on this project. Care should be taken to limit resumes to only those personnel and specialists who will have major project responsibilities. Each resume must include: (a) name of each key person and specialist and his or her title, (b) the project assignment or role which that person will be expected to fulfill in connection with this project, (c) the name of the firm or organization, if any, with whom that individual is presently associated, (d) years of relevant experience with present firm and other firms, (e) the highest academic degree achieved and the discipline covered (if more than one highest degree, such as two Ph.D.'s, list both), the year received and the particular technical/professional discipline which that individual will bring to the project, (f) if registered as an architect, engineer, surveyor, etc., show only the field of registration and the year that such registration was first acquired. If registered in several states, do not list states, and (g) a synopsis of experience, training, or other qualities which reflect individual's potential contribution to this project. Include such data as familiarity with Government or agency procedures, similar type of work performed in the past, management abilities, familiarity with the geographic area, relevant foreign language capabilities, etc. Please limit synopsis of experience to directly relevant information.

8. List up to ten projects which demonstrate the firm's or joint venture's competence to perform work similar to that likely to be required on this project. The more recent such projects, the better. Prime consideration will be given to

projects which illustrate respondent's capability for performing work similar to that being sought. Required information must include: (a) name and location of project, (b) brief description of type and extent of services provided for each project (submissions by joint ventures should indicate which member of the joint venture was the prime on that particular project and what role it played), (c) name and address of the owner of that project (if Government agency, indicate responsible office), (d) completion date (actual when available, otherwise estimated) (e) total construction cost of completed project (or where no construction was involved, the approximate cost of your work) and that portion of the cost of the project for which the named firm was/is responsible.

9. List only those projects which the A-E firm or joint venture, or members of the joint venture are currently performing under direct contract with an agency or department of the Federal Government. Exclude any grant or loan projects being financed by the Federal Government but being performed under contract to other non Federal governmental entities. Information provided under each heading is similar to that requested in the preceding Item 8, except for (d) "Percent Complete." Indicate in this item the percentage of A-E work completed upon filing this form.

10. Through narrative discussion, show reason why the firm or joint venture submitting this questionnaire believes it is especially qualified to undertake the project. Information provided should include, but not be limited to, such data as specialized equipment available for this work, any awards or recognition received by a firm or individuals for similar work, required security clearances, special approaches or concepts developed by the firm relevant to this project, etc. Respondents may say anything they wish in support of their qualifications. When appropriate, respondents may supplement this proposal with graphic material and photographs which best demonstrate design capabilities of the team proposed for this project.

11. Completed forms should be signed by the chief executive officer of the joint venture (thereby attesting to the concurrence and commitment of all members of the joint venture), or by the architect-engineer principal responsible for the conduct of the work in the event it is awarded to the organization submitting this form. Joint ventures selected for subsequent discussions regarding this project must make available a statement of participation signed by a principal of each member of the joint venture. ALL INFORMATION CONTAINED IN THE FORM SHOULD BE CURRENT AND FACTUAL.

Figure 2-10 *(Cont.)*

OMB Approval No. 3090-0029

STANDARD FORM (SF)
255

Architect-Engineer
Related Services
for Specific
Project

1. Project Name / Location for which Firm is Filing:

2a. Commerce Business Daily Announcement Date, if any:

2b. Agency Identification Number, if any:

3. Firm (or Joint-Venture) Name & Address

3a. Name, Title & Telephone Number of Principal to Contact

3b. Address of office to perform work, if different from Item 3

4. Personnel by Discipline: (List each person only once, by primary function.)

___ Administrative	___ Electrical Engineers	___ Oceanographers
___ Architects	___ Estimators	___ Planners: Urban/Regional
___ Chemical Engineers	___ Geologists	___ Sanitary Engineers
___ Civil Engineers	___ Hydrologists	___ Soils Engineers
___ Construction Inspectors	___ Interior Designers	___ Specification Writers
___ Draftsmen	___ Landscape Architects	___ Structural Engineers
___ Ecologists	___ Mechanical Engineers	___ Surveyors
___ Economists	___ Mining Engineers	___ Transportation Engineers

___ Total Personnel

5. If submittal is by JOINT-VENTURE list participating firms and outline specific areas of responsibility (including administrative, technical and financial) for each firm: (Attach SF 254 for each if not on file with Procuring Office.)

5a. Has this Joint-Venture previously worked together? ☐ yes ☐ no

Figure 2-10 *(Cont.)*

6. If respondent is not a joint-venture, list outside key Consultants/Associates anticipated for this project (Attach SF 254 for Consultants/Associates listed, if not already on file with the Contracting Office).

Name & Address	Specialty	Worked with Prime before (Yes or No)
1)		
2)		
3)		
4)		
5)		
6)		
7)		
8)		

Figure 2-10 *(Cont.)*

7. Brief resume of key persons, specialists, and individual consultants anticipated for this project.

a. Name & Title:	a. Name & Title:
b. Project Assignment:	b. Project Assignment:
c. Name of Firm with which associated:	c. Name of Firm with which associated:
d. Years experience: With This Firm _____ With Other Firms _____	d. Years experience: With This Firm _____ With Other Firms _____
e. Education: Degree(s) / Year / Specialization	e. Education: Degree(s) / Years / Specialization
f. Active Registration: Year First Registered/Discipline	f. Active Registration: Year First Registered/Discipline
g. Other Experience and Qualifications relevant to the proposed project:	g. Other Experience and Qualifications relevant to the proposed project:

Figure 2-10 *(Cont.)*

7. Brief resume of key persons, specialists, and individual consultants anticipated for this project.

a. Name & Title:	a. Name & Title:
b. Project Assignment:	b. Project Assignment:
c. Name of Firm with which associated:	c. Name of Firm with which associated:
d. Years experience: With This Firm ___ With Other Firms ___	d. Years experience: With This Firm ___ With Other Firms ___
e. Education: Degree(s) / Year / Specialization	e. Education: Degree(s) / Years / Specialization
f. Active Registration: Year First Registered/Discipline	f. Active Registration: Year First Registered/Discipline
g. Other Experience and Qualifications relevant to the proposed project:	g. Other Experience and Qualifications relevant to the proposed project:

Figure 2-10 *(Cont.)*

7. Brief resume of key persons, specialists, and individual consultants anticipated for this project.

a. Name & Title:	a. Name & Title:
b. Project Assignment:	b. Project Assignment:
c. Name of Firm with which associated:	c. Name of Firm with which associated:
d. Years experience: With This Firm _____ With Other Firms _____	d. Years experience: With This Firm _____ With Other Firms _____
e. Education: Degree(s) / Year / Specialization	e. Education: Degree(s) / Years / Specialization
f. Active Registration: Year First Registered/Discipline	f. Active Registration: Year First Registered/Discipline
g. Other Experience and Qualifications relevant to the proposed project:	g. Other Experience and Qualifications relevant to the proposed project:

Figure 2-10 *(Cont.)*

7. Brief resume of key persons, specialists, and individual consultants anticipated for this project.

a. Name & Title:	a. Name & Title:
b. Project Assignment:	b. Project Assignment:
c. Name of Firm with which associated:	c. Name of Firm with which associated:
d. Years experience: With This Firm ____ With Other Firms ____	d. Years experience: With This Firm ____ With Other Firms ____
e. Education: Degree(s) / Year / Specialization	e. Education: Degree(s) / Years / Specialization
f. Active Registration: Year First Registered/Discipline	f. Active Registration: Year First Registered/Discipline
g. Other Experience and Qualifications relevant to the proposed project:	g. Other Experience and Qualifications relevant to the proposed project:

Figure 2-10 *(Cont.)*

8. Work by firm or joint-venture members which best illustrates current qualifications relevant to this project (list not more than 10 projects).

a. Project Name & Location	b. Nature of Firm's Responsibility	c. Project Owner's Name & Address	d. Completion Date (actual or estimated)	e. Estimated Cost (in thousands)	
				Entire Project	Work for which Firm was/is responsible
(1)					
(2)					
(3)					
(4)					
(5)					
(6)					
(7)					
(8)					
(9)					
(10)					

Figure 2-10 *(Cont.)*

9. All work by firms or joint-venture members currently being performed directly for Federal agencies.

a. Project Name & Location	b. Nature of Firm's Responsibility	c. Agency (Responsible Office) Name & Address	d. Percent complete	e. Estimated Cost (In Thousands)	
				Entire Project	Work for which firm is responsible

Figure 2-10 *(Cont.)*

10. Use this space to provide any additional information or description of resources (including any computer design capabilities) supporting your firm's qualifications for the proposed project.

Date:

11. The foregoing is a statement of facts.

Signature: _____ Typed Name and Title: _____

Figure 2-10 *(Cont.)*

The consultant just needs to know how to use the forms to his or her best advantage.

A few engineers resort to showy or cosmetic techniques like typing with colored ink. Others submit questionnaires as close to the deadline as possible in the belief that their forms will be placed at the top of the list for examination. It is doubtful that such attempts help consulting firms win jobs, though no one seems to know for sure. Basically, it is the quality of the firm's responses to the questionnaires that can impress reviewers most strongly. Further, an appropriate cover letter and photos may enhance the basic data. Follow-up calls and letters are usually not recommended, however, since the number of applicants is often large.

To ensure consideration of your firm, update the forms annually. Agencies will often notify your firm (if you have submitted forms in the past) to fill out a new application; or you can request new forms and update them as desired.

Completing the Federal Forms

Most of the questions on the generalized questionnaire (No. 254) are accompanied by clear explanations and definitions. However, a few suggestions might help you to maximize the benefits you can get from particular items on the form. For example, item 11 asks for examples of representative projects in which the engineering firm has been involved during the past 5 years. (If the firm has been in existence for less than 5 years, you can list projects in which individual members of the firm have been involved.) A maximum number of 30 projects can be listed. What some consultants fail to realize is that they can list the same project more than once to illustrate different services rendered on the particular job.

You can make the most of the application process by evaluating and listing all the services your firm has performed on representative jobs (unless, of course, you can list 30 separate representative projects for the 5-year period). For instance, your consulting firm may have done structural, mechanical, and electrical engineering on a new office building. Using the "experience profile code numbers" provided on the questionnaire, you would therefore list numbers 072 (office buildings, industrial parks); 101 (structural design, special structures); 083 (power generation, transmission, distribution); 043 (heating, ventilating, air conditioning); and 080 (plumbing and piping design). Additional experience could be listed wherever appropriate. Thus, one project yields examples of experience in five different categories.

Too many engineers tend to shortchange themselves by failing to make optimum use of the opportunities provided by the form. Careful consideration of all aspects of each project should actually enable most consulting companies to fill the 30 spaces on item 11. Also, the maximum dollar value—actual or estimated—should be listed for each phase of a project and for each completed project. Jobs should be listed in chronological order.

Give special attention to item 9 to provide an accurate summary of the firm's professional fees for the past 5 years. New firms should list all the fees received to date, if they've been in business less than 5 years.

Form 255, which asks for data relevant to specific projects, is generally considered to be a follow-up to form 254. However, it is often the one that government reviewers check first as they choose between interested firms. If the project for which you are applying is located too far from the offices that the firm lists on the first page of the form, the reviewer might reject the company on the basis of geographic limits alone.

Likewise, item 8 should be answered carefully to display the relevant experience of the firm for the particular project sought. Specificity is vital. For example, if a project will require engineers with experience in building foundations and soils, the firm should note that it has such experience—rather than listing "structural" expertise only. In the final selection, only the company with appropriate experience will be chosen. You should therefore maximize the opportunity for listing all aspects of your experience as it relates to the prospective work. Try to fill in all 10 spaces if possible.

Additional Materials

After the questionnaires are complete, the consulting firm should bind the forms together securely, along with identifying information and additional materials that will support and enhance the basic data provided. You can include a one- or two-page covering letter that highlights particular qualifications of the firm and its members. Project leaders can be profiled briefly, with an explanation of their position, work plans, and schedules for prospective jobs. It may also be helpful to attach photographs of relevant projects on which the company has worked. Last, you may wish to include your firm's brochure.

One caution: Do not overload the government agencies with mounds of inappropriate papers. Some consulting engineer applicants think that quantity will be viewed as a sign of high-quality work. On the contrary, the most important characteristics of your submission will be clarity, thoroughness, conciseness, and evidence that your firm can provide thrifty, effective, and efficient services. Competition for government projects can be keen. Thus the information you provide should be detailed but should not overwhelm the reviewer.

The package of material can be mailed to the government agency in charge of the project, or it can be delivered by hand. Personal delivery is rarely necessary; but in some cases where there may be some doubt about whether the forms will get to the correct people at the right time, it may be worth the added expense of using a messenger or carrying the package yourself. Of course, hand carrying is more feasible when applying for work within your state or local community.

Applying to State and Municipal Agencies

Not all states and municipalities issue questionnaires for application to government engineering projects. Those that do have their own forms usually model them after the federal questionnaires. Therefore, your experience with the federal forms will most likely simplify the task of completing the local government versions.

In localities that do not issue special questionnaires, you can still submit copies of your federal forms. Your responses should be geared toward providing information relevant to projects within the auspices of the localities. A note should be attached to explain your submission of the federal questionnaires. In some cases, an application for state or municipal work will be sufficient if it contains a cover letter, copies of the federal documents, plus the firm's brochure, with supporting photographs and references. (Local references are nearly always helpful, but they are not necessary unless specifically requested.)

Offices of appropriate government agencies should always be contacted for information about application requirements before the consulting firm submits its material.

Overseas Work

The International Federation of Consulting Engineers (FIDIC) has developed a standard registration form for consulting firms (Figure 2-11). Similar to the federal questionnaires, it is oriented specifically toward work in foreign countries. Copies can be obtained from the American Consulting Engineers Council, 1015 15th Street, NW, Washington, D.C. 20005.

Information about overseas work can be obtained from a variety of sources. The sources include: U.S. Department of Commerce, U.S. Agency for International Development (AID), United Nations Develop-

Agency:

Consulting Firm Registration Form

Date (Month, Day, Year)

1. Firm Name (firm to be registered)	Year Estab.	State or Country	Type of Organization			
			Indiv.	Partshp.	Corp.	Other

1a. Affiliated Firms

2. Home Office(s) Business Address(es), Telephone No., Cable Address, Telex No. Officers or Partners to be contacted

3. Former Firm Name(s), if any, and year established

4. Firm ownership

5. Present Branch Office(s) and year established Address, Telephone No., Cable Address, Telex No., Person in Charge

6. Number of personnel in your present organization(s) listed in 1 and 1a

Architects	Civil and Structural Engineers	Electrical Engineers	Mechanical Engineers	Construction Management	Economists	Operations & Management Specialists	Others (Specify)	Others (Specify)

Surveyors	Estimators	Inspectors	Technicians & Draftsmen	Administration and Clerical	Specification Writers	Others (Specify)	Others (Specify)	Total

7. Annual volume of gross fees (last 5 years) in US $ (per 1 and 1a)

Financial rating or bank reference

		Prime Consulting Firm	Associate and Joint Venture	Total
19	$			
19	$			
19	$			
19	$			
19	$			
			Total	

8. Largest projects handled by firm as prime consultant in major fields only
Show name of project, type of service, client reference and engineer's level of effort
Level of effort to be classified as follows:
Class 1 - Fee less than US $ 100,000 or 20 man months.
Class 2 - Fee from US $ 100,000 to US $ 500,000 or from 20 man months to 100 man months
Class 3 - Fee from US $ 500,000 to US $ 1,000,000 or from 100 man months to 200 man months
Class 4 - Fee over US $ 1,000,000 or over 200 man months

9. Partners, directors, officers, and key personnel of firm

Name and Title	Degree(s)	Years with Firm	Year of Birth

Figure 2-11 Consulting firm registration form. *(FIDIC.)*

10. Fields of specialization of permanent full-time staff (Check Appropriate Items)					
A	**Agricultural and natural resources**	**C**	**Public utilities and related fields (continued)**	**D**	**Industry (continued)**
					D-21 Steel and Iron Works
	A-1 Farm Mechanization		C-9 Housing		D-22 Textiles
	A-2 Field and Horticultural Crops		C-10 Industrial Utilities		D-23 Timber Processing
	A-3 Fisheries		C-11 Marina Terminal Facilities		Others (Specify)
	A-4 Forestry and Forest Products		C-12 Mining		D-24
	A-5 Irrigation and Flood Control		C-13 Nuclear Energy		D-25
	A-6 Land and Water Feasibility Studies		C-14 Parking Facilities	**E**	**Architecture and related fields**
			C-15 Power Stations		
	A-7 Land Reclamation and Soil Conservation		C-16 Power Transmission and Distribution		E-1 Conservation
					E-2 Educational Facilities
	A-8 Livestock		C-17 Solid Waste Management		E-3 Factories and Buildings
	A-9 Marketing and Credit		C-18 Buildings		E-4 Parks
	A-10 River Regulation and Control		C-19 Telecommunications		E-5 Social and Low Cost Housing
	A-11 Storage Facilities		C-20 Wastewater Collection, Treatment and Disposal		E-6 Urban Development and City Planning
	Others (Specify)				
	A-12		C-21 Water Supply and Distribution		E-7 Zoos
	A-13		Others (Specify)		Others (Specify)
	A-14		C-22		E-8
B	**Transport**		C-23		E-9
	B-1 Airports and Air Transport		C-24	**F**	**Economic planning and related fields**
	B-2 Bridges		C-25		
	B-3 Bus Transport Facilities		C-26		F-1 Accounting and Auditing
	B-4 Highways and Road Transport		C-27		F-2 Economic Impact Statements
	B-5 Pipelines	**D**	**Industry**		F-3 Economic Policy
	B-6 Public Transport		D-1 Agricultural Products and Food Processing		F-4 Energy Conservation
	B-7 Rapid Transit				F-5 Finance
	B-8 Railroads		D-2 Bricks and Tiles		F-6 Foreign Trade
	B-9 River and Sea Transport		D-3 Cement Works		F-7 Marketing
	B-10 River-, Seaports and Harbors		D-4 Ceramics		F-8 Organization and Management of Public and/or Private Enterprises
	B-11 Subways		D-5 Chemical Plants (including Petrochemicals, Fertilisers, Plastics)		
	B-12 Tunnels				F-9 Regional Development Planning
	Others (Specify)		D-6 Coal		Others (Specify)
	B-13		D-7 Fish Processing		F-10
	B-14		D-8 Foundries	**G**	**Tourism**
	B-15		D-9 Glass Plants		G-1 Hotel Development
C	**Public utilities and related fields**		D-10 Hides and Leather Processing		G-2 Resort Development
			D-11 Industrial Estates		Others (Specify)
	C-1 Air Pollution Control		D-12 Machinery Plants		G-3
	C-2 Cartography		D-13 Materials Handling		G-4
	C-3 Dams		D-14 Metallurgical		
	C-4 Drainage		D-15 Mineralogy		
	C-5 Electrical Installations		D-16 Non-ferrous Processing Plants		
	C-6 Gas Installations and Transmissions		D-17 Oil		
			D-18 Pharmaceutical Plants		
	C-7 Heating, Ventilating, Air-Conditioning		D-19 Pulp and Paper		
			D-20 Rubber		
	C-8 Hospitals				

Date   Name of Firm   Page 4 of 7

Figure 2-11   *(Cont.)*

11. Types of services

A.	Advisory Services	W.	Planning Studies
B.	Aerial Photography	X.	Project Management
C.	Architectural	Y.	Purchasing, Inspection and Testing of Materials
D.	Computer Services		and Equipment
E.	Construction Management	Z.	Resources Surveys
F.	Design of Machinery and Equipment	Z-1	Rate Studies and Appraisals
G.	Economic Studies	Z-2	Sector Studies
H.	Engineering Design, Estimating, Preparation	Z-3	Soils Engineering, Foundation Engineering & Design
	of Contract Documents, Bid Evaluation	Z-4	Supervision of Construction or Equipment
I.	Environmental Studies		Installation Contracts
J.	Farm Extension Services	Z-5	Testing and Inspection
K.	Geological Surveys	Z-6	Topographical and Soil Surveys
L.	Geophysical Surveys	Z-7	Technical Feasibility Studies and Preliminary
M.	Hydro-Geology		Engineering
N.	Hydrological Surveys	Z-8	Traffic Studies
O.	Industrial Process Engineering	Z-9	Value Analysis
P.	Machine Processing of Engineering Data		Others (Specify)
Q.	Management Studies	Z-10	
R.	Market Surveys	Z-11	
S.	Mineral Surveys, Photo Interpretation	Z-12	
T.	Mineral Exploration	Z-13	
U.	Oceanography	Z-14	
V.	Operation and Maintenance	Z-15	

12. Countries in which work performed within past 10 years (Check Appropriate Countries)

Afghanistan	Equatorial Guinea	Lesotho	Sierra Leone
Albania	Ethiopia	Liberia	Singapore
Algeria	Fiji	Libya	Somali (Dem. Rep. of)
Argentina	Finland	Liechtenstein	South Africa
Australia	France	Luxembourg	Spain
Austria	Gabon	Madagascar	Sri Lanka (Ceylon)
Bahamas	Gambia (The)	Malawi	Sudan
Bahrain	Germany (Dem. Rep. of)	Malaysia	Swaziland
Bangladesh	Germany (Fed. Rep. of)	Maldives	Sweden
Barbados	Ghana	Mali	Switzerland
Belgium	Greece	Malta	Syria
Benin (Dahomey)	Grenada	Mauritania	Tanzania
Bhutan	Guatemala	Mauritius	Thailand
Bolivia	Guinea	Mexico	Togo
Botswana	Guinea-Bissau	Monaco	Trinidad and Tobago
Brazil	Guyana	Mongolia	Tunisia
Bulgaria	Haiti	Morocco	Turkey
Burma	Holy See	Nepal	Uganda
Burundi	Honduras	Netherlands	Ukrainian SSR
Byelorussian SSR	Hungary	New Zealand	USSR
Cameroon	Iceland	Nicaragua	United Arab Emirates
Canada	India	Niger	United Kingdom
Central African Rep.	Indonesia	Nigeria	United States
Chad	Iran	Norway	Upper Volta
Chile	Iraq	Oman	Uruguay
China	Ireland	Pakistan	Venezuela
Colombia	Israel	Panama	Vietnam (Rep. of)
Congo (People's Rep. of)	Italy	Paraguay	Western Samoa
Costa Rica	Ivory Coast	Peru	Yemen (Aden)
Cuba	Jamaica	Philippines	Yemen (Sanaa)
Cyprus	Japan	Poland	Yugoslavia
Czechoslovakia	Jordan (The Hashemite Kingdom of)	Portugal	Zaire
Denmark	Kenya	Qatar	Zambia
Dominican Rep.	Korea (Republic of)	Romania	
Ecuador	Kuwait	Rwanda	Others (Specify)
Egypt	Khymer Republic (Cambodia)	San Marino	
El Salvador	Laos	Saudi Arabia	
	Lebanon	Senegal	

Date	Name of Firm	Page 5 of 7

Figure 2-11 *(Cont.)*

13. Narrative Description of firm

(continue on additional page if necessary)

As of this date   the foregoing is a true statement of facts.

Name of firm submitting questionnaire

Typed name and title of person signing   Signature

Date   Name of Firm   Page 6 of 7

14. Typical projects for which consultant services have been furnished during past five years

(Reference sheets may be submitted for as many projects as desired. Sheets should include at least one project in each field of specialization checked in Item 10 and each type of service checked in Item 11).

Name of overall project

Location of overall project

Engineers level of effort
(For classifications see Section 8)

Owner's name and address

Year firm's services completed (indicate if estimated or actual)

Associated firms

Description of project

(firm may submit as many pages as necessary)

Description of services firm provided

List all fields of experience (by symbols from Item 10) for associated aspects of overall project

List all types of services (by symbols from Item 11) for associated aspects of overall project

Date   Name of Firm   Page 7 of 7

Figure 2-11   *(Cont.)*

ment Program, and a variety of international and foreign banks, including the World Bank and the Inter-American Development Bank. However, some engineering firms find that the easiest way to secure foreign work is to apply for projects through the Army Corps of Engineers, the Navy Facilities Engineering Command, and the Department of Defense.

Many overseas projects require commitments of tremendous resources—financial, human, and otherwise. The reason for this outlay is that the foreign nations sometimes lack the basic ingredients for pursuing certain projects unless an engineering firm is willing to provide complete services—design, construction, and even financing. Some consulting engineering firms spend as much as $300,000 setting up an organization in a foreign country before they secure their first contract. Thus, overseas work is best suited to large consulting companies with massive capacity and resources, though some individual consultants do make money on small overseas assignments.

Additional limitations sometimes arise from the need to work with foreign exchange, local help, the peculiarities of different societies' customs, and a variety of other factors. If there is any sort of solution to the complexity of working abroad, it may be to develop steady contacts with other Americans residing in foreign countries. Such individuals may be able to help you establish the foundations for work in particular nations. Alternatively, you can solicit work in joint ventures or consulting with larger firms that already do work overseas. Doing so can enable your firm to expand its experience with minimal risk for financial loss.

CONSULTANT'S PROFIT KEY 4

A good analysis of the requirements of overseas consulting was provided in *Consulting Engineer* magazine by Lee F. Francis while vice president and manager of the International Department of James M. Montgomery, Consulting Engineers, Inc., Pasadena, California. Francis observes:

> For U.S. consulting engineering firms working in the international field, multinational associations and host country professional firm participation are rapidly becoming standard requirements on foreign projects, either by executing and/or lending agency mandate or out of necessity in developing the most economical project management program and effective project staffing. The U.S. consulting firm also faces a period of readjustment in its approaches to business development activities and project planning policies, as developing countries acquire a more substantial controlling influence in technical and cost proposal evaluations and contract awards and negotiations. Primary changes involve a growing practice of multinational consulting firm associations, project country engineering participation, and

the necessity for more productive overseas representation in view of the intense international competition.

PROJECT COUNTRY PARTICIPATION

Many international and regional lending agencies now are encouraging or requiring an association with a project country firm. For example, the Asian Development Bank (ADB) has noted the following for its selection of consultants: "The employment of domestic consultants (either firms or individuals) either alone or in combination with foreign firms, or the collaboration of a consulting firm from a developed member country with consultants from a developing member country, is encouraged where such domestic consultants or consultants from a developing member country are found to be qualified to perform the work. Other things being equal, preference will be given to domestic consultants or combinations of a consulting firm from a developed member country with consultants from a developing member country."

In many of the developing countries, infrastructure projects that 10 years ago were all being financed by bilateral or international funds are now being financed through local funding. In these situations, the participation by U.S. consultants frequently is limited to a percentage of total man-months or percentage of total fees for engineering services. The percentage requirement for local participation varies from 30 to 60 percent, with each country having its own particular stipulations. If a U.S. consulting firm desires to participate in such projects, it is obligated to associate with a local firm, and in most instances both firms of the association must be prequalified and registered with the responsible government agency.

SELECTING AN ASSOCIATE

An inherent drawback to encouraged or required host country participation is that the degree to which a local firm is qualified to participate in an association usually is commensurate with the technical development level of its country. In the smaller developing countries, the local engineering firms, if any, usually are limited in the scope of engineering services they are qualified to perform. Suitable firms may be available to provide only minimal support services such as surveying, limited civil engineering, routine geological exploration, and laboratory testing.

Limited capability is only one problem faced in selecting a local associate. It may be difficult even to find a qualified local firm to associate with. For instance, in some countries a company may be listed by a local government agency as a consulting firm when in actuality the company consists of one engineer who assembles a staff whenever he obtains a project. Frequently this individual is a government employee who is moonlighting on his own while maintaining a civil service position. Although this practice is not frowned upon by the "consultant's" employer, it makes it difficult to utilize his services on a project. There is no guarantee of availability upon project demand.

Dual employment situations generally are encountered in countries that do not have a sufficient number of trained engineers. The number of local engineers graduating from an in-country university or from out-of-country universities may be 10 to 20 per year, whereas the demand is for 300 to 400 per year or more. Consequently, government requirements for trained technical personnel make them the prime employer for all the available engineers.

In the more developed countries, the local engineering firms may be capable of providing general engineering services such as architectural and structural design. Usually, however, these firms cannot provide overall organization and management capabilities or the highly specialized engineering services required for major improvement and expansion projects.

CLIENT AND CONSULTANT BENEFITS

An association with a local firm provides many technical as well as public relations advantages to the client. Among these are: more effective technology transfer and on-the-job training of the local staff; employment opportunities for the local people; reduced foreign currency payments on the project to the benefit of the project country's balance of payments; promotion of more local pride in the project as a result of the local staff involvement; and reduced overall project cost to client due to the lower salary cost for nationals and elimination of transportation, housing, and overseas differential costs for expatriate personnel.

The U.S. consulting firm also can realize many advantages through associating with a local firm: use of local staff to provide technical support services in areas such as mapping, data collection, drilling, and laboratory testing; provision of office space, office equipment, and logistic support; and use of available vehicles and equipment. These items all help to reduce the mobilization costs to the U.S. consultant. In addition, there is a transfer of knowledge from the local professionals to the expatriate engineers on local practices, standards, local construction capability, and local costs, which can serve to avoid embarrassing and expensive mistakes.

However, U.S. consultants must use considerable discretion in selecting a local firm for association on any given project and even more discretion if it is intended that the association be a long-term relationship for more than one specific project. In many countries, even though competent local firms exist, it often is difficult to obtain an objective appraisal from local people as to the local consultant's capability, qualifications, and professional integrity. This problem, in addition to the preexisting cultural differences in acceptable business practices, requires extremely perceptive judgment in evaluating potential associates.

MULTINATIONAL ASSOCIATIONS

The foreign-based engineering firms of other industrialized nations generally are capable of providing engineering services comparable to U.S. performance standards. Many of these consultants are also very competitive with U.S. firms in submitting independent proposals for international projects. These competitors can be highly desirable associates, either under subcontract or in joint venture, when attempting to undertake a large-scale project. Benefits to be derived include: reduced transportation and relocation expenses, since such a selected foreign associate usually is in closer proximity to the project country; and simplified project staffing, where the associate firm has qualified personnel willing to accept single status, short-term or long-term assignments inasmuch as the distance between jobsite and home is only two or three hours by air.

However, U.S. consultants must recognize that the consulting practices, ethics, and engineering approach of consulting firms from other developed countries are sometimes quite different from those in the United States. These differences create a potential for misunderstandings and disagreements, which can hamper efficient execution of the engineering services to be provided by the associated firms. Many potential problems can be anticipated and resolved beforehand by establishing uniform performance standards within the text of a comprehensive joint venture or subcontractual agreement between firms.

BASIS OF ASSOCIATION

An association with a firm or firms from another country can be on a subcontractual basis or on a joint venture basis. It is, however, a general requirement of all lending agencies and clients that in any type of association between firms one firm be designated as the lead firm. In addition, the lead firm should be that one possessing the best qualifications and experience for carrying out the project.

An association with another firm on a subcontractual basis means that the subcontract firm will provide certain services, under the direction of the prime firm, or primary contractor, without responsibility for completion of the overall project. These services usually are provided as a specified number of man-hours or man-months of support or under a lump sum agreement to complete a particular phase of the project.

A joint venture involves an undertaking where there is a "community of interest" in the objective. It is a voluntary contract between two or more parties to commit their money, effects, labor, and skill to a legally binding business agreement with the understanding that there shall be a proportional sharing of the profits or losses between them. The parties of a joint venture have a proportional right to govern the conduct of each other and have a voice in the control and management of the joint venture.

On a large foreign project, a joint venture with another U.S. consulting engineering firm or with a firm from another developed country can work in the best interest of both parties. On a cash flow basis where payments are received in sporadic installments, the burden is reduced for each company, as is the potential risk factor involved in the event of a financial loss. In addition, problems of staffing a large project with qualified senior personnel are shared by the joint venture partners, thus decreasing the high-level manpower commitments of each firm. Staffing for a large overseas project can be a major difficulty, since most foreign clients insist that experienced and well qualified professional engineers be assigned to their projects.

A joint venture with a local firm in a developing country generally is not feasible since few local firms have sufficient monetary reserves to participate on an equal level of financial responsibility. They may, however, have trained personnel who can provide essential support services. Then it is advantageous to enter into a subcontractual agreement.

REPRESENTATION

Achieving productive representation in foreign countries is a complexity in its own right. Many U.S. consultants have attempted to overcome this by establishing regional offices with a roving "ambassador" covering the territory. This is not only a costly position to justify in terms of direct benefits, it is a difficult job to do effectively in areas such as Southeast Asia and South America.

Local representation is generally necessary to maintain contact with public officials and to assay upcoming jobs. These aims can be accomplished through an individual engineer, a local firm, local office of the consultant, or a business agent. Business agents are used frequently in the Middle East countries to act on behalf of the consultant in obtaining work. The agent may work on a retainer basis or on a job basis. The use of an agent, individual engineer, or local firm has the advantage of maintaining close client contact at a minimal cost in comparison to the capital investment required to open and operate a representative office. The desirability of utilizing the consultant's personnel, who are naturally better equipped to promote the company and its qualifications, is far outweighed by the expense incurred in obtaining new business by this method. Overhead costs involved in maintaining a permanent nonproject office and staff in a foreign country are exorbitant.

Discussion of considerations relevant to establishing associations and representations for providing consulting engineering services internationally strongly implies one most important fact. The demise of U.S. and European technological dominance in engineering is within sight. This can be attributed to the increasing economic strength and political influence of the developing world nations, inflationary pressures on engineering project costs, and realization of the end result of effective technology transfer on past and present engineering projects which have involved in-depth technical seminars and on-the-job training programs to project country counterpart personnel. The international consulting engineering services field should continue to flourish as demands for technical training are replaced by organizational and management skills requirements. The need for high-level specialized project staffs to fulfill these goals will enhance the favorability of working within the framework of professional association agreements for international clients. Competition for such projects also will create increased pressures for more active overseas representation.

PROPOSAL PREPARATION

Government agencies and other prospective clients frequently request interested consulting firms to submit proposals that will help them select the right engineers for a particular project under consideration. However, your consulting firm can send unsolicited proposals directly to potential clients. But this practice is relatively rare, except for a few large and well-established companies.

Proposals can lead to substantial amounts of work—often constituting one-fourth of a consultant's business—particularly during the first few years of a practice. Therefore, some of the most successful young firms spend considerable time and effort developing their proposal writing skills. At the same time, they try to maintain a steady flow of new proposals to clients since only a percentage of the proposals will produce major contracts.

Proposals generally must be developed based on materials supplied by the client—such as questionnaires and brief outlines referring to the specific project that is planned. The preparation effort can be both frustrating and rewarding. It usually involves assembling a team of resourceful staff members to sift through the client's materials and through various information sources to produce a broad base of data related to the proposed work. Following the organization of the data, selected members of the team gradually develop a concise, persuasively written presentation that demonstrates the firm's ability to provide expert answers to the client's needs.

Thus the proposal team might consist of a manager, researcher, writer, and additional engineering assistants as required. The manager would also coordinate the plans contained in the proposal with the ongoing organizational and personnel needs of the entire firm.

The effort may become hectic; but a schedule of firm deadlines for the completion of each phase of the proposal task helps minimize confusion. Members of the proposal team should be informed clearly of the deadline for submission of research data and of the dates for completion of the first draft, subsequent versions, and final drafts of the written document. While young consultants may view a 1-week deadline for a final draft as a harsh test of their ability to work under pressure, many experienced firms have learned to prepare proposals in less time—often without anxiety.

One way to help ease the preparation burden is to have the team create a checklist of items that must be covered during the course of the proposal effort. Individual checklists can be developed for separate aspects of the proposal preparation, such as background information about projects similar to the one under consideration; descriptions of the consultant's experience in such work; listings and explanations of the consultant's involvement in projects in the client's geographic area; work scope, plans, and possible schedules for the proposed project; qualifications of staff members who may be engaged in the work; description of organizational plans for doing the work; use of particular equipment and facilities; and financial requirements and contractual arrangements.

The checklists and proposal need not include all the topics just listed. Specific proposal formats and content may vary with each project. However, the following are among the topics that are addressed in most proposals:

1. The firm's experience and capabilities in the type of work needed for the proposed project

2. A discussion of the technical aspects of the proposed work, including information that will evidence the firm's ability to provide high-quality services (whether they will be design, management, or a variety of other engineering tasks)

3. The staffing and organization planned for meeting the client's needs and a description of the management team

4. The background and qualifications of selected staff members who will work on different aspects of the project

5. Discussion of the firm's facilities, its capacity to accomplish the work in the project's geographical area, and advantages or disadvantages that the firm might have in performing the task at the proposed location

6. Field support—arrangements that will be made for the use of additional consultants, contractors, or other professionals from outside the firm

7. Summary of estimated costs as they will be incurred throughout the project's duration

8. Potential impacts that the project will have on the firm if the proposal is accepted, including an examination of current work load and personnel needs to accomplish the work under consideration

Based on the preceding guidelines, the proposal should include a brief review of the firm's previous work of the kind involved, a description of the optimum solution to the client's problem (plus alternative solutions, when appropriate), and the firm's plan to use its resources to achieve the best results. Most proposals contain a schedule of personnel requirements and job assignments, also in summary form.

Cost estimates are an important part of nearly all proposals, although they sometimes get little attention during the preparation effort. A careful estimating of costs requires background investigation of work similar to the proposed project, plus an analysis of the potential job, personnel requirements, overhead, administrative needs, and other factors. Standard fee schedules, like those published by some engineering societies, can also be consulted. Still, many consultants underestimate the importance of cost summaries and thus rely on guesswork.

Some clients do not require a detailed cost summary with the technical proposal, preferring instead to negotiate fees after selecting the consultants for the planned project. But the consulting firm is usually well equipped to estimate costs at the time it prepares the proposal. It is at that time that the proposal team examines the distinct components of the work in detail. Therefore, if the team directs some of its attention to cost estimating, the consultant might avoid potential difficulties during subsequent negotiations. The cost summary—which is of course related to estimates of the time needed to accomplish particular tasks—can thus facilitate proposal preparation, regardless of whether or not the summary appears in the technical proposal.

One method of preliminary cost estimating is to choose a standard fee from the recommended fee schedules that appear in most professional engineering manuals, adding reasonable dollar amounts for administrative expenses and general overhead. That sum would yield a breakdown for direct labor, general and administrative expenses, and overhead. If those quantities are divided by an average dollar value of labor for the project staff, a total number of worker-hours can be approximated. The amount of time can then be weighed against the logistics of the total job to see if it represents a reasonable estimate. Here is an example of the use of this method:

Gross fee		$25,000
Selling and administrative expense at 20% of fee	$ 5,000	
General overhead expense at 40% of fee	$10,000	
Total nonlabor charges		15,000
Balance for direct hourly labor		$10,000
Average hourly labor rate for all types of labor that most likely will be involved using actual payroll-record costs = $15/hour		
Hours available for project = balance for direct hourly labor/average hourly labor rate = $10,000/$15/hour = 667 hours		
Therefore, total worker-hour budget = 667 hours		

Inaccuracies can arise from speculative modes of estimating. But the preceding method—at the very least—enables the consultant to double-check her or his guesswork against standard fee schedules and past experience.

The clients of most consulting engineers tend to be more concerned with obtaining expert, effective service than with getting a job done at the lowest possible cost. But if the consultant produces an extremely low cost estimate as a result of poor guesswork, the client may get the impression that the services are of correspondingly low quality. Likewise, sloppy cost analysis may give rise to excessively high estimates. Either way the firm would run a higher risk of rejection than it would have, had it used a method like the one previously outlined.

A related area of proposal preparation that can also lead to problems if not researched diligently is the discussion of contractual agreements. Certain words and conditions carry special meanings in the context of the contract that some members of the proposal team may not be aware of. To prevent misinterpretation, the discussion of contractual matters should be reviewed by appropriately skilled individuals—including the principals of the firm, accountants, and attorneys—if possible. Thorough editing is necessary for the production of the final, polished draft of the proposal. All possible developments, such as "escalator clauses"—payroll increases that could go into effect during the course of the project—should be addressed in the document.

The general writing task can be approached by developing an outline first. The outline usually indicates briefly the proposal's structure. For example, an outline for a rather short proposal might note the basic data for the introduction, including a statement of the problem, a technical discussion of solutions, a program plan, a description of the consulting firm, a summary, and an appendix listing reference sources used in the proposal preparation.

Following the development of the outline, all relevant information should be collected and reviewed for incorporation in the proposal draft. Once the data is organized, the writers prepare the first draft. Many people find this to be the hardest part of the effort. However, the task seems less foreboding when the initial emphasis is placed on thoroughness, rather than on the writing quality. During revisions, excess material can be deleted while grammar, style, and organization are improved.

In the final-draft form, the proposal generally has a unifying theme. Like an interesting movie, it develops a plot, a cast of characters, and finally provides resolution to a problem. The complete proposal package may include a cover letter, table of contents, and list of illustrations, in addition to the basic text.

Grant Proposals

Grants are an increasingly vital means of funding for academic groups, researchers, nonprofit organizations, and others. Sources of grant awards include agencies of the federal government and corporate foundations. The U.S. government itself funds thousands of grant projects for each year—often as the result of legislation providing for experimental projects involving engineering services.

Federal grants have been awarded to consulting engineers for work involving waste-water treatment; synthetic fuels from municipal garbage; modular housing; prison refurbishment; cost specification and management; plus a seemingly limitless variety of other works. Grant awards can be a boon to a private practice, but considerable background knowledge and acquaintance with government officials is usually required in order to capture the work. Competition for grant proposals is sometimes like a World Series for consultants: The teams train and compete throughout a long season, matching wits and strengths until a few players emerge victorious.

Generally, agencies of state and municipal governments apply for grants directly to the federal government. They, in turn, may ask consultants for assistance in preparing the grant proposals; in other cases, the consultants apply to the agencies themselves.

Much information about grants is available through branch offices of the U.S. Chamber of Commerce, your local chamber of commerce, and individual agencies involved in grant programs. Community planning boards may also be able to provide details on projects in their areas.

Many agencies are initially interested in enhancing their own grant proposals with the technical know-how of consulting engineers. Interested firms can usually obtain data on particular projects from the sponsoring organization; the firms can then review the information to determine whether or not they will be able to participate. Finally, they can contact the agency to offer services in proposal preparation and/or on larger aspects of the overall project.

Additional sources of grant information include libraries and foundation centers. The following are among the best-known regional carriers of grant-related publications:

San Francisco Public Library
Business Branch
530 Kearny Street
San Francisco, Calif. 94100

The Foundation Center
Washington Office
1001 Connecticut Avenue
Washington, D.C. 20036

Atlanta Public Library
126 Carnegie Way NW
Atlanta, Ga. 30300

The Newberry Library
60 West Walton Street
Chicago, Ill. 60600

Associated Foundation of Greater Boston
One Boston Place
Boston, Mass. 02100

The Danforth Foundation
222 South Central Avenue
St. Louis, Mo. 63100

Cleveland Foundation Library
700 National City Bank Building
Cleveland, Ohio 44100

Regional Foundation Library
Hogg Foundation for Mental Health
University of Texas
Austin, Tex. 78700

A variety of books and periodicals are available directly from the headquarters of The Foundation Center at 888 Seventh Avenue, New York, New York 10106.

Directories and Newsletters

The Foundation Directory: Lists the more than 5000 foundations currently providing grants. The latest edition is available from Columbia University Press, 136 Broadway, Irvington-on-Hudson, N.Y. 10533.

The Foundations Center Information Quarterly: Available by subscription from Columbia University Press.

The Foundation News: A bimonthly subscription publication issued by the Council on Foundations, Box 783, Old Chelsea Station, New York, N.Y. 10011.

The Foundation Grants Index: A compilation of grants reported in *Foundation News* for the last year. Available from Columbia University Press.

Books

Andrews, F. Emerson. *Applicants for Grants, Philanthropic Foundations.* Russel Sage Foundation, New York. Available from Basic Books, Inc., Box 400, Scranton, Pa. 18501.

Church, David M. *Seeking Foundation Funds,* Joseph Dermer, ed. National Public Relations Council of Health and Welfare Service, Inc., 419 Park Avenue South, New York, N.Y. 10016.

How to Get Your Fair Share of Foundation Grants. Public Service Materials Center, New York. Available through the Center, located at 104 East 40th Street, New York, N.Y. 10016.

Dermer, Joseph. *How to Raise Funds from Foundations.* Public Service Materials Center, New York.

————. *How to Write Successful Foundation Presentations.* Public Service Materials Center, New York.

Hicks, Tyler G. *Raising Money from Grants—a Complete Kit.* Available from IWS, Inc., Post Office Box 186, Merrick, N.Y. 11566.

Hill, William J. *A Comprehensive Guide to Successful Grantsmanship.* Grant Development Institute, Littleton, Colo. Available through the Institute, located at 2552 Ridge Road, Littleton, Colo. 80120.

Mirkin, Howard T. *The Complete Fund Raising Guide.* Public Service Materials Center, New York.

Nicholas, Ted. *Where the Money Is and How to Get It.* Enterprise Publishing Co., Wilmington, Del. Available from Enterprise Publishing Co., 1000 Oakfield Lane, Wilmington, Del. 19810.

Selection of Proposers

The evaluation of proposals can be a lengthy process. Depending on the size and complexity of the project under consideration, it may take anywhere from a few weeks to several months for a government or foundation selection committee to choose the consulting group that seems best suited for the job.

Specific evaluation criteria may vary among different agencies and projects. Ultimately, most clients are concerned with the engineering firm's overall approach. Clients want to see a demonstration of competence, thoroughness, and clarity in the written description of work scope and plans, personnel qualifications, and technical capability. The staff's prior education and its reputation for successful innovation are also considered important. In some cases, the proposer may also need to report on preliminary testing of specific components of the firm's design, methodology, or work program.

Estimates of costs and fees may also count strongly in the evaluation process but nowadays are often considered to be of less significance. Therefore, fee negotiations usually occur after the majority of proposers have been eliminated from consideration. Clients may invite interested consultants to engage in competitive bidding. Government agencies usually publicize requests for bids in the *Commerce Business Daily*. In 1972 legislation known as the Brooks Law was enacted by the U.S. Congress in an effort to prevent government agencies from selecting proposers merely by bid. The legislation encourages public agencies to evaluate interested firms based on written statements of qualification and/or personal interviews first; then the final selections are made according to the outcome of fee negotiations.

Factors that could cause a proposer to be rejected from consideration early in the selection process include failure to comply with general requirements, standards, or conditions outlined by the sponsoring agency in its official Request for Proposal (commonly referred to as the RFP); lack of experience in the type of work involved, in master planning, in management, or in environmental assessment; lack of financial stability; a lack of qualified permanent staff members for directing the work; and a reputed lack of professionalism or of ethics. Selection committees sometimes tend to favor firms with which they have had successful relations in the past. However, dearth of experience in government projects should certainly not be regarded as a barrier to new work.

The Brooks Law—named for its sponsor, Jack Brooks, a Texas legislator—seeks to reduce the impact of politics on the selection process by promoting fair competition and negotiation among firms. Many states have passed similar laws in recent years. Opinion about such legislation varies throughout the engineering profession. However, experts generally agree that one effect of the laws has been to limit consulting fees to a portion of the total dollar value of government projects, usually about 6 percent.

Since sponsoring agencies sometimes state a price range in their advertisements of new projects, consultants occasionally find it difficult to determine the best fee to quote in their initial negotiations. Many engineers feel that they optimize their prospects for capturing a contract and reasonable fee by basing their initial offer on the higher figure in an advertised price range. They generally do not expect to get that fee but use it as a bargaining tool. By the time the actual contract agreements have been reached, both parties may be satisfied with a fee as low as 6 percent of the middle range of the project's advertised value.

POLITICS AND THE CONSULTING ENGINEER

Opinions about the relationship between politics and government projects vary throughout the engineering profession. Despite legislation encouraging fair competition between consultants seeking government work, questions about the proper role of engineers in the political process remain. Certainly, politics are as important to democracy as physics is to engineering. Full participation in the machinery of our democratic society requires active involvement in political affairs. Such participation is indeed a fundamental characteristic of good citizenship. It may include voting and public support for a particular party and politicians, as well as making contributions to campaigns, lobbying, and fund-raising. But how can engineers practice good citizenship without taking the risk of appearing to engage in political impropriety?

Participation in political affairs generally does not limit a firm's opportunities to acquire government or private-sector work. Active support for the party or candidate of one's choice may help to establish contacts with potential clients. For the beginning consultant who wants to get acquainted with prominent individuals in the community, few routes are more valuable than involvement in party activities. Officials of public agencies may even ask local politicians to recommend engineers with whom the politicians are acquainted.

However, political involvement can be abused. Misuse of relationships with public officials can lead to a breach of professional ethics and long-term damage to a firm's reputation. There are several well-known cases of engineers who were charged with bribing local officials in attempts to curry favor for engagement in government projects. Usually when the charges reached the news media, they erupted into full-blown scandals and further allegations of corruption. One of the more unfortunate outcomes may be that the scandals tend to obscure the honest efforts of the majority of engineering professionals who seek to serve the public interest at the same time that they advance their careers.

Political parties and candidates depend upon contributions from both public and private sources—including engineers—to promote the interests of their constituents. Consultants are not barred from making such contributions; but their professional position often leads to substantial difficulties in this regard. If a consulting firm makes numerous, massive contributions to various campaigns and subsequently acquires contracts through public officials associated with those campaigns, the consultants may be accused of bribery. The same can occur on a smaller scale too. For example, excessive gift giving could be seen as bribery, regardless of its actual intent.

Where do engineers draw the line between proper and improper political behavior? Some may choose to avoid involvement completely. For others, that is not a satisfactory solution. Perhaps the best answer is to provide what we can refer to as *ethical political assistance:* the devotion of reasonable amounts of money and personal effort to support the party and candidates of your choice. Reasonable financial contributions generally range from about $100 at the local level to $1000 at the state level and $5000 at the federal. These figures are variable, of course.

While some campaigns set a suggested minimum amount, no firm is obligated to contribute. Larger contributions (or a series of donations made on a regular basis) may increase the amount of attention that party members give to the consulting firm. However, care must be taken not to exceed legal maximums established for various levels of government. Furthermore, in no case should a donation be viewed as entitling the contributor to either government or private work.

Some consultants use a yardstick of 1 or 2 percent of their total gross sales as representing a reasonable allocation for political contributions. Others use higher percentages. Donations are usually made by check to furnish the engineer with a solid record of all such transactions. The use of checks instead of cash also helps to avoid the appearance of an attempt to make contributions secretly.

Bribery and influence peddling have existed for centuries. But the wise consultant steers clear of situations that may give the slightest impression that he or she is associated with unethical or illegal activities. Political contacts can be helpful as you expand your practice. But involvement in the political arena should not play a dominant role in your business. Properly handled, political activity is one way to establish new relations between your firm and members of the public as you fulfill civic responsibilities.

Other consultants may choose to avoid politics or support different candidates; certainly your staff members have the right to decide for themselves. And you may choose for your company as a whole to assist in a political campaign. Such assistance could enhance the firm's public image. But the firm should not allow its offices to be used for political advertisements.

Another issue that often comes up in discussions about engineers and politics is: What happens if you learn about ongoing bribery or other forms of corruption? Should you report it? These are not easy questions to answer since one might have well-founded fears of retaliation. Some people think that corrupt practices are so pervasive that "blowing the whistle" is not even worth the effort. Not confined to politics, bribery occurs in many forms—in cover-ups of faulty construction, in the favoring of particular manufacturers or contractors for a prospective job, in the granting of shortcuts, and in noncompliance with laws and specifications. Whenever an engineer engages in such actions without valid, legal reasons for doing so, she or he risks harming the reputation of the entire profession. Breaching professional ethics, for the engineer, can mean endangering public safety too. Unfortunately, some engineers do engage in corrupt practices. The only way they can be stopped is if their fellow professionals refuse to involve themselves in similar activities. If necessary, incidents of corruption should be reported to appropriate authorities.

Two significant subclasses of political activity for engineers are: running for office and lobbying. Historically, engineers have been involved in both. Presidents Jimmy Carter and Herbert Hoover both worked as engineers before seeking public office. And voluntary organizations such as the American Consulting Engineers Council and the International Federation of Consulting Engineers represent the interests of consulting professionals by meeting with legislators in various states and in Washington, D.C.

The engineer who runs for office or gains appointment to public positions sometimes encounters situations that could create conflicts of interest with respect to his or her business. If the engineer continues to head the consulting firm, the public office position may actually represent a barrier to many potential projects. One typical example follows:

Ms. X, a consultant in civil engineering, was appointed to serve on the municipal highway commission. Previously, she had done much work designing roadways and traffic lighting for the city. As a member of the highway commission, however, she was involved in decisions affecting the selection of consulting firms for the same kind of projects. This placed her in a conflict of interest between her position as an objective commission member and her role as a civil engineer. Obviously something had to be changed. Rather than give up her business or her official post, Ms. X decided to reorient the focus of her consulting work. Thus she declined engagement in projects related to highway work, even though doing so meant a temporary reduction in her firm's total revenue.

Engineers can run for political office as freely as most citizens can. However, one should thoroughly investigate the impacts that holding an official position might have on one's career—and vice versa—before attempting to enter the political arena.

Lobbying, on the other hand, is a way that engineers can guard and further their professional interests. It generally involves meetings between representatives of one of the major voluntary organizations and elected officials, plus participation by selected members in government committee hearings and the like. The intent of lobbying usually is to influence legislative and other governmental decisions affecting the engineering profession.

The tradition of organized lobbying developed largely from the experience of various groups who sought to influence political decision making during the last 200 years. They found that organized lobbying pro-

AMERICAN CONSULTING ENGINEERS COUNCIL

PROFESSIONAL AND ETHICAL CONDUCT GUIDELINES

PREAMBLE

Consulting engineering is an important and learned profession. The members of the profession recognize that their work has a direct and vital impact on the quality of life for all people. Accordingly, the services provided by consulting engineers require honesty, impartiality, fairness and equity and must be dedicated to the protection of the public health, safety and welfare. In the practice of their profession, consulting engineers must perform under a standard of professional behavior which requires adherence to the highest principles of ethical conduct on behalf of the public, clients, employees and the profession.

I. Fundamental Canons

Consulting engineers, in the fulfillment of their professional duties, shall:

1. Hold paramount the safety, health and welfare of the public in the performance of their professional duties.

2. Perform services only in areas of their competence.

3. Issue public statements only in an objective and truthful manner.

4. Act in professional matters for each client as faithful agents or trustees.

5. Avoid improper solicitation of professional assignments.

II. Rules of Practice

1. Consulting engineers shall hold paramount the safety, health and welfare of the public in the performance of their professional duties.

    a. Consulting engineers shall at all times recognize that their primary obligation is to protect the safety, health, property and welfare of the public. If their professional judgment is overruled under circumstances where the safety, health, property or welfare of the public are endangered, they shall notify their client and such other authority as may be appropriate.

    b. Consulting engineers shall approve only engineering work which, to the best of their knowledge and belief, is safe for public health, property and welfare and in conformity with accepted standards.

    c. Consulting engineers shall not reveal facts, data or information obtained in a professional capacity without the prior consent of the client except as authorized or required by law or these Guidelines.

Figure 2-12 Professional and ethical conduct guidelines. *(American Consulting Engineers Council.)*

d. Consulting engineers shall not permit the use of their name or firm nor associate in business ventures with any person or firm which they have reason to believe is engaging in fraudulent or dishonest business or professional practices.

e. Consulting engineers having knowledge of any alleged violation of these Guidelines shall cooperate with the proper authorities in furnishing such information or assistance as may be required.

2. Consulting engineers shall perform services only in the areas of their competence.

a. Consulting engineers shall undertake assignments only when qualified by education or experience in the specific technical fields involved.

b. Consulting engineers shall not affix their signatures to any plans or documents dealing with subject matter in which they lack competence nor to any plan or document not prepared under their direction and control.

c. Consulting engineers may accept an assignment outside of their fields of competence to the extent that their services are restricted to those phases of the project in which they are qualified and to the extent that they are satisfied that all other phases of such project will be performed by registered or otherwise qualified associates, consultants or employees, in which case they may then sign the documents for the total project.

3. Consulting engineers shall issue public statements only in an objective and truthful manner.

a. Consulting engineers shall be objective and truthful in professional reports, statements or testimony. They shall include all relevant and pertinent information in such reports, statements or testimony.

b. Consulting engineers may express publicly a professional opinion on technical subjects only when that opinion is founded upon adequate knowledge of the facts and competence in the subject matter.

c. Consulting engineers shall issue no statements, criticisms, or arguments on technical matters which are inspired or paid for by interested parties, unless they have prefaced their comments by explicitly identifying the interested parties on whose behalf they are speaking and by revealing the existence of any interest they may have in the matters.

4. Consulting engineers shall act in professional matters for each client as faithful agents or trustees.

a. Consulting engineers shall disclose all known or potential conflicts of interest to their clients by promptly informing them of any business association, interest or other circumstances which could influence or appear to influence their judgment of the quality of their services.

b. Consulting engineers shall not accept compensation, financial or otherwise, from more than one party for services on the same project, or for services pertaining to the same project unless the circumstances are fully disclosed to, and agreed to, by all interested parties.

c. Consulting engineers in public service as members of a governmental body or department shall not participate in decisions with respect to

Figure 2-12 *(Cont.)*

professional services solicited or provided by them or their oganizations in private engineering practices.

d. Consulting engineers shall not solicit or accept a professional contract from a governmental body on which a principal or officer of their organization serves as a member.

5. Consulting engineers shall avoid improper solicitation of professional assignments.

a. Consulting engineers shall not falsify or permit misrepresentation of their, or their associates', academic or professional qualifications. They shall not misrepresent or exaggerate their degree of responsibility in or for the subject matter of prior assignments. Brochures or other presentations incident to the solicitation of assignments shall not misrepresent pertinent facts concerning employees, associates, joint ventures or past accomplishments with the intent and purpose of enhancing their qualifications and their work.

b. Consulting engineers shall not offer, give, solicit or receive, either directly or indirectly, any political contribution in an amount intended to influence the award of a contract by public authority, or which may be reasonably construed by the public of having the effect or intent to influence the award of a contract. They shall not offer any gift or other valuable consideration in order to secure work. They shall not pay a commission, percentage or brokerage fee in order to secure work except to a bona fide employee or bona fide established commercial or marketing agencies retained by them.

Figure 2-12 *(Cont.)*

vided the most effective means of getting their message to the politicians. While opinions on political issues may vary among professionals, many engineers feel they get significant personal and professional rewards from voluntary participation in various organizations. Activities may include public speaking, participation in hearings, meeting legislators, and more.

Finally, some of the professional societies have developed practical guidelines and codes of ethics for consulting engineers. Like the physician's Hippocratic Oath, the codes are viewed as standards for good practice. Figure 2-12 shows the guidelines of the American Consulting Engineers Council.

SUMMARY

1. The successful salesperson is a skilled communicator who has detailed knowledge of the capabilities and goals of his or her firm.

2. Marketing requires skilled salespeople to initiate efforts to acquire new clients. Consultants can rarely, if ever, expect work to come to them unless they first strive to inform potential clients of their availability. Once they do, they must impress upon the clients the consulting firm's commitment to excellence, its expertise, and its reliability.

3. While the public generally seems to view the engineering profession as rather routine and dull, the firms that attract the most clients tend to express enthusiasm, creativity, and confidence through all their actions—in their advertisements, proposals, brochures, personal interviews, phone conversations, and professional relationships.

4. You should not be afraid to reject a potential consulting job if it does not fit within your firm's capacities or desires. Consultants have no obligation to take on every job that is available.

5. Gifts generally do not lead directly to sales. Some consultants feel they can enhance their marketing efforts by giving presents of one kind or another to potential clients; but projects are rarely acquired on this basis.

6. Innumerable marketing research studies have been made to determine what makes people buy. These studies usually find that successful sales are made chiefly on the basis of the client's idea that a particular consulting firm is the best available source of high-quality services.

7. Polite persistence helps to impress the notion that you are serious and determined to perform high-quality work. Close communications with clients should be carried on from the beginning of each potential project, through the follow-up phases.

The following is an example of a professional brochure used by the Mueller Engineering Corporation.

MUELLER

ENGINEERING
CORPORATION

58 SOUTH STREET / WEST HARTFORD, CONNECTICUT 06110 / Telephone: (203) 527-7287

FOREWORD

Since its founding in March of 1950, the Mueller Engineering Corporation has been involved in engineering projects with a total value in excess of $509 million. The requirements of these projects have covered every phase from original evaluation of design to management of construction.

The firm maintains a staff of some 18 specialists who are experts in manufacturing-plant design as well as the more specialized elements of plant engineering. Over the past 32 years the firm has handled over 900 separate projects.

Because of its wide experience in working for the various branches of state and federal government agencies and major industrial firms, the firm is able to bring to any project a broad background of experience. Coupled with this experience is a thorough knoweldge of costs of both labor and materials.

No individual or firm can know everything. Consequently, the Mueller firm has available a list of consultants who, when approved by the client, are retained for their special expertise. With this form of team approach, a service is available to meet the total needs of any project.

SERVICES

General

The Mueller Engineering Coproration services include studies, reports, investigations, cost estimates, and design and management of construction. The firm has in-house staff in electrical, mechanical, structural, and process fields of engineering.

Buildings

Total design capability for manufacturing plants, warehouse buildings, and structures to house special processes.

Electrical

Design of high- and low-voltage, interior and exterior, power distribution. Control, communication, and alarm systems. Energy conservation.

Mechanical

Central HVAC systems. Boiler and refrigeration plants. Heat recovery and specialized exhaust. Clean rooms. Sprinkler and fire protection systems. Control systems. Energy management.

Process

Processes for product manufacturing. Materials handling. Supporting facilities, utilities, and building modifications.

Structural

Building structures, additions, modifications. Machine bases. Roofs, foundations, and renovations.

BUILDINGS

1. **Smith-Gates Corporation.** Expansions of this highly successful electrical products manufacturer required the design and construction of five additions, totaling 89,000 ft^2 of space for manufacturing, warehousing, and shipping facilities.

2. **Top-Notch Office Building.** This 100,000-ft^2 four-floor office building combined a precast concrete exterior with conventional steel frame to provide an office building with maximum tenant facilities. Special foundation design requirements were incorporated, due to its close proximity to the Connecticut River.

3. **Union Station, Hartford, Connecticut.** Renovation and reconstruction of the 1879 structure into a viable transportation center required structural investigation and analysis of the existing building and railroad trestle to provide a basis for structural upgrading and new construction within and adjacent to the existing facility without changing its external appearance, due to its historic classification.

4. **Veterans' Administration Hospital, Manchester, New Hampshire.** Modernization and expansion of the fourth-floor surgical suite required the installation of heavy mechanical equipment and a building on the fourth-floor roof in order to provide a lightweight structure within the load-carrying capacity of the existing structure. A building utilizing the lightweight components of preengineered buildings was designed.

5. *City Garage, North Adams, Massachusetts.* An 11,000-ft^2 city garage for the maintenance and storage of the Public Works Department trucks and equipment, combining existing structures on the site, was designed using preengineered building concepts.

6. *Mite Corporation, Farmington, Connecticut.* A 20,000-ft^2 addition to the manufacturing space incorporating overhead traveling cranes, specialized point spray exhaust and conveyor systems.

7. *Union Carbide, East Hartford, Connecticut.* Renovation and rearrangement of 60,000 ft^2 of space (rearranging and relocating building structure, walls, and other areas for specialized process application).

8. *Submarine Warfare, U.S. Naval Base, New London, Connecticut.* Renovation and rearrangement of existing 5000-ft^2 area for secret computerized simulations including signal shielding of walls and ceilings and clean room atmosphere.

9. *Laboratories, Newington Children's Hospital, Newington, Connecticut.* Renovation of 20,000 ft^2 of space for computerized stick figure Gait Laboratory, raised floor, camera coordination, and pressure sensors. Ophthalmology laboratory with specialized dimmer remote control for vision testing.

2-135

ADDITIONS TOTALLING 100,000

BUILDING REHABILITATION

1. *Clean Room.* Complete rehabilitation of existing manufacturing space with remote air conditioning space and process air systems to create chip manufacturing facility at IBM, East Fishkill. Fully automated. Direct digital control. $1,200,000.

2. *Process Facility.* Complete renovation of existing extrusion area for 12 lines of plastic film production at Union Carbide, East Hartford, Connecticut. Including all process, space, utility, and facility support equipment and energy management. Cost in excess of $2,500,000.

3. *Electrical Distribution.* Renovation, replacement, and retrofit of exterior and interior $1,500,000 substation and distribution of electrical power at Machlett Laboratories in Stamford, Connecticut. Included energy uiltization and monitoring study for energy conservation.

4. *Laboratory Facility.* Complete removal of existing operation and retrofitting to create 20,000 ft^2 of specialized laboratory space at Newington Children's Hospital, Connecticut. Project included all forms of specialized building construction, support utilities, and equipment.

5. **Power Plant.** A $2 million removal and replacement with modern equipment of an operating boiler plant in the East Haven Veterans' Administration Medical Center, Connecticut, without interruption to any ongoing operation of this 1,000,000-ft^2 modern hospital. Emergency power plant also totally replaced.

6. **Production Facility.** Updating, repiping, tank replacement, and enlarged pumping capability; interior and exterior renovation, and new pile foundations. Fuel, test fuel, and cutting oil storage system costing $1,200,000 designed and installed while maintaining constant plant operation.

7. **Pier Renovation.** Retrofit of Pier 15 at a cost of $8 million to accommodate certain overhaul programs at Submarine Base in New London, Connecticut. Included new pier and pile foundations, electrical, sewer, water, air, and potable water system.

8. **Production Facility.** Enlargement and renovation for some $2,800,000 of the manufacturing facility of Smith-Gates in Farmington, Connecticut. Structural, electrical, and mechanical renovations. New building additions including structural, mechanical, and electrical facilities.

…TING, 3000 BTU HTHW GENERATORS
EASTERN CONNECTICUT STATE COLLEGE
WILLIMANTIC, CONNECTICUT
…HEATING, COOLING AND DISPOSAL FACILITY

SPECIALIZED ENGINEERING

1. **Jet Fuel, Avco Lycoming.** Engineering design of 300,000-gal total storage facility for JP-4 and JP-5 with pumped distribution to numerous test cells. $500,000.

2. **Fuel and Cutting Oil, Avco Lycoming.** Engineering design on piles and compacted bases of a 400,000-gal No. 6 oil storage and associated distribution and a 13-tank (10,000 gal each) farm for cutting oil storage and associated distribution. Heated tanks. $850,000.

3. **High- and Low-Voltage Distribution, Machlett Laboratories.** Cost study and design of a 13.5-kV exterior distribution, new transformers and capacitors, and 480-V interior distribution. $600,000.

4. **Facility Redesign, Darworth Company.** Engineering design of piping, facility building, and total new plant electrical distribution to accommodate new paint manufacturing process. $500,000.

5. **5000-psig Air Distribution, Naval Submarine Base, Portsmouth, New Hampshire.** Engineering design of oil-free 5000-psig compressors, controls, tanks, blast-proof building, and 4000 ft of distribution on piers and dry docks. $750,000.

6. **Terminal Power and Control, Mobil Oil Company, New York.** Survey, report, and design of electrical power and control systems at Staten Island, Brooklyn, Port Jefferson, Rochester, and Utica. $3,750,000.

7. **Integrated Circuit Manufacture, Sprague Electric.** Cost analysis and design of 50,000-ft^2 plant retrofit with new interiors, power, lighting, HVAC, exhaust, and air pollution control. $1,200,000.

8. **Power Plant Emission Control, Naval Submarine Base, New London, Connecticut.** Engineering design of stacks, exhaust fans, and air preheater for smokestack pollution control. $900,000.

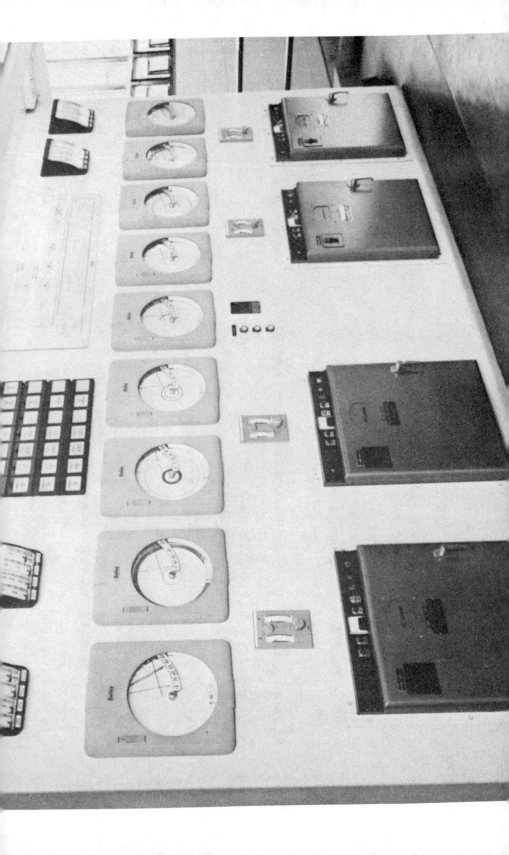

9. **Computer Control, Exxon, New York.** Engineering design support for loading rack and pipeline computer control at Syracuse, New Haven, and Albany tied to remote host computer. $450,000.

10. **Pier Dry Dock Support Services.** Engineering design of electrical power, air, water, and waste-water systems at nuclear dry dock facility. $1,500,000.

11. **Boiler Plant, Veterans' Administration Medical Center.** Engineering and design of a $1,600,000 high-pressure, three-boiler central plant retrofit with 2500 ft of exterior distribution and 200-kW emergency generator capability.

12. **Heating and Cooling Facility, Western Connecticut State College.** A complete $7 million central plant with 200,000,000 Btu/h heating and 8000 tons of cooling. Services delivered as high-temperature hot water and chilled water through 10,000 ft of distribution.

AUTOMATION

1. **Terminal Cardlock Systems.** Some 21 terminals of the Mobil Oil Corporation in New England and New York were retrofitted with total computerized control of fuel dispensing, billing, and access control. All were tied, through modems, to a host computer in Houston, Texas.

2. **Process Control.** Twelve plastic bag manufacturing lines of the East Hartford, Connecticut, facility of Union Carbide, from extrusion through conversion to shipping, were retrofitted to a completely automated mode. Cost exceeded $9 million.

3. **Walk Simulation.** An automated gait laboratory that provides computerized comparisons of actual gaits to stick figures. Can be used to check human structure deficiencies as well as improve sports athletic performance.

4. **Power Plant Automation.** Automated operation of all operational and safety controls of the $1,900,000 West Haven Veterans' Administration Medical Center, Connecticut, power plant with readouts of operating conditions. Automatic emergency power changeover.

5. **Surveillance and Control.** Specialized surveillance, detection, and monitoring systems at the Cheshire Correctional Center utilizing closed-circuit television, pressure and beam sensing alarm devices, and electric current restraint systems with centralized, supervised control.

6. **Building Automation.** Centralized energy management and control of HVAC, fire signaling, and protection systems at the $9,500,000 renovation, addition, and modernization of the Hartford Union Railroad Station, Connecticut. Work involved maintaining historic structure condition.

7. **Sports Arena.** Adaptation of existing mechanical and electrical system in $1,500,000 Springfield Civic Center retrofit to accommodate a system-within-a-system concept permitting the interior system to function both separately and in conjunction with the overall system.

8. **Warfare Simulation.** Automation of related services to accommodate underwater warfare simulation at the Naval Submarine Base in New London, Connecticut. Further details restricted from description or publication.

Planned $5,000,000 Project
Central Heating and Cooling Facility

HEAT RECOVERY

1. *Intensive Care, Newington Children's Hospital, Connecticut.* Due to the requirements of no air recirculation, the 40,000 cfm of exhaust air was investigated on a computerized model, and a coil system recovered 65 percent of the heat exhausted.

2. *Laboratory Building, State Board of Health.* Due to the specialized working environment, some 100,000 cfm of contaminated air was treated and subjected to a heat recovery exchange system that reduced the overall heating cost by 55 percent.

3. *Robertshaw Controls.* Concentrating on the primary source of required makeup air, a heat exchange system was designed to be compatible with the redesigned heat treatment furnace exhaust and plating exhaust. Of the fuel required for makeup air, 50 percent now comes from the recovery system.

4. *YMCA Westport, Connecticut.* The generally accepted requirements of pool, gym, play, and exercise rooms resulted in a greater heat requirement than the available utility gas allocation. A complete restudy of needs, codes, and requirements reduced the heat load by 26 percent and enabled the building to be operated within its allocation.

5. **Hartford Railroad Station Rehabilitation, Connecticut.** This 100,000-ft^2 structure was analyzed and designed as a refurbished structure—all smaller than the original. The fuel savings will pay for the new boilers in less than 9 years.

6. **Top-Notch Office Building.** As is common with many office buildings, a substantial part of the heat available in the winter and the load on the cooling system in summer is lighting. By working with the architect on simple shading and setting up a more intelligent control of heating and air conditioning, a system saving 12 percent in fuel costs was immediately obtained at a very small first cost.

7. **Machlett Laboratories.** Common to many industrial plants is the ever-increasing electrical costs. An analysis indicated that all the possible load clipping would do very little, and an expenditure on this phase was avoided. A very substantial saving was made by shifting certain high-voltage tests to the second shift.

8. **Smith-Gates Corporation.** Since this company makes radiant electric heating products, an electric slab radiant system was installed. Careful analysis of exhaust and ventilation requirements cut total energy requirements by 20 percent.

E.C.S.C. — WILLIMANTIC, CONN.

ELECTRICAL POWER DISTRIBUTION

1. *Machlett Laboratories.* A series of studies and cost data were presented to corporate management on the 10,000-kW connected load distributed at 13.8 and 4.16 kV. A total redesign of the plant distribution system for dual-feed and redesign of 14 transforming locations was made. Cost is estimated to be $600,000.

2. *Naval Submarine Base, New London.* To provide pier services for present and future nuclear attack submarines, a 3750-kVA transformer and 13.8/4.8-kV distribution system were designed, including special all-weather current limiting via 25,000 AIC reactors.

3. *Port Mobil, Staten Island, New York.* A complete analysis and design of a 13.8/4.8 kV explosion-proof distribution system was undertaken and is being funded to the extent of $1,400,000. Some 20 explosion-proof transforming stations are involved.

4. *Loading Rack Computer Control.* At three Exxon terminals—Albany (New York), Syracuse (New York), and New Haven (Connecticut)—a completely computerized fuel dispensing system was designed to interface directly with the central Houston computer.

5. *Process Power.* A series of studies and a distribution alternative design at Union Carbide's East Hartford, Connecticut, plastic bag manufacturing plant provided an increase of 20 percent in power available on the existing 13.8/4.8-kV distribution system.

6. *Power and Signal System.* At the Cheshire Correctional Institute, a new 23.8-kV distribution system was designed for the 600-acre complex. The system also includes infrared and pressure-activated fence security alarms coupled with video camera surveillance. Cost is in excess of $2 million.

7. *Bradley International Airport.* This multi-million-dollar complex of buildings has a new 13.8/4.8-kV distribution system complete with ramp and runway lighting, as well as three strategically located emergency generators for tower and communication facilities.

8. *Eastern Connecticut State College.* A $1 million primary selective spot network underground 23.8-kV distribution system with two primary feeders was designed. Included on the distribution design was the campus communication, telephone, and fire alarm systems.

REFRIGERATION AND AIR CONDITIONING

1. *Hospital, Naval Base.* Refrigeration systems were designed for a number of freezers and walk-in coolers, as well as for the operating room and the autopsy area. Design temperatures for various systems varied from 40°F to −20°F.

2. *Ice Cream Processing Plant.* Design of General Ice Cream Corporation refrigeration systems for cooling systems, batch freezer systems, and hardening and storage rooms. Temperature requirements varied with the individual process from 30°F to −30°F. Design included temperature control of personnel areas.

3. *Chip Manufacturing.* A $500,000 dual space and process system to operate at plus or minus 0.1 percent controls both space and process work in this IBM facility. Specialized run-around heat recovery saved 20 percent of total fuel cost.

4. *Computer Control Area.* The system design was for precise control of temperature and humidity for computer submarine warfare simulation. In addition, special attention was given to lead lining the area's walls for security reasons.

5. **Laboratory Building.** This 60,000-ft^2 state board of health structure houses a variety of pathology, chemistry, and similar suites, each with a need to maintain temperature and humidity control, as well as to avoid cross-contamination.

6. **Central Air Conditioning Plant.** A total of 8000 tons of six central chillers together with 7000 ft of pumped chilled-water piping was designed for Western Connecticut University Campus to handle a total new college campus air conditioning requirement. Included in the design plans were value analysis and life-cycle costing.

7. **Food Service Building.** A complete refrigeration, air conditioning, and heating system was designed for a 4000-student central campus food preparation and service facility for Southern Connecticut University. Design included all kitchen refrigeration and freezer capabilities and total building air conditioning. Facility total cost exceeded $4 million.

8. **Office and Library Complex.** The $3-million Wethersfield Town Hall and Library includes complete energy-intensive air conditioning with individual control of all areas on a totally programmed system. In addition, a separate system controls temperature and humidity for the Town Hall and Library book and document long-term storage areas.

DEAERATOR
W.C.S. — DANBURY

PROCESS PIPING

1. **Stratford Army Engine Plant.** A series of systems totaling more than $1 million were designed for jet-fuel storage and piping, fuel oils 2, 3, 4, 5, and 6 storage and piping, and cutting oil storage and piping, including trichloroethane, cutting oil, hydraulic oil, varsol and corrosion preventive compounds.

2. **Portsmouth Naval Shipyard, *New Hampshire*.** Several unusual compressed-air and vacuum systems were designed, including a 5000-psig compressed-air system and storage in an explosion-proof building with 6400 ft of double extra-heavy stainless steel underground distribution piping.

3. **Western Connecticut Central Plant.** A complete $6 million central heating plant of 200,000,000 Btu/h and refrigeration at 8000 tons distributed as high temperature hot water (HTHW) and chilled water through 10,000 ft of distribution mains. Project included value analysis and life-cycle costing.

4. **Stratford Army Engine Plant.** Fire protection system consisting of fire hydrants, distribution system, sprinkler system in varous buildings, and a form system plus a Halon system in certain flammable-liquid areas.

5. *Exxon Terminal.* Complete analysis of all terminal control and spill prevention systems, including containment of spillage, fuel separation and recovery, and prevention and control systems.

6. *Union Carbide.* Analysis and design of process supplied piping for plastic pellets for extruders to produce plastic film plus associated waste and scrap process recovery systems.

7. *Naval Submarine Base, NLON.* Analysis and design of nuclear submarine pier support facilities, including handling of ship requirements for fresh water, waste water, high-pressure air, and pier fire protection including fire pumps and housing.

8. *Site Utilities, Eastern Connecticut State College.* Analysis and design of high-pressure steam, distribution, drinking water, storm water, and sanitary sewers to cover the total campus.

POLLUTION CONTROL

1. **Power Plant Emission Control.** The 250,000-lb/h steam generating plant at the U.S. Submarine Base had a particulate fallout problem for which the Mueller Engineering Corporation, working with Travelers Research, designed a series of 126-ft stacks on a 60-ft structure with cone jet assist satisfactory to EPA requirements.

2. **Yardney Electric.** The Yardney Plant employs some 300 people in making batteries and other devices. Because of the presence of heavy-metal ions and high alkalinity, an ion exchange system enabled the reuse of the treated water in the production process.

3. **Air Pollution, Western Connecticut State College.** The basic requirement to properly dispense the combustion air was a 150-ft stack. Since soil and other structural conditions precluded a stack higher than 100 ft, a specially designed stack jet cone gave the gas effluent the required 50-ft vertical boost.

4. **Pervel Industries.** An alternative use of spent dye fluids was designed where, through distillation, liquid burning, and storage, the spent material could be reused for heating fuel.

5. *North Adams Solid Waste Disposal.* The result of intensive research, study, and cost analysis was a solid waste shredder and materials-handling system suitable for landfill or incineration at the lowest current cost per ton of any installation in the United States.

6. *State Board of Health Laboratory.* A series of systems was designed to contain such items as tuberculosis germs by negative pressure, and radioactive waste in special containers and exhaust; all totally air conditioned.

7. *Stratford Army Engine Plant.* Complete inviestigation and analysis of state EPA and federal spill control and countermeasure requirements for design of impervious containment area for a 400,000-gal fuel oil storage and a 120,000-gal cutting oil storage facility.

8. *Parsons Manufacturing.* A system designed to contain and treat the waste from 10 paint spray booths so that plant labor could be used and the existing settling lagoon retained with state EPA approval.

ADDITIONS TOTALLING 100,000
SQUARE FEET AT SMITH-GATES
MANUFACTURING — FARMINGTON, CT

PROCESS ENGINEERING

We are process and facility engineers serving companies in the chemical, petrochemical, and machine tool fields.

Currently we are involved in long-term relationships with Union Carbide, Mobil Oil, and the Mite Corporation. To clarify what we do, we will use the Glad garbage bag manufactured by Union Carbide as an example.

For the Glad bag manufacture, we designed the arrangement and installation of all the process systems from the point where plastic pellets arrive to where they are stored, extruded as film, converted into bags, and shipped. We also designed all the structural, electrical, and mechanical support facilities and utilities as well as the building alterations required including energy conservation and pollution control.

We do not presume to be experts in any process at your plant which we have not as yet seen. We are also aware that there must exist a considerable in-house staff doing the work which we have just noted. However, any corporate staff can, and does, get overloaded. That is the situation where we can be of service.

PERSONNEL

Name and Title
Jerome F. Mueller, President

Project Assignment
Principal in Charge

Years Experience
With this firm, 12
With Mueller Associates, 22

Education
BSME, Purdue University
Advanced Management, University of Connecticut

Active Registration

Connecticut
New York
New Hampshire
Florida
Massachusetts
Rhode Island
Vermont

Professional Affiliation

National Society of Professional Engineers; Consulting Engineers Council of U.S.A.; Connecticut Society of Professional Engineers; Connecticut Engineers in Private Practice; American Society of Heating, Refrigeration and Air Conditioning Engineers; Illuminating Engineers Society; Institute of Electrical and Electronic Engineers.

Expertise

Thirty-four years Principal in charge of overall facility, process and utility design. Special expertise in facility design, power plants, process systems, cost and value analysis, electrical and mechanical systems. Responsible for over 900 projects. Author of three technical books published by McGraw-Hill: *Standard Mechanical and Electrical Details, Standard Application of Electrical Details,* and *Standard Application of Mechanical Details.*

Name and Title
Orest B. Budas, Engineer

Project Assignment
Project Manager

Firm with Which Associated
Mueller Engineering Corporation

Years Experience
With this firm, 8
With other firms, 2

Education
BSCE, University of Hartford

Active Registration
Connecticut E.I.T.

Experience and Qualifications
Ten years of experience with this and other consulting firms with rapid advancement and promotion. Knowledgeable in both structural and design engineering. Excellent cost background. Has had extensive experience as project engineer and he has acted as acting Project Manager.

Name and Title

A. E. Saidi, Engineer

Project Assignment

Project Manager

Firm with Which Associated

Mueller Engineering Corporation

Years Experience

With this firm, 6
With other firms, 2

Education

BSCE, Rennsalear Polytechnic Institute

Active Registration

Connecticut PE

Experience and Qualifications

Eight years of experience with this and other consulting firms with rapid advancement and promotion. Knowledgeable of both electrical and structural design. Excellent cost background. Has had extensive experience as both electrical and structural project engineer and has been promoted to Project Manager.

Name and Title
Eric G. Carlson, Engineer

Project Assignment
Project Engineer

Firm with Which Associated
Mueller Engineering Corporation

Years Experience
With this firm, 5
With other firms, 16

Education
University of Hartford

Active Registration
Connecticut E.I.T.

Experience and Qualifications

Twenty-one years of plant and building facilities experience. Excellent knowledge of all forms of piping, ductwork, exhaust, and control systems. Considerable field investigation experience, as well as 5 years

Staffing and Managing the Engineering Office

Whether you start your new consulting engineering practice alone or with one or two colleagues, attention is usually focused first on obtaining your projects, financing, equipment, and other immediate needs—office space, letterheads, etc. Staffing is a secondary concern during the early business life of many professional firms. Yet the principals of the new organization are also responsible for planning for the future. They must prepare to deliver consistent, high-quality services to reap the rewards of a continuing and prosperous business.

At the same time, the principals have the responsibility of seeking additional new work and guiding the firm's growth. Proper staffing becomes a top priority. You, as a principal, must:

1. Determine immediate and future personnel needs
2. Develop a managerial and financial structure for your firm
3. Assign management responsibilities to specific individuals, including marketing tasks, budgeting, accounting, supervision, and hiring
4. Coordinate a personnel policy with the company's short-term and long-term objectives

5. Find potential employees through a variety of sources; interview applicants and make careful hiring decisions
6. Develop and execute effective training programs for new staff members
7. Manage staff members and motivate them to provide superior consulting engineering services to your clients
8. Review employee performance and productivity; consistently and fairly assist or discipline employees who perform unsatisfactorily; grant salary increases, fringe benefits, or other rewards in a program of incentives for excellence

These steps take time. However, they are as vital to the success of the young company as the selection of adequate office space. The new consultant might initially fulfill the roles of owner, manager, engineer, designer, drafter, secretary, and accountant all at once. If the firm prospers, most of these responsibilities will have to be delegated to additional qualified personnel. Therefore, the best time to begin developing an effective system of staffing and management is during the early phases of the new practice.

The National Society of Professional Engineers and other voluntary professional engineering organizations maintain up-to-date listings of job descriptions and qualifications that many consulting firms use as standards in professional practice. Generally, in addition to owners, partners, and officers, personnel are divided into four basic groups: (1) engineers, (2) supervisors, (3) designers, and (4) drafters. These titles comprise the four major areas of responsibility with respect to consulting engineering. Each firm may also use a variety of grades, classifications, and titles to cover these and other specialties.

It is worthwhile for consultants to maintain written job descriptions of the various categories of positions within the company. These job descriptions can help set the foundations of hiring and management that will affect the future of the firm. They usually include salary scale and an outline of individual and group advancement routes.

Here is an example of written job descriptions used by a small consulting engineering firm. You can modify them to suit the needs of most young companies.

Drafter Has thorough knowledge of drafting; makes tracings; performs preset calculations; draws details and plans under supervision; can work in more than one phase of engineering. Salary range: $8000 to $12,000 annually.

Designer Prepares plans and specialized details under moderate to very little supervision; writes specifications with supervision; prepares elements of cost estimates; selects equipment for designs with minor assistance; does original investigation of equipment and materials for design projects; performs project calculations with moderate supervision; can do limited field supervision of construction. Salary range: $12,000 to $17,000 annually.

Engineer   Has a 4-year engineering degree and 4 years of consulting engineering office experience or a 2-year degree and 8 years of consulting engineering office experience, or 12 years of acceptable consulting engineering experience and no degree. Can perform all the tasks of a designer and a drafter with minimum amount of supervision. Can write complete specification for a phase of engineering and do all the work for a cost estimate. Can supervise installation of at least one phase of the design and is knowledgeable in other phases. Can act as assistant project engineer. Salary range: $17,000 to $21,000 annually.

Project Engineer   Meets all the requirements of an engineer, designer, and drafter. Can direct the work of a group of engineers, designers, and drafters. Works by direction and without direct supervision. Can handle all client relations for firm on assigned project. Salary range: $20,000 to $24,000 annually.

Department Head   Is a licensed professional engineer. Meets all the requirements of project engineer, engineer, designer, and drafter. Has the ability to manage a group of project engineers and to set up and maintain a department budget, and possesses the qualifications to become a principal of the firm. Salary range: $24,000 to $30,000 annually.

Secretary   Has thorough knowledge of secretarial and office procedures, types 50 to 70 words per minute, and takes shorthand at 80 to 110 words per minute. Does daily filing and can set up correspondence and job filing systems. Can compile time-card and job-card posting. Can handle the secretarial work of all individuals in the firm. Is punctual, courteous, cooperative, and cheerful. Can operate various office machines, including a blueprint machine. Has good telephone voice and manner. Can delegate or resolve telephone calls in absence of party called. Salary range: $8500 to $13,500.

Besides these basic categories, the engineering firm may use titles such as "Senior Engineer," "Specification Writer," "Cost Estimator," and many others. The number and type of positions within the firm are directly related to the size of the current work load, growth projections, and client's requirements. Ideally, the small firm hires and trains staff members who can master more than one area of responsibility. Large firms can better afford to employ specialists whose skills may not be interchangeable from one job to another.

As the small firm grows, its principals typically perform decreasing amounts of hands-on design work. The principals gradually assume higher managerial roles as they are relieved of the burden of routine tasks. Meanwhile, other staff members may be promoted to positions involving greater responsibility, and new employees can be added.

Many young firms prepare an organization chart to show graphically the staffing and managerial structure of their company. The organization

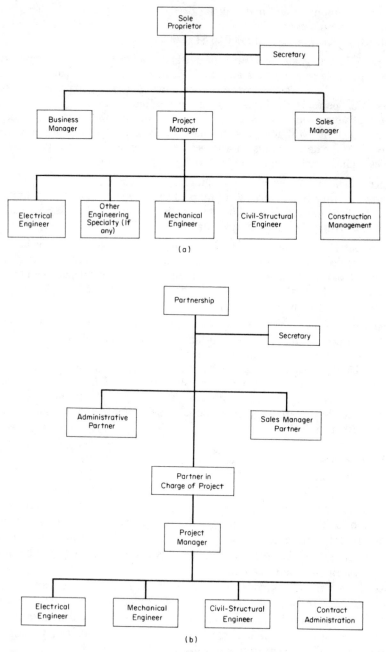

Figure 3-1 Typical organization charts for consulting engineering firms. (*a*) A small firm organized as a sole proprietorship. (*b*) A consulting engineering partnership. (*c*) A consulting engineering firm organized as a corporation.

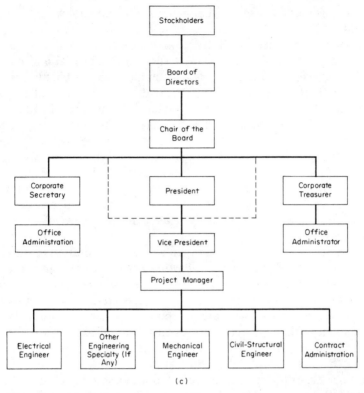

Figure 3-1 (*Cont.*)

chart helps point up areas that lack adequate staffing, even when the firm has only a few members. A chart also assists both the principals and new employees in defining the responsibilities of each member. Figure 3-1 shows several ways of organizing a small consulting engineering firm: sole proprietorship, partnership, and corporation.

A small firm organized as a sole proprietorship (Figure 3-1*a*) is the basic format for today's beginning consulting company. The beginning consulting engineer should recognize that the secretary will probably be a part-timer at the start and the engineer will have to perform many secretarial tasks. The engineer's early goal should be to fill the basic design staff with qualified personnel and not try to be an expert in all fields. It will probably be necessary to use subconsultants (part-timers) in two of the three basic engineering disciplines or not accept work for which the founder is unqualified. Once the engineer has the subconsultants working, the next step is to get a construction manager, followed by a project manager. A qualified secretary can also work as the business manager.

Sales management is usually the owner's job forever—with or without assistants.

A consulting engineering partnership usually begins with two partners, each handling a different discipline—such as mechanical and electrical engineering (see Figure 3-1*b*). One partner also handles clients and is the partner in charge of each project. One or both partners handle supervision of staff. Likewise, one partner handles the running of the office while the other seeks new business for future projects. Your first objective should be to get enough business to hire the three partners plus the project manager. The partner in charge of projects is a highly saleable concept because clients like the idea of having their work handled by a partner. Further, having a partner in charge of administration and another in charge of sales helps a new firm succeed sooner. Remember that including most of the staff on new-client interviews also promotes success in sales activities.

A consulting engineering firm organized as a corporation (Figure 3-1*c*) may be either a *professional corporation* (in which case the principals must be registered professional engineers) or a *nonprofessional corporation* (in which case the principals may be registered but are not required to be). In a small nonprofessional corporation the president is also usually the chairperson and treasurer. A competent office administrator may also serve as the corporate secretary and serve on the board of directors. Also on the board can be the firm's attorney and accountant. The new consulting corporation is often a sole proprietorship in somewhat modified form. Problems may arise when a consulting partnership is converted to a corporation. Then the partners are not equal anymore. An attorney should always be consulted before forming any corporation, so that all laws are followed. For example, some states prohibit the use of the word *engineering* in the name of a corporation unless the principals are registered in the state of incorporation.

A chief goal of conducting an ongoing analysis of the firm's capabilities with the aid of an organization chart is to develop a staff that will provide potential clients with the confidence that the company can deliver superior services of the type the client needs. You can also get good ideas on organizing your firm from an examination of the structure and hiring practices of competing consulting businesses. Fellow consultants can often provide valuable information about their firms' capabilities and former employees.

Another important consideration for the growing company is the ability of its staff members to get along with each other. In the consulting engineering business, personalities play an important role. If those who hire the staff do not select new employees carefully, personal habits, ideas, and emotional quirks may interfere with smooth delivery of professional services. But some personal differences must be minimized or tolerated as a part of normal working life.

DEVELOPING A WRITTEN PERSONNEL POLICY

The development of a written personnel policy is crucial to the proper functioning of a consulting engineering organization. A well-designed policy, enacted during the first few months of the firm's existence, will help minimize personnel problems. It can cover issues as simple as the amount of time allowed for coffee breaks to those as complex as efforts to unionize the engineering staff.

Copies of the policy can be printed and distributed to each staff member in the form of an employee manual. Many firms use loose-leaf binders to hold the pages; such binders facilitate making additions and deletions quickly. Further, some consulting firms ask each employee who receives the personnel manual to sign a document acknowledging that she or he has reviewed the policies and is fully acquainted with all provisions.

Most personnel manuals contain a table of contents plus a foreword briefly describing the firm, its background, and its services. The following list illustrates the wide range of topics that your manual could present in its discussion of personnel policy. (These are by no means the only subjects that could be covered.)

1. **Equal Employment Opportunity (EEO):** Generally, policies state that the consulting firm adheres to all federal and state laws governing EEO programs and does not discriminate on the basis of race, sex, creed, color, religion, or national origin in its hiring and employment practices. Some personnel manuals list the name of the company representative who has been assigned to handle complaints or questions about EEO.

2. **Office Hours and Workweeks:** The firm's daily hours of operation are set forth, including allowances for lunchtime, breaks, etc., and any exceptions to the normal rules. The usual workweek is described, and the number of hours it contains, whether it be a 5-day, 35-hour week, a 4-day week, or some other schedule.

3. **Overtime:** Overtime pay rates, plus the times and the people to whom they apply, are discussed. Professional engineers are sometimes expected to forfeit overtime pay, while clerical and subprofes-

sional employees may be paid time and a half, for example, during weekday overtime, with added pay for Sundays or work on holidays.

4. **Pay Period and Day of Pay:**   Generally, firms that end their week on Friday pay on the following Monday, Tuesday, or Wednesday.

5. **Holiday Schedules:**   Holidays are listed, with indication of whether or not employees will receive pay for those days.

6. **Vacations:**   A clear statement should be provided to inform employees about the amount of paid vacation time they earn for particular durations of employment. Some firms use Labor Day of the preceding year as the qualification point for 1 week's vacation in the first year of employment.

7. **Sick Leave; Leaves of Absence; Maternity, Military, Jury Duty, Funeral, and Sabbatical Leave:**   These are areas of potential conflict with some employees. Therefore, a clear statement of the maximum number of absences that will be allowed—with and without pay— should be provided. Describe each type of absence separately, noting its particular requirements and allowances. (For example, some states have laws governing company policies on maternity leave. These should be researched and adhered to.)

8. **Unexplained Absence and Tardiness:**   Disciplinary measures and limitations are described.

9. **Signing In and Out of Office:**   The normal procedure to be followed by the firm's employees when they enter and leave the office or project site is explained.

10. **Payroll Deductions:**   Any deductions, whether applied automatically by the firm or allowable for the individual member, should be described. Special deductions are listed, along with the customary federal, state, and city taxes and social security withholdings.

11. **Workers' Compensation:**   Describe the policy.

12. **Insurance:**   Describe the rules governing medical or other insurance benefits offered by the firm, if any, and employees' options. (Since this topic may require a lengthy policy discussion, a separate pamphlet or document may be used, with only brief mention in the main policy manual.)

13. **Expense Accounting:**   Describe the policy.

14. **Record Keeping:**   Describe the policy.

15. **Use of Company Cars or Private Cars:**   Most consulting firms do not have company cars but instead pay a fixed sum for each mile that an employee drives in his or her own car while conducting business.

16. **Telephone Use:**   The rules governing use of telephone for personal and business purposes are described.

17. **Correspondence Files; Filing of Plans, Reports, and Specificatons:**   Office procedures are described.

18. **Duplication and Printing of Documents:**   Limitations for personal use and rules governing copying of company materials are explained.

19. **List of Design Plan Holders:**   Rules concerning listing of names of personnel holding particular design plans are explained. (This saves time when newcomers are looking for particular design plans.)

20. **Attendance at Professional Conventions and Other Activities:**   Since some employees seem to be capable of spending more time at various special events than they spend at the office, rules limiting such absences should be described.

21. **Profit-Sharing Plan and Bonuses:**   These special programs should be described at length in separate pamphlets distributed to individuals who participate in profit-sharing and bonus plans.

22. **Probationary Period:**   If the firm has a probationary period for new employees, it should be explained, along with any relation it may have to matters such as eligibility for workers' compensation, benefits, etc.

23. **Employee Performance:**   General expectations of the firm with regard to employee behavior and company goals should be described.

24. **Salary Review and Adjustments:**   Procedures for reviewing performance and making appropriate salary adjustments are explained.

25. **Voluntary Termination and Resignation:**   Procedures to be followed by employees, such as providing advance notice of impending resignation, are described.

26. **Salary Advances:**   Generally, consultants rarely pay employees in advance. Whatever the firm's policy is, it should be described.

27. **Professional Engineering License Registration:**   Requirements for engineering registration for particular tasks are explained. Professional registration of all engineers on the staff should be encouraged.

28. **Outside Employment:**   This is one of the more sensitive issues facing the heads of engineering consulting firms because employee moonlighting and other outside professional work can lead to conflicts of interest and legal liabilities for the employer. Outside employment requirements and limitations should be defined clearly, with the awareness that the firm cannot control activities engaged in by employees after normal working hours.

29. **Construction Engineering Services:**   The duties and limitations of engineers working on-site are set forth in detail, including a discussion of such matters as managerial obligations, use of equipment, etc.

30. **Continuing Education:**   The policy should encourage further education for employees in various positions and describe company programs that promote continuing study.

AFFIRMATIVE ACTION PROGRAMS

Another important element in the personnel policies of many professional firms is an affirmative action hiring program. Government guidelines for affirmative action are in some ways similar to the federal laws governing equal employment opportunities for minority group members, women, veterans, and the handicapped. However, the federal affirmative action requirements state that firms with more than 50 employees must actively seek to hire minorities to be eligible for federal contracts and subcontracts worth $50,000 or more. Additional requirements may apply in some states.

Consultants in private practice should learn as much as they can about the federal and state affirmative action requirements. Failure to comply with rules governing affirmative action programs can result in cancellation of previously awarded government contracts and ineligibility for future awards. So every engineering firm should state its policy in writing. Usually, the affirmative action program is described in a pamphlet or brochure separate from the personnel manual.

The elements of a written statement of affirmative action policy include:

1. Affirmation of EEO policy
2. Description of company procedures for communicating hiring decisions
3. Explanation of administration of EEO program
4. Identification of problem areas such as lack of minority applicants
5. Establishment of goals and schedules to correct deficiencies
6. Development of methodology and execution of the program
7. Description of internal company audits and record keeping
8. Affirmation of support of affirmative action programs

The federal affirmative action program actually encompasses two separate programs: one for members of minorities and one for women. Each has the same basic requirements. These are covered in the Code of Federal Regulations, Title 41, Chapter 60 (sometimes identified as Revised Order 4), for federal contractors or subcontractors. In addition, Title 7 of the Civil Rights Act of 1964 is applicable to businesses employing 15 or more persons. Title 17 makes it unlawful for an employer to take direct or indirect actions that would deprive any employee of individual employment opportunities because of race, creed, sex, color, age, or national origin.

Another federal law, the Rehabilitation Act of 1973, requires businesses to implement affirmative action practices in the hiring of physically handicapped individuals. The law applies, in varying degrees, to employers with federal and federally assisted contracts in excess of $2500. Under the regulations, firms with contracts worth more than $2500 must include a clause in their contract stating that handicapped job applicants will not be discriminated against because of their handicaps. Employers with contracts that will last more than 90 days and are worth up to $50,000 also must prepare a written statement of their affirmative action program, including details on the steps the firm will take to recruit handicapped employees.

Federal regulations like these have attracted criticism from some employers who feel that the laws unnecessarily hamper business. As a result, legislators have started to seek changes in such laws in recent years, with some success. It is therefore advisable to examine the requirements affecting consulting firms, both nationally and locally, and to update your company's policy regularly.

USE OF PERSONNEL FORMS

One of the best-known steps in employee recruiting is the filling out of a job application form by the potential staff member. Figure 3-2 shows a typical employment application form. Such forms can be used for making preliminary evaluations of applicants and should include questions on the applicants' professional and educational background.

Figure 3-3, the applicant data record, is an example of the type of form that can be used to provide documentation of a firm's EEO-affirmative action recruitment efforts. Each applicant for a particular position is asked to complete the form voluntarily so that the forms can be used in periodic government analyses.

Figure 3-4 shows a typical notice of payroll changes. While this form is not usually used in hiring, it provides an effective standardized method for communicating staff changes to affected parts of the firm, such as the payroll department.

APPLICATION FOR EMPLOYMENT
(PLEASE PRINT CLEARLY)

DATE_____

FIRST NAME _____ MIDDLE _____ LAST NAME _____

ADDRESS _____PHONE _____

SOCIAL SECURITY No. _____

APPLIED FOR

TYPE OF WORK; POSITION _____

☐FULL TIME   ☐TEMPORARY   ☐PART TIME (SPECIFY TIMES) _____

EDUCATION

ELEMENTARY	HIGH SCHOOL	COLLEGE	MAJOR	OTHER
☐	☐JUNIOR ☐SENIOR	1 2 3 4		

MILITARY SERVICE

SERVED IN U.S. ARMED FORCES	BRANCH	FROM	TO	RANK AT DISCHARGE
☐YES ☐NO				

EMPLOYMENT HISTORY

EMPLOYER	POSITION	FROM	TO	PAY	PER

REASONS FOR LEAVING LAST JOB _____

_____

ADDITONAL EXPERIENCE _____

_____

SIGNATURE _____ (WE ARE AN EQUAL OPPORTUNITY EMPLOYER)

————————————— FOR OFFICE USE ONLY —————————————

_____

_____

_____

_____ DATE _____

Figure 3-2   Typical employment application. *(Remarkable Products, Inc.)*

Two more forms useful in hiring are the job description form (Figure 3-5) and the applicant rating form (Figure 3-6). Both can help simplify the tasks of interviewing applicants and staffing the firm properly. The job description is used to provide the interviewer with a brief outline of the requirements for each position to be filled, plus its title and the jus-

EMPLOYEE MASTER RECORD

NAME		PAYROLL No.	BIRTHDATE	SEX	DATE HIRED	☐ FULL TIME ☐ PART TIME ☐ TEMPORARY
	ADDRESS	CITY		STATE	ZIP	HOME PHONE
1						
2						

EDUCATION

ELEMENTARY ☐	HIGH SCHOOL ☐ Junior ☐ Senior	COLLEGE 1 2 3 4	MAJOR	OTHER

PREVIOUS EXPERIENCE _____

_____

FAMILY

MARITAL STATUS	MARRIAGE DATE	CHILDREN						
			NAME	SEX	BIRTHDATE		NAME	SEX	BIRTHDATE
		1				2			
NAME OF SPOUSE		3				4			

EMERGENCY CONTACT

	NAME	ADDRESS	CITY	STATE	PHONE	RELATIONSHIP
1						
2						

JOB HISTORY

	DEPARTMENT	POSITION/TITLE	FROM	TO	REMARKS
1					
2					
3					

PAYROLL

SOCIAL SECURITY No.		DATE	RATE	PER		DATE	RATE	PER		DATE	RATE	PER
	1				2				3			
CLOCK No	4				5				6			

INSURANCE RETIREMENT & BENEFITS

	PROGRAM	DATE JOINED	DATE ELIGIBLE	REMARKS
1				
2				
3				

TERMINATION

DATE _____ ☐ RETIREMENT ☐ DISMISSAL ☐ RESIGNATION ☐ OTHER_____

Figure 3-3 Applicant data record. *(Remarkable Products, Inc.)*

tifications for filling it. These descriptions can be written in by the company managers or principals and maintained in a staff file.

Likewise, the applicant rating form can be used by the interviewer as a document listing specific criteria to be used in evaluating individual applicants. Authorized members of the firm list important job criteria on

EMPLOYEE STATUS/PAYROLL CHANGE

EMPLOYEE NAME		LOCATION
SOCIAL SECURITY No.	PAYROLL ID No.	EFFECTIVE DATE

CHANGE OF	FROM	TO	REMARKS
☐ RATE			
☐ STATUS	☐ FULL TIME ☐ PART TIME ☐ TEMPORARY ☐ PERMANENT ☐ DAY ☐ NIGHT	☐ FULL TIME ☐ PART TIME ☐ TEMPORARY ☐ PERMANENT ☐ DAY ☐ NIGHT	
☐ POSITION			
☐ DEPT.			
☐ OTHER			

☐ LEAVE OF ABSENCE   FROM _____ TO _____

☐ TERMINATION   DUE TO: ☐ RETIREMENT   ☐ RESIGNATION   ☐ DISMISSAL

☐ OTHER _____ DATE _____

OTHER REASONS AND EXPLANATION _____

_____

_____

_____

_____

_____

_____

_____

_____

_____

_____

ORIGINATED BY _____ DATE _____ SIGNATURE _____

APPROVED BY _____ DATE _____ SIGNATURE _____

Figure 3-4 Notice of payroll change. *(Remarkable Products, Inc.)*

the form in the spaces provided. Then, the interviewer rates each applicant on the basis of specific requirements, using written comments and perhaps a numerical scale. Criteria can include virtually anything that is relevant to a particular job—from physical appearance to previous job performance—and whatever the firm needs in terms of staff characteristics at a given time.

```
JOB DESCRIPTION

Description of requirements:_____

    _____

    _____

    _____

    _____

    _____


Justification:_____

    _____

    _____

    _____

    _____


Title and/or position in your firm:_____
```

Figure 3-5 Typical job description form.

FINDING QUALIFIED PERSONNEL

Consulting engineers can utilize several sources of information to find potential staff members for their firms. The two most popular ways of recruiting applicants are to place classified advertisements in local newspapers and to solicit through employment agencies. Used effectively, either of these sources can lead to successful staff growth.

Newspaper ads should be worded carefully, describing each job opportunity and its related criteria, with crystal clarity. Most help-wanted ads list the basic requirements for the job being offered, including professional experience and educational background. Opportunities for advancement and starting salary may also be given. However, some firms choose not to state a specific salary, leaving the wages open to negotiation during the actual interview, because they feel that potential applicants might be deterred from responding to an ad when they are seeking a salary higher than the one advertised. However, some employers do list a specific salary in ads, reasoning that they will thus have less difficulty attracting applicants who are well suited to the firm's current economic situation.

CRITERIA SCALE for Rating Applicants

Criterion: Appearance _____

0   1   2   3   4   5   6   7   8   9   10

_____   _____

Comments: _____

_____

_____

Criterion: Personality _____

0   1   2   3   4   5   6   7   8   9   10

_____   _____

Comments: _____

_____

_____

Criterion: Character _____

0   1   2   3   4   5   6   7   8   9   10

_____   _____

Comments: _____

_____

_____

Criterion: Reliability _____

0   1   2   3   4   5   6   7   8   9   10

_____   _____

Comments: _____

_____

_____

Figure 3-6 Applicant rating form. Other criteria, such as experience, ability to communicate, and adaptability, can be substituted for these.

Employment agencies vary greatly from one to another, but they can sometimes be more valuable than advertising. The so-called technical employment agencies—known by some as "body shops"—occasionally act as a handy source of temporary employees. Similarly, secretarial service agencies can provide quick help with typing, telephone answering, dictation, and other tasks when you do not have a full-time secretary or you just need some additional help during an especially busy period.

But there are some disadvantages. The cost of using temporary agency employees is usually considerably higher than what you would pay a full-time staff member for the same work. The reason is that the agency fee normally covers the temporary employee's social security, workers' compensation, and other benefits, and the agency's own profit.

In contrast to the so-called body shops, there are many excellent employment agencies that can provide permanent personnel, usually for a fee that is approximately 20 percent of the first year's salary. Many engineering firms consider such fees quite reasonable and use the agencies almost exclusively in searching for new talent.

In the past, another productive hunting ground for consultants seeking new employees was the government. Through contracts with people in local, state, and federal agencies, consultants could often find highly skilled recruits who were happy to switch to the private sector at an increased salary. Recently the government pay scale has reached—in some cases exceeded—the levels in private consulting. Thus it is now usually harder to woo employees from government jobs.

Still other sources of potential staff members include other consulting firms, staff contacts, colleges, technical schools, friends, business associates, and applicants who walk in "out of the blue." With regard to employees from other firms, recruitment prospects are often substantial. A lot of secret late-night telephone calls between consultants and their competitors' employees occur throughout the profession. Sometimes, deals are made that are satisfactory to all parties concerned; but there are also cases in which ill will and a deterioration in public relations is generated by what some people might view as an attempt to "steal" another firm's talent. Certainly, careful business judgment and ethics are warranted in such situations. The same cautions apply to attempts to recruit new employees from manufacturing firms, sales agencies, contractors, and other companies with which your firm may do business.

Staff contacts are a different matter entirely. They rarely represent a threat to anyone else's business. In fact, members of your staff may be able to introduce you to highly skilled workers who are actively seeking jobs in the types of positions you need to fill. Staff members should therefore be kept informed of job openings with the firm. Likewise, you may get some valuable responses by spreading the word about available positions among your own colleagues, association members, friends, and family.

Some engineers feel that various colleges and advanced technical schools are the best sources of new personnel. Each semester, the schools send out new graduates who are trained, tested, and graded in specific areas of engineering. The graduates are eager to gain experience through employment and to get that experience in firms that can offer rewarding long-term jobs. As an employer, you can ask teachers and school employment services to inform students of your job offerings. You can also benefit by submitting to the schools estimates of your firm's staffing requirements for future projects. Such institutions are usually very willing to help arrange meetings between employers and students.

Besides the preceding sources, consulting firms sometimes use billboard and radio advertising for their job offerings. Finally, there are the applicants who may come to your company seeking work not as the result of any kind of solicitation but on their own accord. Even if you have no need at the moment for applicants who come to you with drafting skills, for example, you should still ask them to complete an application form and leave a copy of their résumé for possible future consideration. In the event you do need a drafter in the future, you may not need to advertise at all. Clearly, the number of sources of new personnel is limited only by the amount of effort one is willing to exert to find the right people.

CONDUCTING EMPLOYEE INTERVIEWS

Once you attract a number of people who are interested in applying for positions within your firm, preparations must be made for interviews of the applicants. A well-planned interviewing strategy is vital to successful hiring. You may wish to ask applicants to submit résumés and completed employment applications on one date, later selecting the individuals whom you will interview on the basis of the written information. Alternatively, you may interview all who submit applications and résumés on the same day that the materials are received.

In either case, the interviewer should be aware that the applicants themselves will be engaging in a selection process. Few applicants are likely to feel they need to work for a particular consulting firm. Therefore, the firm should appoint competent, reliable interviewers who are skilled not only in screening applicants but also in representing your organization with resourcefulness and professional aplomb. When the firm sets up an appointment for an employee interview, the applicant is expected to arrive on time and to be ready to discuss his or her background and interest in the company. The interviewer's own standards should be equally binding.

The start of each interview usually focuses on the data provided in the applicant's résumé and completed employment application. These two documents tend to set the stage for discussions of the prospective employee's general and educational background, interests, professional experience, and relevant skills.

Many other topics can be discussed, such as the applicant's professional interests and career plans, but the interviewer should be fully acquainted with state and federal laws concerning information bearing on equal employment opportunities. For instance, in some areas, questions about a female applicant's childbearing plans may be construed as an attempt to discriminate against women. Information about the laws and their implications can usually be obtained from your state labor department representative or through state and federal commissions appointed to oversee equal employment practices.

Some firms make duplicate copies of employment applications so the interviewer and applicant can both review the documents at the same time. Experienced interviewers can often learn a lot about an applicant's personality by examining the way in which the form was completed and discussing it with the applicant. For example, if you are seeking a designer and one of the job requirements is an ability to do flawless free-hand lettering, you might wish to discuss the applicant's handwriting as it appears on the form. You might even ask the applicant to use free-hand lettering on the form itself.

Another worthwhile topic to cover during the interview is the applicant's employment references. Some applicants provide letters of recommendation or brief references of previous employers along with their résumé; others list these only if the employment application requests them; and some are reluctant to provide any references at all. Generally, the firm can strongly enhance its evaluation of potential employees by securing information for contacting the applicants' former employers and informing the interviewees that the references will be verified.

It is also important that the interviewer does not give the interviewee the impression that the applicant is being policed or interrogated. The best interviews are conducted like a relaxed conversation. The interviewer's manner as well as the physical setting can thus go a long way toward establishing a relaxed atmosphere. Distractions should be kept to a minimum. Ideally, the interview would be held in a private office.

Most interviewers seat themselves behind a desk across from the applicant. While some employment experts feel this arrangement provides the interviewer with an important aura of authority, others say that the participants feel more comfortable sitting in chairs placed side by side. The arrangement of furniture is essentially a personal decision for the interviewer, limited somewhat by the constraints of available space.

Social psychologists and other experts are in agreement about the kinds of behavior and attitude that interviewers can best use to their firms' advantage. The interview is often a tense situation for both sides; therefore, the interviewer should try to put the applicant at ease from the start. Usually, the ice can be broken by injecting some humor into the conversation and discussing mutual backgrounds and interests early in the interview. Then, the more substantive details about the applicant's skills and the job requirements can be approached gradually.

At the same time, the interviewer should convey the clear notion that the conversation is a serious, professional information-gathering session. He or she must control the content and direction of the interview and not allow it to degenerate into a lengthy talk on sports, the weather, or other trivia. At the beginning the applicant should be informed of any time limits or schedule. For example, the interviewer might state that the first 30 minutes will be devoted to talking about the applicant's qualifications and job requirements, while another 30 minutes will be spent touring the firm's offices and meeting another member of the staff. Except in special cases, effective interviews can usually be conducted within an hour's time.

An important part of the interview is the consultant's description of the consulting engineering organization, the job opening, relevant qualifications, salary, benefits, and opportunities for employee advancement. Certainly, applicants deserve a clear-cut explanation of what kind of work your firm does. The characteristics of the available position should be elaborated in sufficient detail to make the job both understandable and attractive to the potential employee. Interviewers who do not plan these steps carefully may lose the interest of desirable applicants.

Most applicants are very much concerned about the details of working hours, time off, etc. In the past, salary was almost invariably thought to be the most important item; but today fringe benefits and vacation time are often considered equally vital issues in the potential employee's choice of whom to work for. The company's policy with respect to such matters should therefore be explained thoroughly by the interviewer. Likewise, the interviewer should be prepared to describe the firm's salary schedule in a straightforward and confident manner.

One caution: It is just as important for the interviewer to listen to the applicant and answer her or his questions as it is for the interviewer to describe the job opening. Good interviews take in a balanced amount of information on both sides. The interviewer may simplify the task in advance by preparing a list of specific questions and criteria to refer to during the interview session. Such a list has the additional virtue of providing a set of standards against which the attributes of several applicants can be measured equally. Figures 3-6 and 3-7 are examples of criteria and questions.

If the applicant appears to become overly anxious or nervous during the session, a short digression from the more weighty topics of job requirements may help ease tension. A short tour of the main engineering section of the firm's offices can also reduce nervousness. Problems rarely

1. Why did you enter the drafting profession?
2. What aspects of drafting do you find most interesting and most challenging?
3. Which aspects of drafting are least interesting and challenging to you?
4. How many days a year—on average—are you unable to get to work because of illness?
5. Do you normally take much time off for personal business?
6. Are you working for any other firm or person in your spare time? If so, how many hours a day do you devote to this work? When and where do you do this work? Would you be willing to stop this work if we employed you?
7. Do you normally get along well with the drafters you work with?
8. What future goals do you have for yourself in terms of your career?
9. Do you plan to further your education in your field? If so, how?
10. What do you look for our consulting firm to do for you and your career? What can you do for our consulting firm that will benefit us?

Figure 3-7 Typical questions prepared in advance for interviews with prospective drafters. (This is only an example; you may wish to develop completely different questions for your firm.)

develop when the interviewer behaves cordially. But a nervous applicant may become emotionally upset if the interviewer does not exercise discretion in his or her comments. There is no real value in bluntly telling an interviewee, "You do not fit in with our staff," for instance. Such a statement could have negative repercussions for the firm after the applicant finds employment with other engineering organizations. While an applicant might make statements that perturb the interviewer, the wise consultant will not engage in verbal duels with potential employees.

When one interview does not seem to provide the firm with adequate information to make a sound choice between applicants, a second session may be desirable. Some consulting firms interview applicants as many as three times, particularly when the candidates are being considered for high positions within the company. Of course, applicants cannot wait forever to find out whether or not they will be hired. Thus the majority of consulting firms consider a single interview sufficient. Additional information can usually be obtained by phoning the applicant afterward.

Another concern of some consultants is how many members of the firm should attend the interview. A number of researchers studied this question, generally concluding that the one-on-one situation is best. Except perhaps in interviews for new partners or principals, the session is not usually helped by placing the interviewee in front of a panel of inquisitors. Too many interviewers spoil the talk.

Besides discussing the potential employee's résumé and application data, some organizations use aptitude tests as part of their procedures for evaluating applicants. The value of such tests is a matter of considerable disagreement in various fields. In consulting engineering, however, it is probably safe to say that attitude and aptitude can be discerned properly through a thorough interview. When aptitude tests are used, they usually must be prepared by outsiders skilled in designing and administering such examinations. This implies an added—and for the most part, unnecessary—expense to the employer.

The interviewer is most interested in learning about the applicant's skills as demonstrated in previous jobs and schooling. Further, the consulting firm should try to discern basic personality traits that will be relevant to employment within the organization. For example, most employers take great care to avoid hiring "floaters"—people who rarely stay at one job for more than a brief period of time. Certainly one does not need to administer aptitude tests to make that determination. The interviewer can normally glean such information from data provided on the applicant's résumé and application. If necessary, the applicant's references can be contacted. Another topic that the interviewer might wish to discuss with the applicant is the reason she or he seeks employment with your firm.

The concluding moments of the interview offer an opportunity to answer any remaining questions and to maximize the positive benefits obtained from the meeting. If the interviewer has decided that the applicant is desirable as a future employee, concluding comments should emphasize the interviewer's pleasure with the meeting, as well as the merits of employment in the consulting firm. Even if the applicant is not highly desirable, he or she should still be left with a positive impression of the company.

Interviewers should constantly be aware that applicants may also be considering taking jobs elsewhere. Therefore, it may be necessary to designate a number of alternate applicants in case the most desirable applicants set their sights on other firms. An organized approach to interviewing—combined with astute judgment of character and a prompt follow-up to meetings—offers the most effective tool for thorough evaluation of potential employees.

SELECTING THE BEST CANDIDATES

The experienced interviewer sets the foundation for a fair evaluation and selection of job candidates chiefly through use of standard criteria and by preventing her or his own personality from dominating the interview. Following the meeting, however, the principals directing hiring may use their own discretion. The final decisions on whom the firm will hire can be strongly influenced by subjective factors. This is especially true when several candidates are considered simultaneously.

Some firms handle this situation by asking the most desirable individuals to return for a second interview. Additional staff members might attend the meeting and review the characteristics of the company and the applicants' qualifications and professional goals with each candidate. Matters such as salary, benefits, and employee responsibilities might also be discussed. The candidates may be introduced to the staff member who would act as their immediate supervisor should they start working with the firm.

At this stage of hiring, a sort of "role reversal" can occur. While the candidates try to sell themselves for the job, the consultants must sell the job to the candidates. Thus the consultants might negotiate with the applicants on issues of salary, vacation time, and the like. Ideally, they impress upon the candidates the notion that working for the organization would be a wise choice. But if an applicant suddenly puts forth special requests—such as a provision for regular absences due to a recurring illness—then the firm should promptly inform the candidate whether the conditions can be met. Some interviewers might feel the request implies a potential problem of excessive absenteeism. If so, the applicant could be rejected from further consideration.

Another issue that often influences personnel selection is the question of how long the candidate is likely to stay with the firm. The business of professional consulting engineering generally requires that employees stay with one company long enough to receive training and to become acquainted with detailed procedures of the particular firm. Only then can staff members begin to work with maximum effectiveness.

Training and adjustment sometimes occur quickly, but in many cases it requires as much as several years. Interviewers therefore have the responsibility of selecting candidates who are willing to make a sufficiently long-term commitment to their jobs. Applicants' professional aspirations count strongly in this determination.

Many firms use legally binding employment contracts to protect themselves against early resignations and other potential problems with employees. Such contracts may include documents, signed by the employees, covering the conditions and duration of employment with the firm. The contracts are usually used only in hiring partners, associates, and employees at high-level positions. Of course, staff members generally cannot be legally retained against their will. In any event, the advice of an attorney skilled in employment issues should be sought before such contracts are prepared.

Following the final selection of candidates, the new employee can be informed of the date to report for work, the contents of the training program, and the direction that his or her efforts will take during the first few weeks. As a precautionary measure, some employers refrain from telling alternate candidates that they are no longer being considered for the job until the individual selected has begun training. The alternates are then informed of the company's decision, usually within a period of 2 weeks.

TRAINING AND PERFORMANCE

An employee training program is an important element of personnel policy for the consulting engineering organization. The provision of a training period helps assure new members that they will be able to make an adjustment to their position in a manner satisfactory both to themselves and to the company. Similarly, it helps guarantee that the firm will be able to maximize the quality of its services and the productivity of its staff. Still, many consulting firms lack adequate plans for conducting training programs. Fearing that on-the-job instruction will cause a costly loss of management time, they try instead to turn the new employee into an immediate production asset by exposing her or him to an onslaught of current projects. Many of these firms consequently develop staffs full of disgruntled workers.

As a service business, it is vital for the consulting engineering organization to monitor and direct the efforts of staff members. People are the main source of production in any consulting firm. So-called nonproductive effort may, indeed, accumulate while training new members. But few business owners would choose to forgo instruction at the cost of service quality and staff morale.

Traditionally, the bulk of training for engineers, designers, and drafters has come through schooling. Consultants rely on a new drafter, for instance, to possess the basic technical skills needed for mechanical or architectural drawing, regardless of whether or not the employee worked for engineering firms before. Part of management's function is to assist staff members in integrating their skills into the current workings of the organization. Also, managers are responsible for guiding employees to contribute to company growth and prosperity.

There is no one best way to accomplish these tasks. Some firms start training a new employee by providing the individual with a written outline describing various tasks that comprise the normal activities of his or her position. For a drafter, this outline might include definitions of tasks such as:

1. Preparation of sketches used to determine a project's equipment needs
2. Preparation of detailed design drawings
3. Use of shop drawings in determining specification requirements
4. Integrating bidding data into the final plans

Such an outline can provide a great deal of information useful in introducing the employee to the firm's procedures.

Unfortunately, some firms fail to supply such an introduction. The owners and principals of the firm may have a clear idea of the kind of performance they expect, but they simply do not take the time to spell out their expectations. Management cannot afford to lead each new employee by the hand. Neither can management assume that staff members will automatically acclimate to new positions without some introductory help.

As part of the training program, the employee may be asked to practice his or her skills on a hypothetical project. Specific tasks should be approached individually, while a supervisor reviews performance. Gradually, the employee can be directed to take on additional challenges to his or her design, drafting, and engineering abilities. Once the individual demonstrates proficiency in particular areas of work, he or she can be assigned to apply the skills to one part of an actual job. Different categories of the employee's discipline can be emphasized through assignments to various types of projects.

A similar method of training can be used for employees in higher-level positions. Lectures, seminars, meetings with other professionals, and additional college-level instruction can also be used to help employees in upgrading their skills. Some firms offer these as part of a systematic benefit plan, combined with the incentives of a salary increase and promotion schedule.

Further, the consulting firm can hold in-house courses. For example, the manager of a particular department can outline the firm's design approach to one phase of a specific project, illustrating the solution to a structural engineering problem, with all its calculations, investigations, conceptual details, and obstacles. Such courses can be held during normal working hours or on weekends and weekday evenings. Some organizations consider the in-house seminar to be mandatory for all engineers, designers, and drafters. It is generally believed that the courses bring about an overall increase in the staff's knowledge, thus strengthening the entire firm.

An added virtue of such a training program is that it tends to increase communication and cohesiveness among members of the consulting organization. Without some form of instruction, employees may have little opportunity for contact with company principals. The most effective training programs encourage such contact, making use of the idea that thorough communication motivates employees to perform superior work.

Regular meetings between the new employee and his or her manager are essential to evaluating the employee's performance. Following the completion of the training program, the performance review may be used as the basis for decisions on salary increases and promotions. Under these conditions, the importance of communication and monitoring becomes clearly visible. Managers must make certain that staff members know the requirements of their jobs, and they must observe the staff's performance to see how each member works to meet those requirements.

Most consulting firms consider employees for raises at least twice each year. Generally, the heads of the organization base their decisions on a manager's evaluation of the progress an employee has made toward meeting specific requirements. These requirements can be discussed in regular employee meetings, then recorded in the employee's file. During the performance reviews, a manager can use a simple rating sheet to list her or his comments with respect to the employee's work on each objective. A weekly or monthly review of this type may produce a vivid performance record. But even if the review is conducted less frequently, it will provide a foundation for discussion about promotions, salary increases, and other actions that may be warranted for particular staff members.

EFFECTIVE MANAGEMENT FOR MAXIMUM PRODUCTIVITY

The consulting engineer must learn a variety of management techniques to maximize productivity and profitability. During the early phases of operation in a small consulting firm, management responsibilities are relatively simple: The individual aspects of a project are attended to by a few engineers, designers, and drafters, while the firm's principals oversee the office work, budgeting, marketing, and planning tasks.

Some firms succeed in growing without changing their management practices substantially. But many managers and principals find themselves in charge of increasingly complex organizations, for which their original style of management is no longer suitable. If proper management techniques are not learned and adopted at the appropriate stages of a firm's evolution, the business usually stagnates or fails. Time schedules become ineffective, budgets are not met, the quality of the firm's service diminishes and its reputation declines.

The special needs of management in consulting engineering stem from the nature of the engineering organization itself. It is a hybrid form of practice, combining the characteristics of a professional service organization with those of a product-producing business. The consulting engineer has the qualities of a manufacturer, a contractor, and an engineer all at once. Managing a consulting firm—as well as working as an employee in one—thus requires a high degree of flexibility (and the ability to simultaneously integrate the work of several disciplines.)

The consultant's professional engineering expertise acts as the "top layer" of decision making in the firm. Beyond that layer lies the managerial foundation that guides the organization's performance and provides the firm with the strength it needs to succeed.

Most engineering firms are regarded as "line-and-staff organizations"; that is, they are operated in such a way that the principals of the firm oversee the securing of work; then they pass the work out to a group of people within the company who perform the actual jobs that are required, usually with the assistance of supervisors appointed by the principals.

Generally, problems in the functioning of line-and-staff organizations can be traced to poor or inappropriate management. For example, when there are a number of projects to be completed and the responsibilities of the various individuals assigned to do the work overlap excessively or become confused, or the schedule of tasks is unclear, the difficulties most likely have arisen from a lack of management communication with regard to task definitions, job assignments, scheduling priorities, and the like.

In an attempt to avoid such potential problems, some engineering firms have paid high-level staff members to attend management training seminars and business schools. Unfortunately, most consultants find this to be only a partial solution since academic programs historically lack management courses specifically suited to consulting engineers. In the 1970s, however, some schools began to tailor programs that better fit the requirements of engineering managers. We can expect to see further growth in this area of study in the years ahead.

In the meantime, what techniques can you use to effectively manage your firm? To answer this question, we shall examine some of the basic functions of business management in the context of the professional practice of consulting engineering.

Job Task Definition

Among the chief responsibilities of managers are determination of the work requirements for each project, definition of the tasks necessary to meet those requirements, and delegation of specific tasks to staff members who have the appropriate expertise. These responsibilities as a whole can be referred to as "job task definition."

In engineering firms, job task definition usually occurs along the lines of the four major engineering disciplines: civil, chemical, electrical, and mechanical. Of course, each discipline contains a variety of specialities. Thus a manager might assign one civil engineer to work on structural aspects of a given project, while appointing another civil engineer to take part in the design of sanitary and environmental phases of the job. Further, the responsibilities of employees from different disciplines may occasionally overlap, such as when an electrical engineer coordinates a system design with the structural requirements of a particular building. Maximum productivity can be achieved only when each staff member knows his or her range of responsibilities and any limitations—such as deadlines and budget restrictions—relevant to carrying them out. It is management's function to make certain that this information is communicated and understood.

When the consulting firm does not possess a staff with sufficient capabilities to carry out certain parts of a project, management must determine whether some of the work will be farmed out to subconsultants or whether the engagement might have to be rejected entirely. Each potential project must be analyzed carefully by the firm's principals beforehand to determine whether or not it is within the organization's capabilities.

Management by Project

Once a consulting organization decides to undertake a new project, a specific plan is outlined. The plan may include development of a budget, schedules, equipment specification preparation, and job task definitions. Several individuals may be involved in planning one project, but most engagements require only one project manager or project engineer. It is that individual's duty to direct the activities that are necessary to carry out the project plan.

Most consulting firms having several projects in process find it is not feasible for a single project manager to handle the direction of all jobs at once. Therefore, many organizations designate a group of employees as project managers, appointing each to take charge of a particular project that suits her or his expertise. This allows the principals of the firm to maintain control over the volume and direction of work. Moreover, it provides for a distribution of authority within the firm, which often gives rise to enhanced employee satisfaction. But the rotation of authority can sometimes lead to a situation in which certain staff members experience an inflated notion of their own power, while others feel slighted.

A number of firms solve this problem by appointing selected members to act as project managers on smaller jobs while reserving the larger, more complex projects for management by one of the firm's principals or by a senior staff member who holds the permanent title of "Project Engineer." Care is taken to make certain each employee understands whether his or her appointment is temporary or implies a permanent elevation in status. Once the appointments are made, the project manager is in a good position to undertake the challenge of controlling the project in a way that fulfills the plan and meets the requirements of the client.

Management by Discipline

In the past, many consulting firms organized themselves into relatively independent departments, according to discipline. The heads of each department had the responsibility of managing individual projects or parts of a project relevant to the department's field. At first, this would seem to be a rather simple way of delegating authority. But organizations of this kind usually run into difficulties in properly managing a large volume of work.

The principals or top-level managers of the firm hand down one set of priorities to the departmental managers, while the departments each develop their own priorities. Often, the priorities conflict. For example, the mechanical engineers might be struggling to meet a deadline on a project confined to their department, while at the same time the electrical and structural departments require coordination with the mechanical engineering aspects of a different project, also under deadline pressure. Such circumstances usually lead to an unsettling of priorities, not to mention the displeasure of the key personnel involved. As a result, this system of management has largely given way to other, less troubled forms of administration.

This does not mean it is counterproductive for employees representing different disciplines to assist each other or to exchange roles occasionally. On the contrary, it is probably more counterproductive for managers of an engineering firm to try to restrict a particular type of work to one group or individual. An interchange of roles may help managers and staff members to better understand each other's problems. Mechanical engineers and electrical engineers should sometimes be encouraged to observe or get involved in structural and civil engineering, for instance. Likewise, it may be of benefit for the firm to encourage project engineers and managers to assume the roles of designers and drafters when circumstances permit.

Another area where greater communication and understanding is often needed is in the relationships between designers and engineers. For many consulting firms, the two positions are not as different as implied by the simple dictionary definitions of the titles. Many designers have an engineering background, but they are usually supervised by engineers in their organization. This condition sometimes results in conflicts between the supervising engineers (who generally direct the designers to follow a particular approach to a project) and the designers (who may want to use a different approach).

One way to prevent this from happening is to hold a preliminary design conference on each new project, asking both the designer and the engineer to attend. The goal of the conference is to allow the key personnel who are to work on the project to discuss various approaches to solving the client's problem and to determine the objectives for the preliminary design. During the meeting, either the project engineer or the firm's principals outline the broad scope of the engagement, the proposed approach to the project, and the reasons behind that approach. For example, the project engineer might explain why she or he prefers to equip a building with a circulating warm-air heating system, rather than a hot-water or steam system. Once all key personnel understand the logic behind the approach, there should be little or no disagreement over the development of the actual design.

If other sorts of conflicts develop, staff members should be able to discuss them with other administrators in the firm, such as the project engineer, a managing partner, or company principal.

MANAGERIAL FRAMEWORK FOR LARGER JOBS

In Section 1 we noted that most successful consulting engineering firms try to maintain a balanced mixture of projects. In most cases, the goal is for approximately half the total income to derive from one or two major projects yearly, while the other half comes from an assortment of small- and medium-sized jobs. Those lucky enough to attain such a balance find themselves with a tremendous profit potential.

To ensure that major projects are completed according to schedule and budget requirements, experienced consultants make certain that a principal of the firm is placed in charge of each major project. The principal, who thus serves as chief manager, generally appoints a project engineer to report directly to him or her. The project engineer, in turn, provides a direct line of communication between the principal and all other personnel involved in the project.

A group of chief engineers, representing the various disciplines, reports directly to the project engineer. Finally, specific teams would be assigned to perform different phases of the actual work. Rarely will the members of the teams work on any other project before their responsibilities on the major job are fulfilled.

The resulting managerial framework is usually highly effective. But things can get hard to control if the firm suddenly acquires additional engagements. While the managers of the firm ought to exercise some flexibility for work assignments and scheduling, they also need to direct the efforts of the staff according to a known set of priorities. One of the most important objectives of any project is that it be completed on time. Therefore, a top priority is to maintain all important projects on a strict schedule. The potential success of a major job will be threatened if an engineering team is fragmented and a schedule is interrupted in an effort to take on more work than the firm is ready to handle.

SHOP DRAWINGS AND FIELD SUPERVISION

The handling of shop drawings can become a problem for both small and large organizations. The drawings, supplied by equipment manufacturers to show how their equipment can be used on a particular project, must be examined by someone who is thoroughly acquainted with the project design and specifications. It thus seems logical to assign the project designer to this task. Unfortunately, by the time the shop drawings are received, the designer frequently has been engaged in other projects.

When this happens, the managers of the engineering firm have three options:

1. Take the designer off other projects so she or he can check the shop drawings.
2. Familiarize another employee with the design and have that person do the checking.
3. Appoint a field supervisor or one of the firm's principals to review the drawings.

The approval of shop drawings is generally a managerial task since the firm's managers know most of the ramifications of a given design. They also know the problems that can arise from use of certain types of equipment.

Field supervisors have the added perspective of being able to see how the proposed equipment will fit into the overall project site. Most consulting engineering firms hire a full-time field supervisor at some early stage of their growth. But who supervises the field supervisor?

The most workable arrangement is to have the field supervisor report directly to the project engineer responsible for managing the design work. The project engineer and field supervisor can then work together to resolve any problems in the design, shop drawings, or proposed equipment and to make the changes that may be required.

SUPERVISING NONSTAFF PERSONNEL

Consulting firms sometimes hire outside consultants and part-time employees to handle increased amounts of work. The work of such nonstaff assistants should be managed with the same (if not more) diligence and attention to detail as is used in the supervision of staff members. Ideally, the work of the nonstaff people should be integrated into the overall system of project management. When this is not feasible, a manager may be appointed to oversee part-timers' work. If the firm employs a small group of part-time designers and drafters during the evenings, for example, it is part of management's function to see that a member of the firm is available to oversee their work.

MANAGEMENT BY OBJECTIVE

In the United States, there are more than 1000 management consulting firms helping a variety of businesses to install effective management systems. The system that these firms use most often in professional practice can be referred to as "management by objective." This system is utilized under different names depending on the type of organization, but it is based on a single principle that is applicable to virtually all consulting engineering businesses. The principle states that each firm has access to an almost unlimited amount of internally generated data than can help guide management to optimum utilization of labor. Generally, this data is collected and recorded in the form of managerial accounting reports on current and future work. (See Figures 3-8 through 3-17.)

The specific objectives of the reports are myriad. They can provide managers or company principals with an overview of the firm's current work; estimates of the percentage of completion of individual projects;

PROJECT DESIGN CRITERIA

PROJECT NAME _____

CLIENT _____

DATE STARTED _____ 19 ____ DATE DUE ____ 19 ____ BUDGET $ _____

CRITERIA _____

_____

_____

_____

_____

_____

_____

_____

_____

_____

_____

_____

_____

_____

_____

_____

_____

_____

_____

_____

_____

Figure 3-8 Project design criteria form.

PROJECT DESIGN CRITERIA

PROJECT NAME _Gait Lab - Curtiss Bldg - Newington Child Hosp_

CLIENT _Newington Children's Hospital, Newington, CT_

DATE STARTED _9/12_ 19 - - DATE DUE _12/15_ 19 - - BUDGET $ _110,000 - 130,000_

CRITERIA _The space shall be arranged to suit the needs_
of the United Technology computer and camera requirements and
the physical and medical constraints of Newington Hospital.
The floor shall be a special computer application floor
raised 8" off existing floor throughout. The clear height
throughout shall be a minimum of 8'-0". The walls
separating the lab from other areas of the building
shall be lightweight concrete or cinder block. All interior
space partitions shall be steel joist, dry wall, painted with hard
finish, washable surfaces. Doors from outside the lab
to the lab shall be 3'-6" x 7'-0" solid. Hardware in
all cases shall be keyed to present master key system.
Ceilings shall be 2x4' lay-in tile — quality materials;
2 hr fire rating. Windows on exterior walls shall be
bricked up. Rooms shall have sprinkler system and
smoke alarms tied to present system. Present ductwork
shall be raised tight to present ceiling for clearance.
The Gait Lab office ceiling may be lower than 8'-0"
if moving ducts in that one area proves not
feasible. The present stairwell doors and possibly the
hose cabinet may have to be relocated (door to opposite side)
to secure proper stair access. The heating and air

Figure 3-9 Filled-out project design criteria form.

PROJECT DESIGN CRITERIA

PROJECT NAME _Gait Lab - Curtiss BLDG - Newington Child Hosp_

CLIENT _Newington Children's Hospital, Newington, CT_

DATE STARTED _9/2_ 19.-- DATE DUE _12/5_ 19.-- BUDGET $ _110,000 - 130,000_

CRITERIA _conditioning shall be separate air systems with unit located outside in area adjacent to present exterior exit door. Ductwork shall be below raised floor and above new dropped ceiling. One thermostat centrally located. Control shall be day-nite, auto-manual, weekend and holiday skip-time clock. Electrical supply shall include special isolating transformer to guarantee no voltage variation. Temperature shall be 76° in summer, 72° in winter. Computer is primary concern. Special concrete slab required to allow installation of two forced pressure-sensing devices (see UT data). Bid date probably 1/15/--. Construction must be completed 5/15/--. Careful take of all existing systems required as some duct work, sprinklers and smoke alarms will be relocated and access must be maintained to items covered by new dropped ceilings._

J. Mueller

Figure 3-9 (*Cont.*)

```
┌─────────────────────────────────────────────────────────────────────┐
│                                                                       │
│                      PROJECT PROGRESS REPORT                          │
│                                                                       │
│   PROJECT NAME _____WEEK ENDING _____ 19___│
│   PROJECT MANAGER _____│
│   PERSONNEL EMPLOYED FOR THIS REPORTING PERIOD _____│
│   _____│
│   _____│
│                                                                       │
│   WORK COMPLETED FOR THIS REPORTING PERIOD_____ │
│   _____│
│   _____│
│   _____│
│   _____│
│                                                                       │
│   WORK PLANNED FOR COMING WEEK _____ │
│   _____│
│   _____│
│   _____│
│   _____│
│                                                                       │
│   PROJECT STARTED_____ 19____COMPLETION REQUIRED BY_____ 19_____│
│   % COMPLETED AS REPORTED LAST WEEK ___%   % COMPLETED FOR THIS REPORT ___%│
│   REMARKS _____ │
│   _____│
│   _____│
│   _____│
│                                                                       │
│                              SIGNED_____                 │
│                                                                       │
└─────────────────────────────────────────────────────────────────────┘
```

Figure 3-10   Project progress report form.

the use of time by various personnel; direct and indirect costs; summaries of output and efficiency; fee breakdowns; and other vital information.

The size and structure of the firm largely determines what records are needed, who will maintain them, and how frequently they should be updated. In many cases, the person in charge of the reporting for a particular project is the project engineer assigned to that engagement. As noted earlier, the project engineer may be someone who holds another, permanent position within the firm, such as a designer, supervisor, or even a company partner.

A project manager reports on the activities of the various project leaders and communicates with the client and any outside consultants that

Monthly Output Summary

Name of project	Dates		Project manager	Worker-hour totals		Percentages of		Remarks
	Started	Due		Budgeted	Actual	Work completed	Budgeted worker-hour	

Figure 3-11 Monthly output summary form.

Project name _____ Month of _____

Projected Fee Allocation

Total estimated fee $ _____

Less consultants retained estimated costs:

Architectural $ _____
Civil-structural   _____
Mechanical   _____
Electrical   _____
Site   _____
Survey   _____
Boring   _____
Other   _____

Total   _____

Net estimated fee $_____

Net Estimated Fee Allocation

Item	Net fee,%	Value
Profit	_____	$ _____
Sales effort	_____	_____
Administrative	_____	_____
Overhead	_____	_____
Direct labor	_____	_____

Estimated Direct Labor Worker-hours

Estimated average hourly rate of employees assigned to project $_____/hour

Estimated hours which project must be budgeted to complete equals
the total dollars available in the estimate value above for direct
labor divided by the estimated hourly rate of all employees who
will be assigned to the project.

Figure 3-12 Projected fee allocation form.

```
Project name _____     Month of _____

                    Actual Fee Breakdown

Total fee received                                    $ _____

Less consultants paid:

Architectural                    $ _____
Civil-structural                   _____
Mechanical                         _____
Electrical                         _____
Site                               _____
Survey                             _____
Borings                            _____
Other                              _____

Total                                                  _____

Net total fee for firm                                $ _____

                    Net Fee Allocation

      Item                Value               Total Net Fee,%

Profit              $ _____              _____
Sales effort          _____              _____
Administrative        _____              _____
Overhead              _____              _____
Direct labor          _____              _____

Total               $ _____              _____



Note:  You cannot have more than 100% as a value of total net fee
       in a normal situation.  If the totals for direct labor and
       overhead equal more than 100%, which can occur, then sales
       effort, administrative, and profit items are negative
       values.
```

Figure 3-13   Actual fee breakdown form.

might be retained. In a small consulting firm, one or more principals usually act as project managers.

Examples of forms used in managerial accounting appear throughout this section. These forms include reports of project design criteria, project progress, monthly output, staff time utilization, and projected fee allocations.

The criteria report (Figures 3-8 and 3-9) is usually completed by the project manager, who works directly with the client. It provides details and specifications relative to the design requirements of the job and may be fairly extensive. Figure 3-9 shows a completed form.

The project progress report (Figure 3-10) is completed by the project engineer or by the project manager if no leader has been appointed specifically to the project being covered. The progress report can be used to give a weekly or monthly synopsis of work completion. It is a valuable

Project name_____   Month of_____

Financial Analysis of Project Performance

Employee name	Hours	Rate/hour	Total	Budgeted	Budgeted Spent,%
_____	_____	$_____	$_____	$_____	_____
_____	_____	_____	_____	_____	_____
_____	_____	_____	_____	_____	_____
_____	_____	_____	_____	_____	_____
_____	_____	_____	_____	_____	_____
_____	_____	_____	_____	_____	_____
_____	_____	_____	_____	_____	_____
Totals	_____	$_____	$_____	$_____	_____

Net Fee Allocation

Item	Net Fee,%	Budgeted	Actual	Profit (Loss)
Direct labor	_____	$_____	$_____	$_____
Overhead	_____	_____	_____	_____
Administrative	_____	_____	_____	_____
Sales expense	_____	_____	_____	_____
Profit (loss)	_____	_____	_____	_____
Total	100	$_____	$_____	$_____

Figure 3-14   Financial analysis of project performance.

tool for managing staff time on current and future jobs. The monthly output summary (Figure 3-11) is used to provide owners, partners, or principals of the firm with a wide-ranging record of all recent projects, their starting and ending dates, percentage completion, and budgeted and actual employee-hours.

Such data can be translated into dollar values for the firm's comptroller, auditors, and owners. This dollar-value translation is the purpose of the projected and estimated fee allocation forms (Figure 3-12); the actual fee breakdown and allocation forms (Figure 3-13); the financial analysis of project performance (Figure 3-14); and the accrued-cash-flow report (Figure 3-15). The projected and estimated fee allocations are listed during the project planning phases, whereas an actual fee breakdown is prepared after project completion. All these reports, properly controlled and audited, can be used to show the organization's financial status. Further, they can help management pinpoint areas needing improvement.

The financial analysis report examines the time spent on a project,

Accrued Cash Flow

Month of_____ 19_____

Project Name	% Completed This Month	Fee Value Total	Completed Value
_____	_____	$_____	$_____
_____	_____	_____	_____
_____	_____	_____	_____
_____	_____	_____	_____
_____	_____	_____	_____
_____	_____	_____	_____
_____	_____	_____	_____
_____	_____	_____	_____
_____	_____	_____	_____

Total completed value   _____

Total cash disbursed except for consultants   _____

  Profit (loss) $_____

Figure 3-15   Accrued cash-flow record.

converts it to dollar values, and then compares the expenditures to the project's budget. Cost overruns can be detected simply by comparing the project progress report to the financial analysis. If, for example, the financial report shows that 70 percent of the budgeted direct labor has been spent and the project engineer's progress report shows 50 percent completion, then the firm is either losing money or there is something wrong with the engineer's estimate. This could indicate a serious management problem, requiring swift action by the consulting company principals.

The accrued-cash-flow report highlights the financial status of a group of projects on an accrued basis. It gives the value of completed work, the amount of fees earned and disbursed, and the resulting profit or loss. This report contrasts with cash-accounting reports in that it shows the amount of money earned rather than the amount received. For example, if you had five projects, each with a value of $20,000 in fee income and you completed 10 percent of the work on each in 1 month, your accrued income would be: 5 times 10 percent of $20,000, or $10,000. Thus, if your

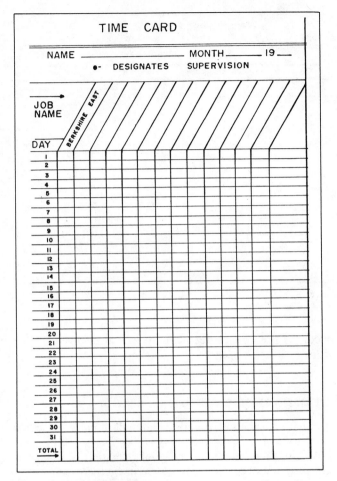

Figure 3-16   Monthly time report.

monthly expense was $12,000, then you lost $2000 (no matter what your cash receipts show).

Figure 3-16 shows a monthly time report that gives a daily or weekly record of the hours worked by a staff member on one or more projects. The monthly report is easier to file than weekly time cards and can be useful for monthly billing purposes, especially where hourly billing is involved. A weekly time report, of the type that might be used by company principals and personnel involved in administrative and sales efforts, appears in Figure 3-17. The form can be modified to suit the indi-

Weekly Time Report

JONES & JONES
CONSULTING ENGINEERS

WEEK ENDING _____
EMPLOYEE NAME _____
EMPLOYEE NUMBER _____

PROJECT NAME	CODE	PROJECT NO.		MON.	TUES.	WED.	THURS	FRI.	SAT.	SUN.	TOTAL
			ST								
			OT								
			ST								
			OT								
			ST								
			OT								
			ST								
			OT								
			ST								
			OT								
			ST								
ADMINISTRATION			OT								
TECHNICAL DEVELOPMENT			ST								
			OT								
BUSINESS DEVELOPMENT			ST								
			OT								
SICK LEAVE			ST								
UNASSIGNED			ST								
HOLIDAY			ST								
VACATION			ST								
		TOTAL									

APPROVED _____   RECORDED _____

Figure 3-17 Weekly time report.

vidual needs of the consulting firm by including the number of hours budgeted for each type of task.

CONSULTANT'S PROFIT KEY 5

Another popular system to keep projects on budget and on schedule while generating useful control data and reports is known as the Jobtrax method. David Burstein, PE, Vice President and Manager of Operations at Engineering-Science, Atlanta, Georgia, described use of the system in *Consulting Engineer:*

In recent years there has been a growing trend toward the use of CPM (Critical Path Method of Schedule), PERT (Program Evaluation and Review Technique), or some combination of the two as a tool for planning design and construction projects and monitoring their progress. While these approaches generally provide more care and accurate planning, they may result in a reduced emphasis on timely and accurate monitoring of actual project progress. This is due to three inherent weaknesses in most CPM/PERT methods of scheduling:

• The time required to update CPM/PERT diagrams often results in late or erratic progress reporting.

• Most CPM/PERT methods concentrate on critical path activities, which certainly are important, but ignore overall schedule status.

• Most CPM/PERT methods do not include a procedure for monitoring both budget status and schedule status.

Computerized CPM systems can be used to overcome some of these deficiencies for large, complex projects (over 1000 tasks). However, such methods tend to be too cumbersome to use on smaller projects. But there is a method, known as the JOBTRAX® system, that can be used either independently or in conjunction with CPM/PERT systems for projects of almost any size. It is based on the use of independent measurements of progress, expenditures, and calendar-time to obtain both schedule and budget status estimates.

PLANNING THE PROJECT

The first step in this procedure is to prepare a task outline for the project. The level of detail of the outline depends on the project's magnitude and complexity. A listing of only the major tasks generally is sufficient for small projects, while a breakdown into subtasks and sub-subtasks is required for larger projects. The example of a simple task outline presented below is used as the basis for the example computations used here to explain the JOBTRAX system:

• Review topography, utilities, zoning, etc.
• Finalize preliminary design.
• Prepare mechanical drawings.
• Prepare equipment specifications.
• Prepare civil drawings.
• Prepare structural drawings.
• Prepare electrical drawings.
• Project management.

Having completed the task outline, it is necessary to schedule these tasks. This can be done with a bar chart, CPM chart, PERT diagram, or any other method that the project manager deems appropriate for the particular project. The JOBTRAX system imposes no constraints on the type of schedule to be used. The only requirement is that a single set of start and finish dates be selected for monitoring progress. In the case of a CPM chart, these usually would be the early start and early finish dates. For a PERT schedule, they would be the most probable start and stop dates.

The next step is to assign a total budget amount to each activity listed in the outline. Again, the JOBTRAX system imposes no constraints on the method used to assign task budgets except that the total of the task budgets must equal the total project budget (excluding contingencies and profit). The results of the task scheduling and budgeting can be represented in a simplified format as shown in Fig. 3-18.

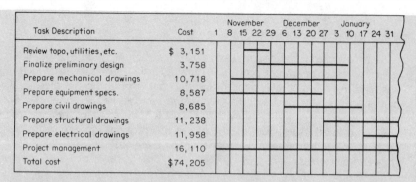

Figure 3-18   Task outline, budget, and schedule.

These results provide the necessary input for the final step required for project planning—the expenditure projection. This projection is derived by apportioning each task budget into the scheduled time frame for the corresponding activity, as shown in Fig. 3-19. These values then can be totaled for each month to obtain an estimate of that month's expenditures and summed to estimate the cumulative expenditures throughout the project duration. The cumulative expenditures are plotted onto a graph, as shown in Fig. 3-21.

It is important to remember that this curve represents not only the projected expenditures, but also the estimated rate of progress for the project. This concept can be displayed graphically by defining the total project budget to equal 100 percent completion and then establishing a project progress scale on the right side of the page (see Fig. 3-21). This graph serves as the baseline against which the schedule status of the project is measured periodically.

MONITORING SCHEDULE AND BUDGET STATUS

A periodic estimate of the overall project schedule and budget status is important to assure completion within preestablished goals. In this method the first step is to prepare a realistic estimate of the progress on each task in terms of percent completion of that task. Progress is defined as work actually accomplished and is independent of either budget expenditures or calendar time spent on the activity. For example, if a task consists of pouring 4000 cubic yards of concrete and a total of 3000 have been poured, then that task is 75 percent complete regardless of how much time or money it took to accomplish it. Estimating the progress of each activity is the single most important (and difficult) step, since it is nonmathematical and relies solely on the experience and judgment of the project manager and staff. One commonly used statistical method to reduce potential for errors is to preestablish various milestones for each activity. This reduces the range in which the progress estimate must be made.

Task Description	Budget	November				
		1	8	15	22	29
Review topo, utilities, etc.	$ 3,151			1,575	1,576	
Finalize preliminary design	3,758				537	537
Prepare mechanical drawings	10,178		1,191	1,191	1,191	1,191
Prepare equipment specs.	8,587	1,073	1,073	1,073	1,073	1,073
Prepare civil drawings	8,685					
Prepare structural drawings	11,238					
Prepare electrical drawings	11,958					
Project management	16,110	732	732	732	732	732
Totals	$74,205	1,805	2,996	4,571	5,109	3,533
Cumulative totals	—	1,805	4,801	9,327	14,481	18,014

Task Description	Budget	December			
		6	13	20	27
Review topo, utilities, etc.	$ 3,151				
Finalize preliminary design	3,758	537	537	537	537
Prepare mechanical drawings	10,718	1,191	1,191	1,191	1,191
Prepare equipment specs.	8,587	1,073	1,073	1,076	
Prepare civil drawings	8,685	1,447	1,448	1,447	1,448
Prepare structual drawings	11,238				1,215
Prepare electrical drawings	11,958				
Project management	16,110	732	732	732	732
Totals	$74,205	4,980	4,981	4,983	5,123
Cumulative totals	—	22,974	27,975	32,958	38,081

Figure 3-19   Projected expenses.

After the percent completion has been determined for each task, these percentages are multiplied by the corresponding task budgets, as shown in Fig. 3-20. These products are totaled and divided into the total project budget to obtain a weighted average of total project progress. This value represents the actual amount of progress made to date on the total project.

Task Descriptions	Budget ($/Task)	% Done	Budget Spent
Review topo, etc.	$ 3,151	100	$ 3,151
Final prelim. design	3,758	55	2,067
Prepare mech. drawings	10,178	35	3,751
Prepare equip. specs.	8,587	40	3,435
Prepare civil drawings	8,685	35	3,040
Prepare struc. drawings	11,238	20	2,248
Prepare elec. drawings	11,958	10	1,196
Project mgt.	16,110	35	5,639
Totals	$74,205		$24,527

Overall project progress = $24,257 − 74,205 = 0.331 = 33.1%

Figure 3-20 Overall project progress.

The total percent completion for each update period is plotted onto the same graph as the projected expenditures and the actual expenditures for the corresponding update periods, as shown in Fig. 3-21. The overall schedule status can be determined by comparing the estimated progress curve with the projected expenditure curve (distance A in Fig. 3-21). The overall budget status is obtained by comparing the estimated progress curve with the actual expenditure curve (distance B in Fig. 3-21).

A common mistake is to compare the projected expenditures with the actual expenditures in an attempt to determine budget status. If this comparison is made in the example project (Fig. 3-21), it can be seen that the actual expenditures are less than the projected expenditures. This leads to the erroneous conclusion that the project is under-budget when, in reality, this clearly is not the case. It can be seen that comparing projected expenditures with actual expenditures has no validity in terms of either budget or schedule status and may tend to lull the project manager into a false sense of security.

If the results of this analysis indicate that the project is behind schedule, or even if the overall project is ahead of schedule, it is desirable to know which activities are in trouble. This can be learned by filling in the lines on the project schedule to a distance corresponding to the estimated progress for that task. For example, if a task is scheduled to take four months and is estimated to be 25 percent complete, then the first month of the line for that task would be filled in. The end of the filled-in portion then is compared with the date that the update is prepared. If the update is beyond the filled-in schedule line, the task is behind schedule by an amount corresponding to the distance between these points. Similarly, the task is ahead of schedule if the filled-in schedule line extends beyond the date of the update. An example of this graphical analysis is shown in Fig. 3-22. (Tasks 1 and 8 are on schedule; tasks 2, 3, 4, and 5 are behind schedule; and tasks 6 and 7 are ahead of schedule.) Although the example uses a simple bar chart, the same analysis can be done as easily using a CPM or PERT diagram.

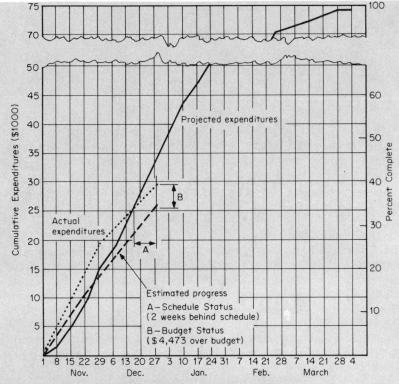

Figure 3-21   Progress curves of budget and schedule status.

Monitoring the budget status of each individual task is not as simple as the schedule status and requires independent cost accounting for each task that is to be monitored. This usually results in increased paperwork and policing the project team to ensure that everyone is charging his time and expenses to the proper task code. The project manager must determine on

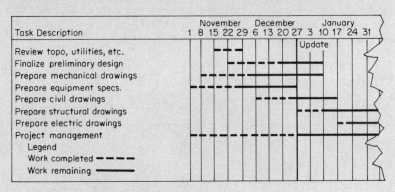

Figure 3-22   Schedule status of each task.

a case-by-case basis whether the additional information is worth the effort it takes to maintain it.

COMPUTERIZED JOBTRAX METHOD

The JOBTRAX method is ideally suited for computer application. The computer is used for three basic functions, which are:

- Improving the accuracy of the projected expenditure calculation by considering the effects of inflation on a weekly or monthly basis and using these escalated rates to compute expenditures for various tasks.
- Storing the results of project planning computations for use in updating project schedule status reports.
- Using the input of labor manhours to calculate manpower requirements.

Unlike most computerized CPM/PERT systems, the computerized JOBTRAX system is easy to use, usually requiring less than 30 minutes to compute the initial input forms which generate project planning output. These forms ask for such information as labor rate multiplier, other direct costs multiplier, labor escalation rate, and initial labor rates and labor categories. The project may be divided into as many as 200 tasks. The input form required to obtain an update of budget and schedule status as well as an updated manpower projection is even easier to complete, generally requiring less than 15 minutes. This form generates output that summarizes the project status.

PROS AND CONS

Many scheduling and budgeting methods are available to assist project managers in properly planning various types of projects, but most treat the need to monitor progress as a second priority. The JOBTRAX method uses other planning tools for quickly and accurately determining the budget and schedule status and for anticipating manpower requirements. This approach is now in use on many projects of various sizes and has met with considerable support not only from project managers, but also from clients.

The main advantages of the JOBTRAX system are:

- It is easy to use for setting up new projects.
- It is easy to use for monitoring progress.
- It provides accurate assessment of both budget and schedule status.
- It provides manpower requirement.

Its principal disadvantages are that it:

- Must be revised whenever there is a contract modification that changes the scope of work, budget, or schedule.

- Does not provide the level of control that CPM or PERT provides for multi $100 million projects.

- The information software for the computerized JOBTRAX system was developed by Don Fry of Engineering-Science, Atlanta, Georgia.

CONSULTANT'S PROFIT KEY 6

Large projects, such as chemical and petroleum plants, require careful planning to develop a project schedule. Arthur E. Kerridge, Project Manager, M. W. Kellogg Co., Houston, Texas, presents here a technique for developing a project schedule for such jobs. First presented in *Hydrocarbon Processing,* this method is both practical and useful. As the author says, "Early in its life, a project may be too ill-defined to list activities in detail or estimate their durations. Consequently, a 'top-down' rather than a 'bottom up' approach is needed." Here it is.

Before developing a project schedule, some things must be known. Fig. 3-23 lists some key information which should be available. Additional information may be useful. But do not get into the trap of asking for too much information before starting a schedule. Too much information may merely confuse. It is possible to create a project schedule with even less information than listed, because much information can be developed statistically or by pro-rating.

A Sample Project. For illustration purposes, the following typical project is assumed.

- *Total Installed Cost.* (engineering/procurement/construction)—$100 million

- *Home Office Manhours.* (at 10 percent at $35 per hour)—300,000 hours

- *Direct Field Labor Hours.* (at 20 percent at $15 per hour)—1,300,000 hours

- *Ratio of Field Labor to Home Office Hours.* 4.3 to 1

- *Type of Project.* Refinery hydro-desulfurizer for a major oil company

- *Operating Conditions.* Maximum temperature 1,200°F, maximum pressure 600 psig

- *Number of Equipment Items.* 250 (approximate hours per piece—1,200)

- *Equipment Types.* Exchangers, pressure vessels, compressors, furnace, towers

- *Long Delivery Items.* Furnace tubes, compressors, and alloy exchangers

- *Materials of Construction.* Mainly carbon steel, but some alloy

- *Piping.* Large bore pipe, and also alloy pipe

- *Structural Steel.* Major plant structure required

- Approximate Total Installed Cost—Engineer/Procure/Construct
- Approximate Home Office Manhours
- Approximate Direct Field Labor Manhours
- General Description and Type of Project
- Source of Process Design and Engineering Design
- Maximum Temperature/Pressure Conditions
- Number and Type of Equipment Items
- Identification of Special Long Delivery Items
- Materials of Construction
- Maximum Size and Extent of Piping
- Major Structures Required
- Location Where Plant Will Be Built
- Site Conditions—Approximate Area of Plot
- Labor Availability (Union/Open Shop)
- Major Field Fabrication & Subcontract Items
- Procurement Policy and Material Sources—Local/Worldwide
- Approvals Required
- Schedule Constraints
    —date of award of work/start of engineering
    —date funds released for purchase
    —date site available
    —date of permits/approvals to build

Figure 3-23 Procurement specifics.

- **Plant Location.** To be built on the U.S. gulf coast. Could be union or open shop labor
- **Site Conditions.** Level and clear/approximate plot area—400 ft $\times$ 400 ft
- **Field Fabricated/Erected Items.** Furnace and main tower
- **Procurement.** Worldwide purchase based upon competitive price and delivery
- **Approvals.** Client approves all major drawings
- **Schedule Constraints.**

    Process design completed as a package to be released to engineering contractor three months after award

    Start of engineering in three months

    Release of funds to commit purchase in six months

    Construction jobsite available in nine months

Major Elements. The major elements or activities in any schedule are engineering, procurement, and construction. The first step is to develop independent subschedules for each of these major activities.

- *Engineering.* Process design, conceptual/analytical engineering, production design, specifications, requisitions and drawings
- *Procurement.* Inquiries, bid evaluations, purchase orders, expediting, inspection and delivery to jobsite
- *Construction.* Temporary facilities, material receipt and erection (civil, structures, equipment, piping, instruments, electrical, paint and insulation)

Schedule development can start with any of the above elements. A start should be made with either procurement or construction. Leave engineering until last. The reason is that engineering activities are the most flexible and adaptable. The engineering approach may well be modified or governed by schedule constraints arising in the procurement or construction areas. Construction logic is governed by the required building sequence. Procurement durations may be dictated by third parties.

PROCUREMENT

Procurement starts with an inquiry requisition and ends when material is received. The logic is simple. There are a large number of similar independent parallel activities. For each of these there are a fixed number of sequential subactivities.

The procurement schedule is influenced mainly by outside sources. Technical data and technical reviews are controlled by engineering. Approvals are governed by the clients and/or management. Quotation durations, drawing and material deliveries are set by vendors. Inspection and transportation are also set by others. The durations directly controlled by the procurement group are a small part of the total time span.

Procurement Cycle. The first step in the development of the procurement schedule is to establish a "procurement cycle." It lists the subactivities which must be performed from the inquiry requisition up to purchase order placement and receipt of vendor drawings. All subactivities should be included even though each one may only take a few days. Cumulatively, they can add substantially to the procurement schedule.

Fig. 3-24 illustrates a chart used to establish the procurement cycle. The left hand column lists the major sequential activities. The chart is completed by assessing the number of working days required for each of the subactivities. Working days are used in place of weeks to allow for the effect of weekends. Overall calendar weeks are determined by dividing the cumulative working days by five.

Since durations differ for major complex equipment items, standard equipment items, and bulk materials, three columns are included in the chart for three separate procurement cycles: 1. major complex equipment,

REF	Activities NOTE Durations are <u>working days</u> ( ) are cumulative working days	Durations Complex Equipment Item A	Standard Equipment Item B	Bulk Material Item C
1	Requisition Ready for Inquiry	0 (0)	0 (0)	0 (0)
		8	5	5
2	Client Approval Received	(8)	(5)	(5)
		7	5	3
3	Issue Inquiry to Vendors	(15)	(10)	(8)
		30	20	15
4	Receive Quotations	(45)	(30)	(23)
		20	15	7
5	Complete Bid Evaluation	(65)	(45)	(30)
		10	10	5
6	Client Approval Received	(75)	(55)	(35)
		5	5	5
7	Place Purchase Order	(80)	(60)	(40)
—	TOTAL EQUIVALENT WEEKS	16	12	8
		20	15	10
8	Receive Preliminary Vendor Prints	(100)	(75)	(50)
		20	15	10
9	Receive Final Vendor Prints		(90)	(60)
—	ADDED EQUIVALENT WEEKS	8	6	4

Figure 3-24 Procurement cycle chart.

2. standard equipment and 3. bulk materials (such as piping, electrical, instruments, structural and civil).

When establishing activity durations, be realistic. Do not unreasonably shorten vendor bid times. Bids will merely be late or incomplete, which will require more rebid or follow-up time.

Having established the procurement cycle, which sets the time required up to the purchase order, add delivery and shipment times for each material category. This can be done with a chart as shown in Fig. 3-25.

Item NOTE: Delivery Promise Start Point P.O. = Purchase Order A.D. = Approved Drawing	Cycle Type From Fig. 2	Delivery Promise (Weeks)	Delivery Promise Start Point	Added Time From P.O.	Total Delivery (Weeks)	Shipment Time (Weeks)
EQUIPMENT						
Furnace—Structural Steel	A	14–18	A.D.	8	22–26	3
—Tubes	A	38–46	A.D.	8	46–54	3
Towers—Alloy	A	36–44	A.D.	8	44–52	4
—C.S. Heavy Wall	B	24–36	A.D.	6	30–42	4
Towers Field Erected—Materials	A	20–24	A.D.	8	28–32	3
—Erection	-	16–20	-	0	16–20	3
Exchangers S & T—Alloy	A	30–38	A.D.	8	38–46	4
S & T—C.S.	A	20–24	A.D.	8	28–32	4
Air Fin Exchangers	B	20–24	P.O.	6	26–30	3
Vessels—Alloy	A	36–42	A.D.	8	44–50	4
—C.S. Heavy Wall	A	24–36	A.D.	8	32–44	4
—C.S. Light Wall	B	16–32	A.D.	6	22–38	4
Tanks—Shop Fabricated	B	20–28	A.D.	6	26–34	4
Tanks Field Erected—Materials	A	16–24	A.D.	8	24–32	3
—Erection	-	16–20	-	0	16–20	3
Pumps Process—Alloy	A	32–44	P.O.	0	32–44	3
—C.S.	A	24–32	P.O.	0	24–32	3
Compressors—Large Process	A	48–60	P.O.	0	48–60	4
—Light Utility	B	30–40	P.O.	0	30–40	3
Special Equipment—Packaged	A	32–52	P.O.	0	32–52	4
STRUCTURAL STEEL						
Major Plant Structures	C	12–20	A.D.	4	16–24	3
PIPING						
Large Bore—Alloy	C	24–40	P.O.	0	24–40	3
—C.S.	C	12–24	P.O.	0	12–24	3
Large Valves—Alloy	C	32–48	P.O.	0	32–48	3
—C.S.	C	16–24	P.O.	0	16–24	3
Pipe Fabrication—Alloy	C	16–24	A.D.	4	20–28	3
—C.S.	C	8–12	A.D.	4	12–16	3
INSTRUMENTS						
Control Valves	C	32–48	P.O.	0	32–48	3
Consoles/Panels	A	24–48	A.D.	8	32–56	4
ELECTRICAL						
Transformers	B	24–36	P.O.	0	24–36	3
Major Switchgear	B	24–36	P.O.	0	24–36	3
Motor Control Centers	B	16–24	P.O.	0	16–24	3
Power Cable	B	20–32	P.O.	0	20–32	3

Figure 3-25 Materials delivery chart.

Equipment/Materials List.   The first step is to list the major equipment and material categories required. Durations are then added from experience or by telephone contact with leading vendors. Vendors' telephone quoted deliveries are usually their best and most optimistic. Frequently, a vendor's quoted delivery starts from receipt of approved drawings released for fabrication, not necessarily from receipt of purchase order. Also, promised delivery may be the day it is completed in the shop without allowance for inspection, witnessed performance test, or transportation arrangements.

Fig. 3-25 takes these factors into account by noting the procurement cycle type from Fig. 3-24, the delivery promise in weeks and the probable start point for the delivery promise which may be purchase order or approved drawings.

The chart also allows for transportation/shipment to the jobsite. Assessment of this duration may be difficult if the source is unknown, and could be overseas. However, it is best to be conservative. Shipment times can be the most unpredictable, varying from a few days to several weeks. If a conservative allowance is put in here, it can be a buffer to cover inspection delays, rejections or other transportation holdups.

CONSTRUCTION

Before construction can start, drawings and materials must arrive at the jobsite. Thereafter, construction proceeds in a sequence governed by the work's physical nature. Excavation precedes foundations, which are followed by steel and equipment erection. Piping, electrical and instruments then commence in sequence, finishing with insulation and paint.

Sequences.   In general, construction activities have a predetermined logic. They must follow one another in a set order like building blocks. The field has little opportunity to be flexible or adaptable. Drawings and materials must be delivered to the jobsite in the right sequence ahead of the time when they will be installed.

If it can be assumed that drawing and material deliveries will not present a problem, then how long should the construction work take? It becomes a matter of resource. How large a labor force is available and how many workers can be economically applied to the job? Too few and the job will take too long. Too many may mean lower productivity and inefficient working. There could also be problems of inefficient use of construction equipment and costly short term supervisory peaks.

Manhours.   Fig. 3-26 shows a series of typical overall construction curves for one million labor manhours spread over construction durations of 6 to 36 months. Although the curve shapes are the same, the peak manpower requirements, as shown by the dotted line, increase exponentially as durations are shortened.

If manpower availability is not limiting, a method of determining the economic construction manpower peak is to evaluate the maximum density

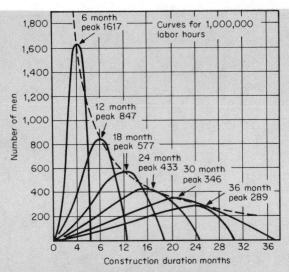

Figure 3-26 Effect of duration on manpower peak.

of labor per plot area. For a typical process plant, a construction worker requires an average working area between 150 and 250 square feet. For preliminary planning purposes, 200 square feet per man is a good number to use. This does not apply to offsite areas or to areas where pipe and equipment are more spread out. In these areas, the amount of work available may not support a labor density of one man per 200 square feet.

For the project under consideration, the plot area is 400 × 400 feet which equals 160,000 square feet. The economic peak manpower, therefore, is 160,000/200 = 800 men.

Field Labor. Fig. 3-27 (top) shows a series of curves for a range of total field labor manhours in which the manpower peaks are plotted against construction durations. These curves are derived by the method shown in Fig. 3-26. For an economic construction peak of 800 men and field labor hours of 1.3 million the construction duration is approximately 18 months.

Fig. 3-27 (bottom) shows a typical field labor curve. The horizontal scale represents the construction duration shown in 10% increments. The vertical scale shows percent progress achieved for each 10% increment in construction duration. The maximum or peak percent achieved per 10 percent of calendar time is 18 percent. This curve can be used to compute manpower peaks for a known duration in the following manner.

Let economic manpower peak = P (unknown)
Total construction duration = M months
Ten percent of construction duration = M × 0.1 months
Hours worked per week = H hours
Hours worked per month = H × 4.33 hours
Total construction labor hours = T hours

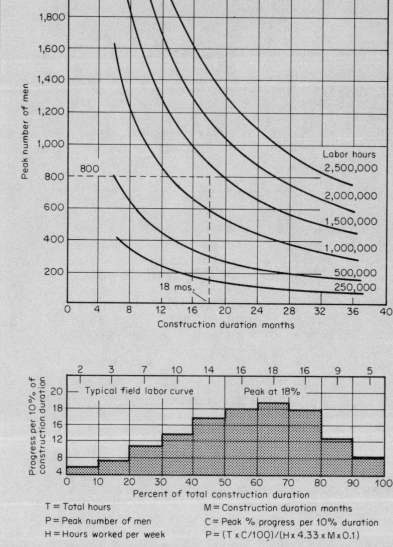

Figure 3-27 Upper: Peak manpower by construction duration. Lower: Percent of total construction duration.

CRAFT	APPROX % OF TOTAL LABOR HOURS	% OF OVERALL CONSTRUCTION DURATION			PEAK PROGRESS PER 10% OF CRAFT DURATION	BAR CHART SCHEDULE
		START %	FINISH %	DURATION %		10 20 30 40 50 60 70 80 90
CIVIL	20	0	85	85	13	
STRUCTURAL	8	15	75	60	13	
EQUIPMENT	8	15	85	70	13	
PIPING	40	15	95	80	16	
ELECTRICAL	8	25	100	75	18	
INSTRUMENTS	7	35	100	65	18	
INSULATION & PAINT	9	50	100	50	16	
TOTAL CONSTRUCTION	100	0	100	100	18	

Figure 3-28   Craft timing and distribution.

Maximum percent achieved per 10% duration increment = C%

Hours expended in peak 10% period for C% achievement = $T \times C/100$ hours

Hours expended in peak 10% period for P men = $P \times H \times 4.33 \times M \times 0.1$ hours

Then $T \times C/100 = P \times H \times 4.33 \times M \times 0.1$

Therefore, $P = (T \times C/100)/(H \times 4.33 \times M \times 0.1)$

Substituting known values

$T = 1,300,000, M = 18, H = 40, C = 18$

Then $P = 750$, which is close to the figure shown in the chart

For $P = 800$, M becomes 17 months

Labor Peaks.   The next step is to look at the individual peaks for each labor craft. Fig. 3-28 shows typical craft timing and distribution. This chart shows the percentage of total hours for each craft, the start, finish, and duration time and the peak progress achieved for 10 percent of the craft duration. The previous formula can also be used to calculate the craft peak.

Taking the piping craft, Tp piping hours = $T \times 40\% = 520,000$ hours

Mp piping duration = $M \times 0.8 = 14.4$ months

Cp peak piping progress for 10% duration = 16%

Therefore, $Pp = (520,000 \times 16/100)/(40 \times 4.33 \times 14.4 \times 0.1) = 334$

This gives the peak piping labor force required for the project.

Material Deliveries.   The final step is to determine the required material deliveries. All materials are required ahead of the craft schedule periods

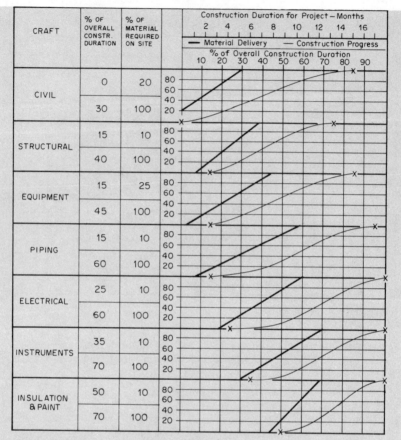

Figure 3-29   Construction materials delivery requirements.

shown in Fig. 3-28. Fig. 3-29 plots the typical percentage progress completion curves for each craft over their duration. Material delivery requirements are added in ahead of these curves with adequate lead times as
shown. For smooth construction work each craft requires a minimum backlog of material.

Across the top of the curves the actual construction duration for the project as determined previously is superimposed. This now positively identifies material delivery requirements in relation to the established construction duration.

ENGINEERING

Engineering has been left until last because it is the most difficult to schedule. For many reasons, the sequence of engineering activities may not follow typical logic. It comes down to priorities, preferences, and discipline.

If minimum schedule is the priority, many engineering shortcuts can be taken based upon conservative, empirical assumptions. Equipment and material can be "sole sourced" to save time. The result may be a less than optimum design, with some added costs. If design optimization and/or minimum cost is the priority, then engineering may go through many recycles with consequent effect upon the schedule and manhours.

Orderliness. Even though engineering activity sequences and durations may be radically altered or rearranged by priority dictate, engineering development must proceed in an orderly progression. Secondary engineering or detail design must not be permitted to start until primary decisions have been made. Nothing is gained in the schedule by working out of sequence; resources are wasted and time is lost from recycle.

The fundamental steps in engineering are:

* *Process Design.* Process flow diagrams; heat and material balances; temperature, pressure, physical, chemical and flow conditions; process equipment specifications and data sheets; utility systems balance diagrams and environmental systems concepts and specifications.

* *Basic Engineering.* Piping and instrument diagrams (process and utility); general engineering specifications; equipment mechanical specifications, data sheets and requisitions; plot/site plans (block model); plant arrangement studies (planning model) and single line electrical diagrams.

* *Production Engineering.* Final site and plot plans, production design model, civil and structural fabrication/erection drawings, final certified vendor drawings, piping key plans and isometrics, instrument and electrical field installation drawings and bulk material lists and requisitions.

In general, engineering development must proceed in the order of the above steps. A certain amount of overlap may be permitted, and on larger projects, specific areas or systems may be released ahead of others.

Spread Charts. This fundamental staging of engineering work coupled with the fact that there is a typical interrelationship between the manhours and timing of each engineering discipline makes it possible to develop prototype engineering discipline spread charts for typical project conditions. Fig. 3-30 is a typical chart for an EPC project initiated with a process design package which is a good starting point to develop an engineering schedule. This figure shows typically: the distribution of hours between disciplines, the start/finish point for each discipline relative to the overall project duration and the normal progress curve for each discipline.

A prototype chart of this type applies fairly consistently for the majority of process units. The unknowns are the realistic durations for each discipline's time span.

Quantify. The next step is to come up with some quantitative estimates of the work to be done by the key disciplines, by statistical methods.

Total home office manhours = 300,000

CASE-EPC Project with Process Design Supplied Note Bar Charts show scheduled time to 90%					ENGINEERING DURATION TO 90% POINT FOR PROJECT (MONTHS) 2 4 6 8 10 12 14						
DISCIPLINE	% OF TOTAL HOURS	START %	90% POINT %	CURVE TYPE	% CALENDAR TIME TO MECHANICAL COMPLETION						
					10	20	30	40	50	60	70
PROJECT ENG. & ADMIN.	12	0	70	2	6	15	28	44	59	75	90
P&ID ENGINEERING	16	0	35	2	15	44	75	90			
VESSELS	6	0	35	1	20	53	80	90			
EQUIPMENT	5	0	35	1	20	53	80	90			
PRINT LAYOUT & PIPING DESIGN	33	5	65	2	3	13	27	45	63	81	90
CIVIL & STRUCTURAL	13	20	50	3			16	50	90		
INSTRUMENTS	10	15	70	3		3	12	26	46	64	90
ELECTRICAL	5	20	70	3			7	22	42	67	90
TOTAL ENGINEERING	100	0	65	-	6	19	37	46	71	84	90

Basic Engineering 40%   Production Engineering 60%

Figure 3-30   Prototype engineering discipline spread chart.

Total engineering manhours at 75% = 225,000
Individual discipline hours may be determined from Fig. 3-30.
Number of equipment items = 250
Number of P&ID lines = 250 × 6.5 = 1625
Number of P&ID's = 250/3 = 83
Number of model tables/planning areas = site area/10,000 = 160,000/ 10,000 = 16 (10,000 sq ft of site = a model table of 2½ × 4 ft @ ⅜ in. = 1 ft)
Number of piping key plans = site area/4,000 = 40 (4,000 sq ft of site area = a drawing area of 19 in. × 29 in. @ ⅜ in. = 1 ft)
Number of isometrics = number of lines = 1625

Having quantified the major categories of work, the next question is: How long should they take? The critical durations in basic engineering are: P&ID development time to production release, equipment design time to requisition issue, receipt of proprietary equipment vendor drawings and development of planning arrangement studies to production release.

An analysis of the work scope, the manpower availability in each discipline and matching these against the durations and progress curves shown in Fig. 3-30 suggest a realistic time span of six to seven months for basic engineering.

Critical Durations.   In production design these are piping and instrument production. A balancing of key piping manpower availability against the work scope of the 16 piping areas/40 key plans suggest an overall duration for piping production in the range of nine months. Coupling the two together gives an overall engineering period in the region of 15 to 16 months. This time scale can now be added to Fig. 3-30 above the total engineering duration, as shown. We now have the backbone of an overall engineering bar chart which shows start and finish points for each discipline and a typical progress curve. Dates for specific activities can now be determined from the individual discipline progress curves. Specific activity completion points can be related to specific progress completion percentages for that discipline which can now be related to a schedule date from Fig. 3-30.

THE TOTAL PROJECT

The last stage in the development of the project schedule is to integrate the three independent subschedules for engineering, procurement, and construction to produce an overall master project schedule. This may require a series of iterations. Start with a standard master project schedule layout, and make a first assumption for a realistic (not optimistic) project schedule duration. There are two early indicators which can be used to make the first project duration assumption.

Engineering (to the 90% point) normally should not exceed 65 percent of the total project duration.

Therefore, if engineering = 15 to 16 months, then the total project duration = say $15.5 \times 100/65 = 23.8$ months.

Project construction (excluding pre-project early site development work) should normally never start earlier than 30 percent of the total project duration.

Therefore, if construction = 17 to 18 months, then the total project duration = say $17.5 \times 100/70 = 25$ months.

Master Schedule. For a first pass, set up a master project schedule format assuming a project duration of 25 months from release of the process design package to mechanical completion. Starting from the front, lay in the engineering durations and requisition dates for equipment and bulk materials based upon the engineering schedule. Add onto these the material delivery times from the procurement schedule to give equipment and material field delivery dates.

Starting from the back end, lay in the construction craft durations and the construction material required dates from the construction schedule. Now look for "float" or "overlap." Float will show up if material deliveries are ahead of the construction material required dates. Overlap will show up if drawing or material deliveries are later than the construction required dates. The overall schedule may then be compressed or elongated to eliminate float or to accommodate overlap. Usually, the second pass will achieve a reasonable fit and produce a realistic balanced schedule.

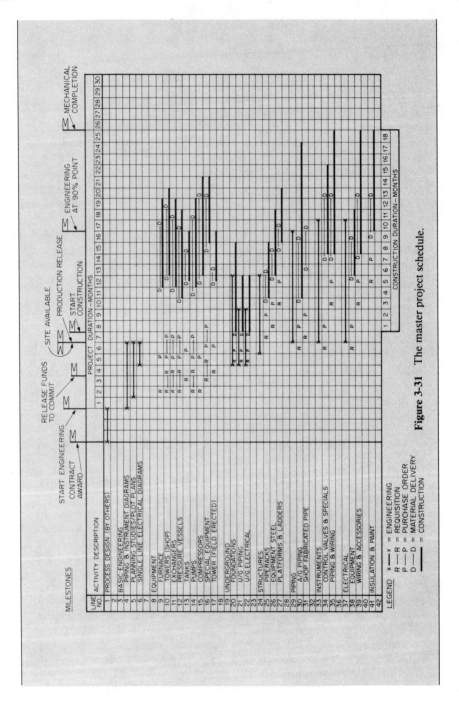

Figure 3-31 The master project schedule.

Fig. 3-31 illustrates a typical master project schedule which integrates and balances the three separate project schedules of engineering (Fig. 3-30), procurement (Figs. 3-25 amd 3-26), and construction (Fig. 3-29).

Comparative Analysis. One final check is to look at the relationship between the progress curves generated from the master project schedule for engineering, material delivery, and construction. The relative positions and lead times between these curves must always fall within certain limits for all projects. If the progress curves for the project in question differ to a marked degree from the typical curves, then something is seriously wrong and we do not have a satisfactory preliminary master project schedule. It will be necessary to take another look at the points where the project progress curves deviate from the typical progress curves. Fig. 3-32 illustrates the shape and relationship between the engineering, material delivery, and construction curves for a typical project.

Make Adjustments. Finally, the "realistic" schedule produced by this method (Fig. 3-31) is not necessarily the minimum schedule. If, after this exercise, the durations produced are unacceptable, then compress the overall duration to the desired point. Look at the overlaps to determine what positive steps must be taken to shorten what otherwise would be considered a normal duration. Extra priorities in engineering to make earlier decisions? More pressure in procurement for shorter cycle times and accelerated deliveries? More manpower peaking in the field? These are all options that can be considered.

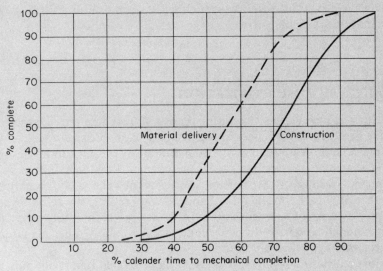

Figure 3-32 Typical project progress curves.

The schedule format will show up "pinch" points at an early date so that schedule critical areas can be identified. This means that specific action plans and priorities can be directed to these areas from the project initiation if minimum schedule is an overriding priority.

Realistic Milestones. A master project schedule developed by this "top-down" method provides realistic milestones. These can subsequently be used for the development of detailed schedules and networks within each of the engineering, procurement, and construction disciplines.

Another new development in the scheduling and control of projects is the use of microcomputers for this task. A variety of project control programs are now available for scheduling and controlling engineering projects of many types. Applications programs, such as "The Harvard Project Manager," "Milestone," "LisaProject," and others, shorten the time for project scheduling while reducing the labor input. With the wider acceptance of microcomputers by consulting engineers in all fields, these scheduling programs will find ever greater use. And—with computer-aided design also available on these small computers—the consultant's office of the future will be much different from what it is today. See later sections of this Handbook for more information on computer-aided design (CAD).

Now we'll examine some ways such reports can be utilized in a system of management by objective.

ANALYZING TIME UTILIZATION

The various managerial accounting reports shown here provide the consulting firm with several possible approaches to analyzing employees' use of time. The project progress report and monthly output summaries represent a general perspective on the performance of the staff and team members working on a particular engagement. The daily, weekly, or monthly time card, on the other hand, directly reflects the activities and the productivity of each employee. Incidentally, the monthly time card has the added advantage of being comparable to many other managerial reports, most of which are monthly.

Some consulting organizations use time cards that report not only the amount of time that a staff member has worked on each project but also the amount of time spent in meetings, instruction, and other so-called nonproductive effort. Vacation time, breaks, sick leave, and the like may also be covered. Accurate reporting of such items can help managers to pinpoint the sources of delays and cost overruns on virtually any job. While a certain amount of nonproductive effort is included in every employee's workday, it is management's duty to control these factors. Further, the examination of time utilization on a continuing basis builds a foundation for making time and budget projections on future engagements.

TIME PROJECTIONS

Reliable time projections help simplify the general task of management. Employees tend to work more efficiently when their superiors set forth specific deadlines; otherwise, effort tends to be spread out and productivity shrinks. For most consulting firms, records of earlier projects are considered to be the only valid source of information for making time projections and relevant deadlines.

For example, the time cards, progress reports, and output summary of the electrical engineering phase of a 50,000-ft^2 (4645-m^2) elementary school project might show that such a job requires approximately 100 workhours to complete. This would be a reasonable time projection for another school project of comparable size. Of course, other variables should be taken into account in extrapolating figures from one job to another.

One of the only alternatives to using time cards and managerial records for time projections is to base the projections on the total consulting fee. Thus, if the total consulting fee for a project is $10,000 and the net fee available from the total for direct labor is about $3600, the value of the direct labor can be divided by the hourly payroll cost per employee to give a time projection. For example, if the average hourly payroll cost is about $9, then there are about 400 workhours ($3600 ÷ $9) that can be allocated if all other times such as overhead and profit are to be maintained at current levels.

This method of time projection can be used effectively to double-check estimates based on managerial reports and vice versa. Used together, the two techniques can help management to develop production standards for specific types of projects and tasks; these standards can then be used as part of a systematic approach to setting budget requirements and improving the overall productivity of the firm.

DEFINING THE "NORMAL" OUTPUT

With experience, the managers of a consulting firm can get a fairly good idea of what sort of output is "normal" for employees in different positions within the organization. The data supplied by managerial accounting is invaluable in this regard.

So many variables exist in the operation of a successful consulting engineering practice that it is almost impossible to define the normal output of a designer, drafter, or engineer in a way that would be applicable to the profession in general. The managerial and financial reports of the individual organization, however, can show you whether your employees are producing the work at a profitable rate. In the final analysis, the normal output of any firm is whatever output allows the business to enjoy a profit while its principals and staff members earn a reasonable salary.

INSPECTING DESIGNERS' WORK

Nearly every successful business has procedures for inspecting (or auditing) its financial and managerial reports to ensure accuracy and prevent misuse. Every consulting engineering firm should implement effective internal audit measures. Further, the engineer's role in work that can affect public safety demands that the consulting firm's designs be inspected carefully before they are used in any project. Insurance companies offering errors-and-omissions insurance often insist that a firm have a system of inspection procedures and suitable checklists before they will provide errors-and-omissions coverage.

This system requires the following steps:

1. The firm must develop appropriate lists of items to inspect on designs from each engineering discipline (i.e., checklists).

2. One or more staff members must be appointed to inspect each item on the checklist before any design is sent to individuals outside the firm or used outside of the firm's offices. These "checkers" should be fully capable of determining whether the design meets all relevant requirements.

3. In most cases, the work of the checkers should be reviewed by additional managers or consultants.

Generally, the amount of time devoted to inspecting the design on any given project should be proportional to the project's total value. The checkers' responsibility is large, but their efforts should not be allowed to cause the firm to exceed budget requirements or delay the proper completion of an engagement, except in cases where serious problems are discovered in the design.

Designs should be inspected three times, if possible: (1) once during the preliminary phase; (2) once in the basic drawing stage, when most of the design has been finalized and the concept requires details and expansion to be fully understandable to contractors who will use the drawings; and (3) once before the design leaves the office, to make certain that all specifications, budgetary needs, and other requirements have been met and that there are no errors.

The checker then signs the final set of drawings to indicate that they have been inspected. Usually, the title block (Figure 3-33) of engineer's drawings has a space for this purpose. Unfortunately, some engineers leave the checking task to architects or government representatives who are involved in the use of the design. So it is management's duty to see that the appropriate inspector reviews each design and signs the drawings before the firm sends them out.

Ideally, the design inspector does not work directly on developing the design itself. The person assigned to check a particular design should be someone who can evaluate that design with objectivity. Thus the design inspector might be a project manager or an engineer who is working on a separate project. It is best to rotate inspectors at least twice to avoid placing the burden of evaluation on any one individual for too long a time. Rotating checkers also helps to prevent friction between those doing the checking and those whose designs are being inspected.

Who checks the checkers? This responsibility usually falls into the hands of a company principal or engineering associate. Only in large consulting firms can some principals afford to be detached from project management and design. However, some small- and medium-sized firms appoint an associate to supervise design inspection and project management, thus enabling the principals to devote more effort toward expan-

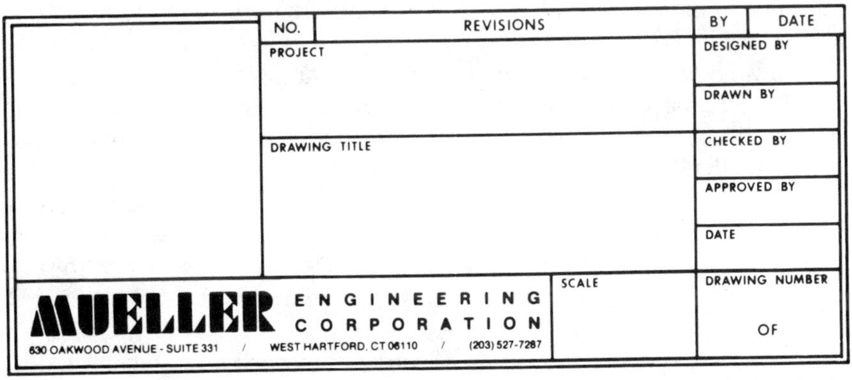

Figure 3-33 Typical consulting engineer's drawing title block showing spaces for the checker's and approver's initials.

sion and planning for the future. An associate who assumes this role works with the project managers and acts as design assistant, inspector, and perhaps even as a production assistant on certain engagements.

TWO KINDS OF CRITICISM

Nearly everyone taking part in the management of a consulting engineering business—a supervisor, inspector, project engineer, or other—uses criticism in managing the organization, at least on occasion. But criticism is a means to an end, not an end in itself. What effect it has on staff performance is determined largely by the form the criticism takes. Will it motivate employees to work more efficiently? Will it steer them away from errors made in the past? Or will it restrain or discourage staff members from putting their "all" into their work?

The effects—experts agree—stem from two different kinds of criticism: productive and nonproductive. The former tends to enhance an employee's ability to perform a task at the same time that it inspires the individual to excel in general. The latter tends to degrade the staff member, usually resulting in little or no improvement in performance and no inspiration to excel. Negative criticism, destructive management—call it what you will—does little, if anything, to benefit the consulting firm.

One way in which nonproductive criticism finds its way into management in a professional engineering practice is when a manager orders the staff to perform certain tasks, giving little explanation and expecting blind obedience by the employees. Engineers with many years of experience sometimes make this mistake because they forget that they are more familiar with certain tasks than are their employees. While such a management style may have some merits in a prison or military organization, it has little place in a consulting firm. Orders must be given calmly and logically so the entire staff is able to envision the project and the way in which specific tasks fit within it.

A similar cause of nonproductive criticism and wasted effort is the tendency that many engineers have to continually restudy a problem or design. It is not surprising that many managers do this, considering the emphasis that engineers are taught to place on detail and excellence. But it can be overdone. Commonly, excessive scrutiny by managers leads to a lackadaisical attitude among employees, who begin to feel that their efforts are not valued. By contrast, staff members who are trusted to produce work that needs only a brief but thorough inspection feel motivated to work harder and better.

It goes without saying that an effective manager knows when to stop restudying an employee's work, when to suggest changes, and when to commend a member for a job well done. These are techniques that most professional consultants learn from experience as they build their private practice into a successful engineering organization. Unfortunately, many beginning consultants find it harder to offer productive criticism than nonproductive comments. Once the firm becomes established, however, the wise consultant allows the staff to contribute ideas on the company's operations and in this way forges a cohesive, positive organizational structure.

INCENTIVES FOR EXCELLENCE

Strong motivational tools include salary increases, promotions in status, pay bonuses, and assorted fringe benefits (sometimes referred to as "perks"). Another valuable perk staff members appreciate is that of having some flexibility in work hours, when personal obligations require time off. Management should lay out a policy explaining the provisions for taking time off from work, as well as any requirements for compensating for this time. Loyal staff members are usually happy to work after normal working hours or at other special times to repay their firm for time taken off for personal reasons. Being allowed to do this is a valuable perk to most staff members.

Used properly, bonuses are an effective tool for rewarding good employees and encouraging a continued pursuit of excellence. But fluctuations in a company's financial circumstances sometimes make it unwise to dispense bonuses, even though they may be well deserved at the time. Special care should be taken to award bonuses in a judicious and equitable manner.

Perks, such as company credit cards, expense accounts, paid membership in professional societies, and automobile use are additional ways many businesses reward hardworking staff members. Further, your firm might issue occasional in-house bulletins to provide special recognition to employees for their achievements. Some very large companies even send news releases to local media representatives to publicize newsworthy promotions. And any manager trying to succeed will meet with employees from time to time and acknowledge the members' contributions to the group.

The manager, principal, or associate who strives to maintain personal communication with the staff generally commands respect from all members, whereas an aloof administrative team tends to have more difficulty managing an organization properly.

Still other forms of employee recognition include the so-called status symbols—a drafting or design table located near the office of a company principal, for instance, or a personal nameplate, personalized business cards, letterheads, or even a new drafting stool. The resourceful manager can think of numerous incentives to impress staff members with the esteem derived from the pursuit of excellence. But managers should exercise discretion in offering such rewards; to be beneficial, they must not be dispensed frivolously or in ways that would diminish the importance of other employees.

Professional discretion is vital in selecting staff members to participate in research projects, in meetings with clients and contractors, and in design inspections or equipment selection. Each of these duties carries a certain amount of prestige for the person assigned to it. They also carry some big responsibilities. Thus, delegating the responsibilities to employees unaccustomed to such tasks brings a degree of risk into the company's activities. Occasionally, such assignments may result in unexpected rewards for the consulting firm, however. For example, the appointment of a designer to a team engaged in field investigations may add a whole new perspective to the team's efforts; the designer may help the field investigators to unearth problems that had previously been overlooked because of the team's familiarity with its customary investigation routine.

All the organization's incentives and rewards should be monitored carefully to ensure consistency and prevent overuse. The importance of monitoring is particularly vivid with regard to such items as expense accounts and the use of company automobiles. Both can be used as effective incentives, especially for assignments requiring travel and entertainment of clients. Used to excess, however, they can drain the firm's finances to a dizzying level.

INSURANCE PLANS AND BENEFIT PROGRAMS

As noted earlier, it is part of the function of the consulting firm's administrators to provide staff members with an explanation of fringe benefit and insurance programs in the organization's written personnel policy. We shall now examine a few benefit programs in greater detail.

Workers' Compensation

Most employers are required by law to pay between 1 and 5 percent of each employee's salary to the state workers' compensation fund. In addition, the employer is responsible for a portion of the state's unemployment insurance. If an employee has reason to collect workers' "comp" or unemployment insurance—due to a debilitating work accident or layoff, for example—the costs of these items to the employer usually increase.

Blue Cross, Blue Shield

The rates for these health insurance premiums generally amount to between 4 and 8 percent of an employee's annual salary. Many consulting engineering firms nowadays pay the entire premium as part of their benefit package, although some companies request that employees pay a portion of the total amount.

Major medical coverage is also provided in many firms. Generally, insurance companies require that the major medical policy be accompanied by life insurance with some form of payment in the event of normal or accidental death. The cost of this type of insurance varies, but the usual premium rate runs between 2 and 4 percent of the employee's salary. Many large corporations also include a dental plan. However, the majority of small consulting engineering firms do not offer a dental program.

Based on the preceding insurance plans alone, a consulting firm might pay 10 to 15 percent of its staff's salaries in the form of insurance benefits. Further, most engineering practices provide a limited number of paid vacation days and sick days, not to mention time off for doctors' appointments and the like.

While some consultants ponder the possibility of having an additional insurance policy to cover sick leave, most consider such a policy to be an unnecessary expense, except in cases of serious and prolonged illness. Thus, some firms do provide coverage for extreme conditions. Fortunately, such illnesses are rare; the cost of the policy can be minimized by providing coverage only for illnesses lasting more than 1 or 2 months.

The cost of all these items can be quite substantial. Providing 2 or 3 weeks of paid sick days would require the firm to allocate about 5 percent of the employee's annual pay for this purpose. When you add in the cost of the insurance premium for prolonged absence, the allocation rises to about 10 percent. The following table lists the various fringe benefit programs provided by many consulting firms and gives the approximate percentages of employee salary that you could expect to pay for each type of program.

Program	Percentage of Employee's Annual Salary
Workers' compensation	3
Blue Cross, Blue Shield	3
Major medical	4
Time off for all medical reasons	5
Sickness insurance	5
Time off for personal reasons	2
Total	22

Further, there are three more categories of benefits provided by some organizations: (1) retirement and pension plans, (2) bonuses or profit sharing, and (3) stock options. A retirement and pension plan could cost the firm between 1 and 5 percent of an employee's annual pay (for our purposes, we shall assume an average of about 3 percent). Bonuses, profit sharing, and stock options could cost an additional 5 or even 10 percent. Thus the total cost of the entire benefit package, including all items in the preceding table, comes to 35 percent. Add on 2 or 3 weeks of paid vacation time, and your package will cost approximately 40 percent of staff members' annual salaries.

This summary reveals some of the factors that influence the overhead costs of most consulting firms. Overhead costs may appear inordinately high to the uninitiated; but a brief reference to the preceding items should help put any discussion of overhead into the proper perspective.

SUBCONSULTANT ARRANGEMENTS WITH EMPLOYEES

Another type of program that some consultants view as an employee benefit is an arrangement for employees to change to a subconsultant status in certain situations. Conditions that might call for such an arrangement include extreme financial stress within the firm, a lack of suitable advancement opportunities for senior staff members, and the desire of an older employee to retire from daily in-house work.

Subconsultants do not collect a salary but usually work for a fixed fee acceptable to both the firm and the subconsultant. For example, if an electrical designer has been earning $20,000 a year and desires to advance to a higher-level position at an increased salary, the firm may choose to transfer the designer to a subconsultant role if it does not have the financial resources or capacity to promote him or her to a higher, salaried position.

For most employees, the change in status to subconsultant is an opportunity to increase their income while gaining greater independence. Thus, the electrical designer might charge a fee of $1000 to do part of the firm's electrical work, perform it in 2 weeks, and average $500 a week, or $26,000 each year. At the same time, both the subconsultant and the engineering firm could be bringing in new projects. Similarly, a retired employee might be retained on a subconsultant or part-time basis to the mutual benefit of the firm and the individual. Such arrangements can give rise to a variety of complications, however. They should therefore be approached cautiously.

The guidance of attorneys and accountants skilled in subconsultant arrangements can be an invaluable aid. These professionals should also be consulted if the firm incorporates and considers providing an investment program for employees to buy company stock or "earn" stock in lieu of salary increases.

THE VALUE OF CONTINUING EDUCATION

A major goal of most consulting engineering firms is the development of a large group of consultants possessing professional engineering registrations. To achieve this goal, management must motivate staff members to continue their engineering education. An education program can include in-house courses for drafters, designers, and individuals who hold college engineering degrees but who have not yet secured their professional licenses. The courses can be taught by experienced professionals within the firm or by instructors from nearby colleges and universities.

The consulting firm usually pays part or all of the cost of continuing education for staff members. While there is only a limited amount of home study available to engineers, the firm might pay the tuition fees for employees attending night courses at local universities. Employees who are seriously interested in completing their bachelor degree work view such programs as a strong performance incentive. Others should be strongly encouraged to undertake continuing education. To prevent abuse, some firms pay only for courses necessary to the attainment of a professional engineering license. Frequently, the employee is required to maintain her or his grades above a certain level. Eventually, a continuing-education program will benefit both the consulting group and the individual member. Not only will it enhance the organization's capability and status, but it will also help the firm to keep abreast of the newest advances throughout the engineering disciplines.

When an employee obtains his or her professional registration, the license should be framed (at the company's expense) and displayed prominently in the firm's office, preferably in a place where all members' licenses can be seen by visitors. A collection of licenses, plaques, and certificates of membership in professional organizations is a mark of prestige that consultants should not underrate.

Likewise, employees should also be encouraged to participate in activities of various professional societies and associations. Many consulting businesses allow staff members to devote several working days to these activities each year. As an added incentive, some firms provide cash awards to employees for the attainment of a professional certificate, license, membership, or publication of an article in an engineering journal.

Another type of continuing-education opportunity is the professional research assignment. The assignment might derive from a contract acquired by the firm; or it could involve an independent project, requiring the employee to take a leave of absence from her or his regular job. Participation in such projects generally enhances staff members' knowledge of recent developments in their field, while it increases the firm's overall standing.

Whenever possible, research assignments should be acquired by the consulting firm first, then offered to selected employees. Of course, the conditions of employment and how the employees will be affected by the research assignment should be explained in detail before such arrangements are made. Managers must be prepared to dispense research assignments without disrupting the overall operation of the consulting business. In particular, management must be aware that an employee's successful completion of special assignments and educational programs may be accompanied by requests for considerable increases in salary. All these factors should be taken into account when considering participation in research projects, continuing-education, or other programs for enhancing employees' professional status.

JOB SECURITY

One of the fundamental incentives for good performance among employees is the sense that their jobs provide protection against financial hardship. Rare indeed is the staff member who does not strive toward maximum job security, particularly during periods of widespread economic uncertainty. However, most employees feel disinclined to put much effort into jobs that lack a stable future.

The private practice of consulting engineering is subject to a wide array of market influences and economic fluctuations. One month a firm might have several projects that generally require electrical engineering, for example, while a few months later the firm may have only mechanical consulting work. At another time, the efforts of consultants from all the major engineering disciplines may be required for one major project throughout an entire summer.

The nature of the consulting engineering business thus presents management with an interesting question: How to provide job security in the context of an ever-changing market? The answer is by no means simple. In fact, there may be no real solution since the consultant cannot control the nation's economy. However, there are several steps that management can take to instill employees with the confidence that their efforts are all worthwhile.

Aside from satisfying clients' expectations and conducting a successful marketing program, managers must constantly "put their best foot forward" in relations with employees. The principals of the firm could seriously lower staff morale if they were to reveal every problem, every loss of a contract, or every work shortage to employees. Yet it is management's responsibility to inform the staff of the organization's objectives and to explain its activities aimed at meeting those objectives.

Staff members, for their part, must understand that their job security depends primarily on the quality of their performance. Management should not imply that the stability of the consulting firm hinges solely on securing a prospective engagement. While a little worry can sometimes motivate people to work harder or faster, managers can usually use a positive attitude more effectively than a negative one.

When the bulk of the firm's work is temporarily in the electrical field, for instance, management should explain that the positions of staff members in other disciplines will not be permanently affected. Engineers with mechanical or civil expertise might be directed to assist the electrical engineers for the duration of the work; alternatively, they could be assigned to marketing tasks, research, or some type of educational program. In this way, management protects the business and its employees on all sides—the firm bends to meet the requirements of the current market and it "reinvests" the talents and expertise of the staff, distributing responsibilities according to the organization's changing needs.

Principals and associates should meet with staff for about an hour each month to discuss the overall activities of the firm, what sort of progress has been made toward specific objectives, and plans for the future. Changes in disciplinary rules or organizational structure should also be explained during these meetings.

Proper communication from management will effectively negate misunderstandings that commonly arise as a firm grows. Without this communication, a seemingly minor change in procedure, such as appointing someone new to sign the "checked by" position on the title block of a design, may have disruptive repercussions. Employees with high levels of seniority, particularly, deserve to be kept informed of changes affecting their jobs.

SUMMARY

The managers of a consulting engineering practice are responsible for implementing a wide array of measures to ensure the firm's prosperity. Among management's chief duties are: staffing the organization, controlling the work, and directing employees to provide high-quality consulting services as efficiently and effectively as possible.

Specific efforts toward fulfilling these responsibilities include: finding, screening, and training competent staff members for a variety of positions; developing a comprehensive personnel policy; providing general supervision and management suited to individual projects; and defining each member's tasks as they relate to the consulting engineering operations. Further, managers generally use selected managerial accounting reports to help the firm achieve planned financial and professional objectives. General objectives may include increased output and efficiency with respect to particular consulting services.

Also, management must provide a system of incentives and rewards to motivate employees to do exemplary work. To remain satisfied with their positions in the firm, engineering and other personnel need more than just a good salary. They need to be presented with interesting challenges, and they must have a sense of security in their jobs. In this context, it is vital that the firm offer an appropriate system of employee benefits, combined with some form of continuing education.

Managers must constantly be aware of the expectations of the consulting firm's clients in order to direct personnel accordingly. If a client's opinion on the best solution to his or her problem differs from the opinion of a staff member, the employee should not be permitted to argue with the client but should satisfy the client's needs in accordance with sound engineering principles. While the consultant may be allowed some discretion with regard to engineering solutions, the client has the freedom to choose among various options as long as they will not ruin the effectiveness of the final result.

Generally, the client expects to deal with one or two people as her or his source of information. The source may include management personnel. However, the source should be involved directly in the work being done for the client. It is imperative that managers communicate with the individuals involved in each engagement and that they coordinate all information necessary to completing the project on time. Maintaining work on schedule and meeting budget requirements are probably the two most basic measures of the effectiveness of any consulting engineering organization.

Earning Maximum Profit from Your Consulting Practice

So far we have examined several of the tasks involved in opening a consulting engineering office: (1) finding a suitable location, (2) selecting adequate space, and (3) installing the furniture and equipment needed to build and maintain a successful private practice. We also discussed techniques for staffing the firm, maximizing sales, and managing employees. No handbook of professional engineering would be complete, however, were it not to focus on the details of these topics as they relate to the practical aspects of daily office operation. We shall now turn to these details.

OFFICE ORGANIZATION

The way you organize and manage your firm's offices is determined partly by the types of consulting work you do and partly by your own personal preferences or "style." If your firm engages chiefly in design and supervision of construction, for example, its needs will vary considerably from those of a firm engaging in specialized research studies or forensic engineering. Likewise, your idea of the proper use of managerial accounting reports or other office systems may be considered uniquely innovative—or clumsy and ineffective—by someone else.

Despite professional differences, there is a common denominator that seems to distinguish the more effective consulting managers from less effective ones. Successful consultants keep themselves involved in the office projects and oversee their firms' progress toward meeting specific objectives on each engagement that is acquired.

While the atmosphere in the office of a successful consulting engineering firm may be hectic, it still is organized. By contrast one frequently finds the employees of less successful firms commenting, "Nobody knows what's really going on around here."

As discussed in Section 3, managerial accounting reports comprise one of the basic "tools" of the effective manager. Further, the consulting engineer must know how to supervise the four types of personnel who produce the engineering work: (1) drafters, (2) designers, (3) engineers, and (4) project managers. The responsibilities of individuals in these four positions may vary substantially from one firm to another, and each title may have several subdivisions or levels. Yet most small consulting engineering organizations categorize about 20 percent of their staff members (not including clerical personnel) as engineers. The remainder are usually designers and drafters. Thus the basic corps of personnel that is directly involved in producing the engineering work consists largely of drafters and designers.

To manage the organization properly, the consultant must attempt to balance the supervision these employees require with the freedom they need to do their best work. The consultant must have a system for presenting each engineering problem and its proposed solution to the design and drafting personnel and must be able to utilize their input to tailor the result to the client's desired solution. This means that the designers and drafters must be allowed to contribute actively to the overall consulting efforts of the firm.

It is essential for the manager to establish a rapport with staff members as well as an atmosphere of mutual respect. Probably the only way to achieve this is for the consultant to be involved in the day-to-day activities of the staff and to make his or her responsibility known to all personnel. Above all, the consultant should be able to answer questions that might arise dealing with one or more engineering disciplines as they relate to the design solutions for individual projects.

To be fully responsible for a project requiring more than one discipline, the consultant must have mastered skills in each of the relevant fields. Alternatively, the work must be managed by another member or principal skilled in the appropriate engineering disciplines.

Unfortunately, some consulting firms engage in projects involving disciplines for which none of the managers or principals are properly trained or licensed. Those who do so run the risk of being held legally responsible for work for which they are not professionally qualified. If a serious error develops in a particular job, it no longer matters who commited the actual error. The consultant whose stamp is placed on the drawing is responsible regardless of the original source of the problem. Thus it is mandatory that the consultant staff her or his firm with individuals possessing the appropriate licenses and skills in all the disciplines in which the company will practice.

Further, the consultant must seek out and hire highly skilled, well-organized individuals to perform the routine office chores—including typing, filing, shorthand, bookkeeping, and the like. The importance of spending the necessary amount of money to screen and employ the people best suited to these tasks should not be underestimated. The consultant must focus an even greater amount of care on the selection of outside professional assistants such as cost estimators, shop-drawing experts, specification writers, accountants, attorneys, and others. These people should be selected and retained in the most cost-effective manner possible.

Research shows that principals and project managers function most effectively while directly supervising five or fewer employees. In cases where a firm has a substantially greater number of employees than its principal or principals can directly manage, one or more of the most knowledgeable and reliable staff members should be put in charge of particular projects and/or particular activities of the firm. This form of management creates what is in effect a combination of a line-and-staff organization and a project-by-project organization. The "line" portion of the firm is the principal, while the staff would be the key employees who are capable of working relatively independently to supervise the rest of the organization. Key employees serve as "partners in charge" of certain projects, and the principal continues to control their work, plus other projects.

The principal's chief tool in managing the key employees is the managerial accounting system discussed in Section 3. This system forms the vital link between the principal and the projects not directly under his or her supervision, while affording the key employees independence and enabling the principals to manage their other projects. The key employees—whether they hold the title "Project Manager" or something else—should be required to provide weekly reports on the progress of their work. In particular, careful attention must be paid to time schedules and budgets so the principal can make certain that all projects are completed on time and within budgetary requirements.

Some firms post charts in the drafting room or other parts of the office showing the percent of project completion and the amount of time spent on each project, versus budgeted employee-hours. This has the beneficial effect of keeping personnel up to date on the status of individual projects. However, some experts feel the positive value of the charts is overshadowed by the negative reactions from those employees who see the charts as a heavy-handed management attempt to dictate schedules or force competition. Indeed, few if any employees want to feel they are working according to a "racing form" or sweatshop rule.

Nonetheless, the principal must constantly analyze the number of projects and tasks planned for completion in the immediate future and must determine the total number of employee-hours the work will require as well as the total number of hours available from staff. Such analyses should be conducted at least once a month. As mentioned, daily and weekly time and progress reports of ongoing work are also essential for the successful management of the private engineering practice.

LEADERSHIP POSITIONS

One of the advantages of delegating some management responsibilities to a few key employees is that doing so helps prepare certain staff members for leadership positions. As the firm grows, the management skills learned by these key individuals will be highly valuable.

Basically, leadership positions are developed by giving selected employees increasing amounts of responsibility. The principal must supervise the employees as they assume higher levels of responsibility to provide direction and prevent errors during the training phase.

Those selected for leadership training must have certain qualities of personality if they are to become effective managers. The key employees should be able to get along well with fellow staff members, must be able to produce their own engineering work efficiently, and must gain the respect of those whom they may someday supervise. The leadership "trainees" should also be free of personal habits or mannerisms that may interfere with the smooth functioning of the organization.

SPECIAL ASSISTANCE

Certain managerial responsibilities should not be fulfilled by key employees in leadership positions. Vital tasks such as record keeping, billing, and payroll dispensing must be handled by the consulting firm's principals or by qualified specialists.

The importance of such efforts is sometimes underestimated by the consulting engineer, who often sees the tasks as so routine as to be unworthy of special attention. Actually, the way these "routine" responsibilities are fulfilled may determine the success or failure of the business. Deficiencies in any one of these areas may seriously undermine the efficiency of the organization.

Unfortunately, many engineering firms do lack personnel with expert capabilities in the so-called routine areas. The principals of these firms find themselves struggling to avoid being overwhelmed, especially during periods of greater work loads. Thus they might decide to seek additional partners or principals to help with the day-to-day managerial functions. But the work-load fluctuations inherent in the consulting business often make this solution unsatisfactory.

Large engineering firms can usually afford to employ full-time specialists—for specification handling, for cost estimating, and for many other tasks. Since most small firms of 15 or fewer members find this an excessively costly option, the best solution for many is to hire certain specialists on a temporary or part-time basis. Before deciding to do this, the principal should analyze the time and costs incurred by doing the work within the firm and compare them to the cost of hiring outside experts. If the part-time specialists can perform the same work at a rate below that of the usual selling rate for the principal's services, the hiring of the outsider could save the firm a lot of unnecessary expense. Otherwise the principal may have to reexamine the tasks and attempt to delegate them to a partner or key employee.

The control of specifications, in particular, must be constantly maintained by the firm's management since some manufacturers and suppliers may seek to influence the firm's specification writing for their own benefit. Most manufacturers and suppliers would like to see their products specified on almost all engineering projects. Their representatives consequently offer consultants all sorts of arrangements to this end, some legitimate and some illegitimate. Regardless of the nature of the deal offered, however, the consulting firm's specifications are crucial to the success of its engineering solutions. Any tampering with the specification writing may pose a risk to the success of both the firm and its work. Therefore, it is advisable to seek outside professional assistance—especially for the firm's first several projects—and avoid excessive commercial influence.

Less risky but equally time-consuming for the consulting manager is the task of signing checks—employee payroll checks or payment for yesterday's business lunch. In most small firms, the principal signs most of the checks in the midst of many other activities. Studies show, however, that a substantial portion of the manager's week—as much as one-half to two-thirds—is spent on this rather routine chore. Certainly every firm must maintain control over its payments and check signing; but most principals could make more profitable use of their time were they freed of the check writing chore.

Several options are open to the manager who wants to reduce his or her check signing burdens. Payroll can be done by computer, for instance, and you can have a bank-approved signature stamp made. Also, there are a variety of computer services available, some of which will handle all bookkeeping, calculation, and check distribution tasks you and your staff usually perform. Generally, the costs of such services are negligible when compared to the amount of valuable time lost by doing the work yourself. Of course, every consulting firm should also retain a qualified accountant to help manage its record-keeping systems and ensure the most cost-effective use of all financial procedures.

One aspect of office management that usually does not require outside assistance but which some small firms perform inadequately, is the collection of fees from clients. Small firms tend to send out their bills haphazardly. Further, some lack a system for collecting overdue payments.

Frequently the underlying reason for such poor billing procedures is the consultants' fear of offending their precious few clients. But no business can succeed without a proper system of billing and collection. Fortunately, it is not difficult to install an effective system.

One secretary or other member of the firm can easily be instructed to send out bills and payment requests according to a standard schedule. Generally, bills are mailed monthly. All bills should accurately indicate the amount owed and the date payment is due. Additional notices should be mailed monthly (or more frequently) to request overdue items. Clients who continue to avoid payment for more than 6 months should be notified that they will be referred to a collection agency and/or an attorney if problems persist.

Good collection procedures are likely to reduce the need for loans to help finance your business. Effective procedures can also help to increase your financial standing with your bank, thus making it easier to obtain monetary assistance when necessary.

Overdue bills add an unnecessary burden to the consulting firm since most banks will not accept bills as collateral when the payments are more than 3 months overdue. Further, the overdue bills themselves may force you to seek loans to continue operating, possibly causing your firm to lose as much as 25 percent of its fees to interest payments and other added expenses. Clearly, training your staff in the proper techniques of soliciting regular and overdue payments is an effort that is well worthwhile.

OFFICE MANAGEMENT

As noted, the organization and management of your consulting practice will be influenced by a mixture of professional requirements and personal tastes. The arrangement of your office and equipment will be determined by similar factors. Yet most successful consultants find that a few basic considerations guide decisions in office arrangement. For example, the arrangement of tables and other equipment for the design staff must be for maximum efficiency and productivity. Thus designers should be located where they can be supervised by management and where they can conveniently communicate with the firm's principals.

Each staff member must have an adequate working space. Since the design drawing is an essential part of the language of engineering, nearly all personnel involved with any phase of design drawings should have a drafting table and a "throw" space for handling their work. This applies to drafters, designers, plus most engineers and project managers. The following are reasonable space allocations for individuals in these positions:

1. Drafters: 70 ft^2 (6.5 m^2)
2. Designers: 100 ft^2 (9.3 m^2)
3. Engineers: 120 ft^2 (11.1 m^2)
4. Project managers: 120 ft^2 (11.1 m^2)

Drafters should have at least as much space for laying down their drawings (throw space) as their drafting tables occupy. Designers, however, need more: their throw space should be double or triple the amount of space occupied by their drawing boards.

Secretaries or other clerical personnel also need to have a permanent area of the office that they may utilize—about 75 ft^2 (7.0 m^2).

In addition, several other office areas must be considered in the overall plant. These include private offices for the firm's principals [150 to 250 ft^2 (4.6 to 18.6 m^2)]; plan areas with enough room for a blueprint machine [about 120 ft^2 (11.1 m^2) or more].

Again, the focus of good office design is optimum efficiency. Potential distractions should be minimized whenever possible. Generally this can be accomplished by providing barriers to visual disruptions and curbing conversations between employees. Too many consulting firms fail to recognize the value of privacy and instead set up their drafting area as if it were a sea of tables. Many large organizations use such a setup in an attempt to save space, reasoning that the average drafter needs nothing more than a drafting board, throw space, and stool to do the firm's work. It is, in fact, possible to fit a drafter in a space of 50 ft^2 (4.6 m^2) or less.

But as soon as somebody walks into such a room, the staff's concentration is broken. Further, staff members will be tempted to waste time talking, especially if they are seated close together. Some firms solve these problems by installing movable partitions or other visual restrictions between desks. Partitions that are 5 ft (1.5 m) high, or higher, can significantly increase work efficiency by improving employees' ability to concentrate on the work at hand.

The chief guideline for all aspects of office arrangement is utility. Your office design should be cost effective and consistently conducive to high-quality, efficient work. Aesthetics also play a role in office design, but some firms tend to overemphasize looks to impress visitors. Pursued to extremes, the emphasis on aesthetics reduces cost effectiveness. Few clients actually walk into the consulting engineer's office, and the few who do visit do not expect the engineering office to look like the executive suite of some huge corporation. By emphasizing utility first, you will be better able to impress clients with your excellent services. Then there will be little need for expensive furnishings.

Other items to consider in the office arrangement include files, bookshelves, and special equipment such as copying machines and computers. Let's examine these items to note how the consulting engineer can utilize them for maximum effectiveness.

FILES

Every consulting engineering firm must have an organized filing system to properly control and record its activities. The basic filing system of most firms consists of:

1. A correspondence file
2. A billing and contract file
3. An original tracing file
4. A plan and specification file, including contracts
5. A blueprint file
6. Some form of work-in-progress filing system (may be merely a space where the drawings can be placed for easy reference)
7. A catalog file or shelves for arranging manufacturers' and suppliers' catalogs—either in alphabetical order, by discipline, or by type of product

The ordering of most of these files is determined by their content. For instance, correspondence files are usually filed alphabetically, with the most recent letters near the front of the binder or file drawer. The same principles apply to billing files, although some firms prefer to order these alphabetically by name of client.

BUILDING YOUR OWN SPECIFICATIONS FILE

While most consulting firms produce somewhere between 10 and 20 sets of specifications each year, the task of organizing specifications is not always simple. Over a few years, the average organization accumulates vast amounts of specification information. Properly filed, the data can become a valuable reference source for details concerning past, current, and future projects.

The standard data for a particular project or type of project should be typed single-spaced on 8½- by 11-in pages. Figure 4-1 shows such a page, this one for a boiler. By recording data from a series of projects involving boilers, an engineer could develop his or her own standards for different boilers applicable in various situations. For example, there currently are at least four or five good-quality boilers available—made of cast iron or steel and fired by oil, gas, even coal—whose characteristics are fairly common. Each is suited to particular types of projects.

How does one keep track of all the details of the different kinds of boilers, not to mention the plethora of the other heating, air conditioning, ventilation, and electrical equipment? The standard specification file should contain an index listing every item that is put in any specification. The items and their corresponding descriptions can then be listed alphabetically and arranged accordingly in the file by type of equipment, material, construction process, and so on. For example, under "Heating" you might list: boilers, burners, expansion tanks, etc. Under "Electrical," lighting fixtures, panel-boards, power distribution equipment, service equipment, etc.

The file itself can consist of a series of ring binders or file folders containing the master copies of standard specifications. Thus arranged, these prized bits of information may soon fill two or three drawers of a file cabinet. Incidentally, some consultants create an index listing the different specifications for each project in the chronological order in which the specs were originally developed. This system has the added advantage of helping staff to "walk through" the sequence of events on each project— beginning, for example, with the source of power or heat and continuing to the termination of heat, air conditioning, ventilation, exhaust, or other requirements.

The engineering profession has developed all sorts of standard specifications for general building construction and architecture that can be added to your firm's spec file. However, few valuable specifications are available covering mechanical and electrical phases of construction. Several books—including *The Construction Specification Handbook,* Second Edition, by Hans W. Meier, FCSI, published by Prentice-Hall, Inc.—can assist the small firm in building its specification file. But even these books

1. BOILERS: (Cast Iron)

A. Furnish and install where shown _____ ( ) cast iron low pressure (hot water) (steam) boilers designed in accordance with ASME standards for (oil) (gas) fired (hot water) (steam) heating. Boilers have a net IBR rating of _____ square feet of (170 degree hot water) (steam) radiation each with standard rear smoke outlet and designed for front firing.

B. Boilers shall be equal to H. B. Smith #_____ and shall be complete with all operating and safety controls and shall include one flue brush and one scraper.

C. Concrete supporting base will be installed under another division of the specification. The front steel plate for mounting the burner will be furnished and installed by this contractor. Anchor bolts will be furnished by this contractor, but set in the base as outlined under another division of the specification.

D. Boilers shall (each) be furnished with a smoke hood with an induced draft fan mounted at rear of boiler. Smoke hoods and ID fans shall be furnished by the boiler manufacturer as a package part of the boilers. See paragraph #_____ this division for ID fan specification. Smoke hood shall be furnished with shutter type motor-operated dampers furnished by the boiler manufacturer and installed in each boiler outlet elbow on the suction side of the induced draft fan. Damper motor and linkage shall be interlocked with the oil burner controls and shall be automatically operated in conjunction with the ID fans. All control wiring shall be furnished and installed by the oil burner contractor.

2. BOILERS: (Steel)

A. Furnish and install where shown on the plans _____ ( ) forced draft boiler-burner units of the fire tube scotch marine type with rotary atomizing direct drive, modulating type (oil) (gas) burner. Each unit shall be completely integrated, factory wired, factory insulated with jacket, and mounted on skids.

B. Each unit shall be fitted with a smoke box, side opening tube access doors, rear combustion chamber including complete refractory lining fitted with observation port and access door, saddles with extended base for mounting (oil) (gas) burner assembly, (water trimmings for low pressure, boiler altitude gauge, thermometer,) (steam trimmings consisting of steam gauge with siphon and cock, M, M # 191 combination water column, pump control and M, M #91 low-water cut-off and required number of (water) (steam) relief valves to meet ASME code requirements and as recommended by the Hartford Steam Boiler Company.

C. Each boiler shall be provided with all standard boiler openings with _____ in outlet (with dip tube and vent tap), _____ in return inlet (with internal flow baffle) and with 2 in drain.

Figure 4-1 Typical specification page for boilers.

generally do not focus on the mechanical and electrical areas, which can easily comprise 40 percent or more of a project's total construction cost.

Thus, for most engineers, writing mechanical and electrical specifications becomes a task of borrowing publicly available data from other engineers, then refining and redefining their data to suit their projects' specific requirements.

A variety of specifications can be obtained from materials released by the Army Corps of Engineers, the Navy Facilities Engineering Command, and the Veterans' Administration. Many are written from a "materials and performance" standpoint. The only disadvantage in using these specifications is that they are often extremely general and may allow the use of products of lesser quality than your firm desires. The specifications must therefore be molded to the work required for your project. At the very least, the specifications can help you describe different grades of installation—from quick, low-cost designs to work involving the "Cadillac" of materials.

Specifications may sometimes be provided by manufacturers and their agents. However, the engineer must be cautious to avoid using them "as is" since the manufacturer's shop drawings and descriptions are generally designed to favor their own products.

The many sources of available data will certainly enable you to develop a series of standard specifications for your firm's files. As long as the specifications are carefully researched, tailored to your firm's needs, and regularly updated, you should be able to use them effectively. Staff members must, of course, be instructed in the use of the specification files. Usually, no deviations are allowed to appear on projects leaving the office unless prior approval has been obtained from the principals or others in charge of specification writing.

Today many large consulting engineering firms are storing their standard specifications on word processors. When a specification is to be prepared for a new job, suitable changes are entered in the word processor. Then the entire specification is printed by the word processor, including the changes made for the particular job. Smaller consulting engineering firms are also finding that a word processor will save both time and money in specification preparation.

FILING PLANS DURING DESIGN AND AFTER

Consultants sometimes forget that a drawing being prepared by even their lowest-paid drafter may represent hundreds or even thousands of dollars' worth of work. Hence, something must be done to protect the firm's drawings from damage or loss. Yet the frequent use of plans during design makes filing them a repetitive and sometimes annoying chore. You cannot put away every tracing each night and take it out the next morning without spending a considerable portion of time on effort alone.

The alternative to tedious filing procedures and risky nonfiling is to store plans in flat files 40 to 50 in (102 to 127 cm) long by 30 to 40 in (75 to 102 cm) deep. These files can be used as temporary storage during design, with interchangeable nameplates noting the contents of each drawer. Since even a small firm may have six or seven different projects in the active design state at one time, it is logical to secure at least 10 file drawers, to contain both active project plans and original tracings for engagements recently or partially completed.

Original design data, including partial prints and notes marked by designers, staff engineers, outside consultants, public officials, and others, can be a valuable part of the consulting firm's permanent records. If a project develops any kind of problems, these notes and scribblings may save the firm a lot of trouble when it needs to locate background data.

However, original plan data is hard to store in regular file cabinets due to its bulkiness. Some consultants set aside an area in the office for storing the materials in cardboard tubes. Most of the tubes used for this purpose are 3 ft (1 m) long and about 4 in (10 cm) in diameter. Each tube is labeled to indicate its contents.

Shop drawings and approval materials should also be filed with the original notes to allow convenient access to details on equipment, approvals, and other information—before, during, and after project completion. Shop drawings may also be filed chronologically by discipline.

While many organizations keep their original tracings in flat file drawers, sooner or later most consultants reach a point where their permanent plan file takes up large amounts of space. Some choose to place the materials on microfilm. But microfilming has its own drawbacks: It costs more, and if you ever need to make additional prints from the film, the process of reducing and enlarging the material may produce a less accurate print. In most cases, it is best to file the original documents and reprint them directly when needed.

To prevent loss or damage of documents, the consulting engineer must maintain every filing system in proper order. Materials for a single project should be removed from a file as a whole, not separately. Inform your staff about your rules pertaining to removal and replacement of project materials. Otherwise chaos may result, with current projects getting mixed up with completed projects, drawings being lost, etc.

Management must be vigilant and make sure that file drawers are not overloaded to the point where documents may tear when removed. Generally, file drawers should not be more than half full since packing down the tracings invites damage.

Consulting engineers have tried all sorts of systems to control their files, including issuing "library" cards to users. Still the simplest and perhaps most effective method is to allow only certain personnel to remove and replace original documents. It is also good to appoint one or more staff members to review the files at least twice a month to keep the papers from getting creased, torn, or disorganized.

STANDARD DESIGN DETAILS

Consulting engineers throughout the world know that the use of standardized details for repetitive designs helps to increase efficiency and reduce design errors. (Figure 4-2 shows several typical standard design details used in consulting firms.) Hardly anything is more frustrating than

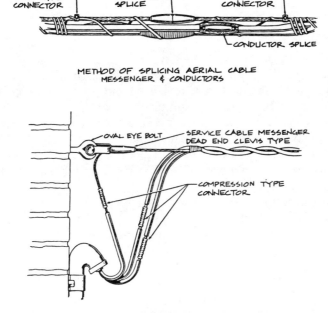

Figure 4-2 Typical standard details developed by a consulting firm. *(J. Mueller, Standard Mechanical and Electrical Details, McGraw-Hill, New York, 1980.)*

being unable to find a standard design detail when needed and having to waste 1 or 2 hours reconstructing it in a drawing. To ensure that standard details are conveniently available at all times, the firm should file all loose design papers carefully and have at least one copy of any book that applies to the kinds of standard design details the firm uses. Three such books, all published by McGraw-Hill, are Mueller, *Standard Mechanical and Electrical Details;* Newman, *Standard Structural Details for Building Construction;* and Callendar, *Time-Saver Standards for Architectural Design Data.*

Copies of the standard design details should be placed in your firm's engineering and drafting areas. If possible, a duplicate set should be kept in the principal's office as well.

Remember it is part of management's function to instruct the engineering staff on the use of various design details if the firm wishes to standardize certain design practices. Some firms insist that particular design details be used in every applicable drawing. Also, some consultants provide their staff with guidelines on the use of standard notes, legends, and symbols for the documents prepared for each job.

CATALOGS

Every engineering office needs to have complete, up-to-date detailed information on all the products that it may use in designs. The most common sources are manufacturers' catalogs and catalog supplements. Unfortunately, the beginning consulting firm sometimes has difficulty obtaining these publications because manufacturers may be reluctant to provide expensive catalogs to companies that are not certain to become regular customers. The new consultant may even have to pay for certain catalogs; but once the engineering firm becomes established, catalogs seem to pour in from every manufacturer.

A standard technique of acquiring catalogs for a new firm is to tell manufacturers' representatives that you will not be able to consider using their products until you have received the companies' publications. This approach is a valid one which will produce the desired catalogs on your desk.

The best general source of catalogs is *Sweet's Catalog File,* available from McGraw-Hill. It may be difficult to obtain since sales are restricted to firms which have performed particular types of projects and specified dollar amounts of work. Many architectural organizations and large engineering firms receive new copies of *Sweet's Catalog File* each year. A judicious inquiry may enable you to acquire older versions before a firm disposes of them.

While a 1-year-old set does not provide complete up-to-date information, it still represents a wealth of useful data, especially for the beginning consultant.

Most manufacturers will supply catalogs or other information to engineers who specify their products. All you need to do is call or write the company to request the data. The average practicing consultant finds it fairly easy to get all the information needed for various projects by combining a set of *Sweet's Catalog File* with data provided by the manufacturers and advice from other engineering professionals.

Another quick way for beginning engineering consultants to obtain the catalogs she or he seeks is by requesting them through professional magazine reader service cards. These cards regularly appear in most professional magazines (see Figure 4-3). All you need to do is circle the number of the catalog you want and send the card to the publication. *Consulting Engineer* magazine has an annual issue devoted exclusively just to catalogs. Using this issue alone should provide you with a large percentage of the catalogs you need.

Besides the catalogs themselves, you may benefit by maintaining a list of additional publications that the firm may someday need. Most professional organizations such as the American Society of Heating, Refrigeration and Air Conditioning Engineers (ASHRAE) will provide, upon request, information on various manufacturers and their products as well as where to send for catalogs applicable to the disciplines in which you practice.

Even if you do not have many catalogs when you organize your office, it is a good idea to allocate space for those you may eventually accumulate. The common catalog file consists of four shelves, each about 3 ft (1 m) wide. Ideally, the shelves should be centrally located so that all staff members can get to them without interrupting others. One staff member should not have to lean over another member's table to reach a catalog.

Many well-established consulting firms eventually require as many as 10 of the four-shelf units. To make effective use of space, some consultants position the units as aisle dividers. Doing so creates a barrier to visual distractions and eliminates the need for additional partitions at each work station. It also allows easy access to the books. Figure 4-4 shows one way of arranging catalog shelves.

There is no ideal system for organizing catalogs, but they are generally shelved according to type of product and usage. Arranging the books alphabetically by name of manufacturer leads to a haphazard mixing of catalogs on such diverse items as boilers with ones on lighting fixtures and sanitation equipment; such an arrangement should be avoided, if possible. Instead, you can develop an alphabetical card file for the manufacturers' names, cross-indexed by type of product, and coordinate the

Electronics
Reader Service

Electronics 1983 ARTICLE INDEX

If the cards below have already been used, you may obtain the needed information by writing directly to the manufacturer, or by sending your name and address, plus the Reader Service number and issue date, to Electronics Reader Service Department, P.O. Box No. 2530, Clinton, Iowa 52735, U.S.A.

To obtain a free copy of the complete index of articles published in Electronics from January 13, 1983 to December 15, 1983, circle Reader Service No. 475.

Electronics

February 23, 1984 This reader service card expires May 23, 1984

NAME _____ TITLE _____

PHONE (_____) _____ COMPANY _____

STREET ADDRESS (Company ☐ or home ☐ check one) _____

CITY _____ STATE _____ ZIP _____

Was This Magazine Personally Addressed to You? ☐ Yes ☐ No

Industry classification (check one):
a ☐ Computer & Related Equipment
b ☐ Communications Equipment & Systems
c ☐ Navigation, Guidance or Control Systems
d ☐ Aerospace, Underseas Ground Support

e ☐ Test & Measuring Equipment
f ☐ Consumer Products
g ☐ Industrial Controls & Equipment
h ☐ Components & Subassemblies

5 Source of Inquiry—DOMESTIC
j ☐ Independent R&D Organizations
k ☐ Government

Your design function (check each letter that applies):
x ☐ I do electronic design or development engineering work.
y ☐ I supervise electronic design or development engineering work.
z ☐ I set standards for, or evaluate electronic components, systems and materials.

Your principal job responsibility (check one)
t ☐ Management
v ☐ Engineering Management
r ☐ Engineering

Estimate number of employees (at this location): **1.** ☐ under 20 **2.** ☐ 20-99 **3.** ☐ 100-999 **4.** ☐ over 1000

1 16 31 46	61 76 91 106	121 136 151 166	181 196 211 226	241 256 271 348	363 378 393 408	423 438 453 468	483 498 703 718
2 17 32 47	62 77 92 107	122 137 152 167	182 197 212 227	242 257 272 349	364 379 394 409	424 439 454 469	484 499 704 719
3 18 33 48	63 78 93 108	123 138 153 168	183 198 213 228	243 258 273 350	365 380 395 410	425 440 455 470	485 500 705 720
4 19 34 49	64 79 94 109	124 139 154 169	184 199 214 229	244 259 274 351	366 381 396 411	426 441 456 471	486 501 706 900
5 20 35 50	65 80 95 110	125 140 155 170	185 200 215 230	245 260 275 352	367 382 397 412	427 442 457 472	487 502 707 901
6 21 36 51	66 81 96 111	126 141 156 171	186 201 216 231	246 261 338 353	368 383 398 413	428 443 458 473	488 503 708 902
7 22 37 52	67 82 97 112	127 142 157 172	187 202 217 232	247 262 339 354	369 384 399 414	429 444 459 474	489 504 709 951
8 23 38 53	68 83 98 113	128 143 158 173	188 203 218 233	248 263 340 355	370 385 400 415	430 445 460 475	490 505 710 952
9 24 39 54	69 84 99 114	129 144 159 174	189 204 219 234	249 264 341 356	371 386 401 416	431 446 461 476	491 506 711 953
10 25 40 55	70 85 100 115	130 145 160 175	190 205 220 235	250 265 342 357	372 387 402 417	432 447 462 477	492 507 712 954
11 26 41 56	71 86 101 116	131 146 161 176	191 206 221 236	251 266 343 358	373 388 403 418	433 448 463 478	493 508 713 956
12 27 42 57	72 87 102 117	132 147 162 177	192 207 222 237	252 267 344 359	374 389 404 419	434 449 464 479	494 509 714 957
13 28 43 58	73 88 103 118	133 148 163 178	193 208 223 238	253 268 345 360	375 390 405 420	435 450 465 480	495 510 715 958
14 29 44 59	74 89 104 119	134 149 164 179	194 209 224 239	254 269 346 361	376 391 406 421	436 451 466 481	496 701 716 959
15 30 45 60	75 90 105 120	135 150 165 180	195 210 225 240	255 270 347 362	377 392 407 422	437 452 467 482	497 702 717 960

Affix
Postage
Here

Electronics

P.O. Box No. 2530
Clinton, Iowa 52735

Figure 4-3 A reader service card from a technical magazine can be helpful in obtaining catalogs.

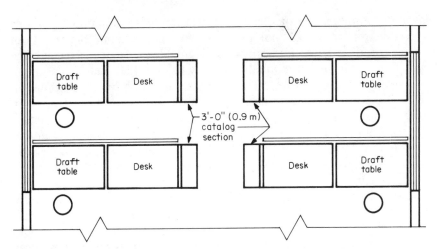

Figure 4-4 Catalog shelving makes reference to information sources easier and quicker.

card listing with the shelves. The business card of the current representative of each manufacturer should be permanently affixed to the inside front cover of the company's catalog so that anyone using it can find the person's name and telephone number at a glance.

Label the edge of each shelf with the categories of catalogs contained on it. For example, on a shelf labeled "Indoor Lighting Fixtures," one could expect to find all the manufacturers' publications that deal with the topic.

Most catalogs come in ring binders. Supplements and loose pages should be inserted in the appropriate binders when they are received. Since more than one member may need to use a particular catalog at the same time, you may want to advise employees to photocopy the required pages so the catalog can be returned to the file for later use. For short-term use, however, photocopying is usually not necessary. Another guideline for controlling catalog use is to instruct employees to keep the materials in the office at all times. No books or parts of books should be removed to the private files of any member, except in the form of photocopies or other duplicate copies.

OTHER REFERENCE MATERIALS

Besides filing catalogs, plans, standard specifications, shop drawings, and other materials, the consulting engineer intending to keep abreast of the latest developments in the field must be aware of the many publications available from technical book publishers and professional societies. Both sources produce books that can be of invaluable assistance to the average consultant, and they continually add new titles to their lists.

One important item for the engineering firm to have is a manual on cost estimation. Many excellent books on the topic can be obtained from R. S. Means Company and McGraw-Hill, as well as from other publishers of professional and technical reference books. Cost-estimating manuals can be used constantly by the consultant preparing cost estimates and budgets for his or her firm. Most of these books are updated annually for greater accuracy.

Considering the variety of printed materials available to the consultant for improving the business, it is easy to imagine that the average firm's library might consist of 20 to 30 ft (6 to 9 m) of shelving devoted to publications on consulting engineering topics. Larger firms sometimes set up a separate library containing two or more copies of the most important books. Obviously, the consultant must maintain control over the use of the materials. The very largest firms create a central filing system, which contains records of all its standard details, plans, specifications, catalogs, price lists, reference books, and other materials. They may even appoint a member to act as librarian in charge of the circulation, tracking, and retrieval of the items. The average consulting firm may not need to go to these lengths to maintain a quality library; but it should at least keep records of the use and location of its most important publications.

The same requirements apply to journals and magazines acquired by the firm. Ideally, at least one member of the organization should be responsible for checking the magazines and selecting articles that are relevant to the company's activities. This individual may also copy (with suitable permission) and distribute the articles to appropriate members.

The Consultant's Reference Shelf, located at the end of this section, lists suggested book, magazine, and journal titles for a basic library for the beginning consulting engineer. While not all the books and publications may be suitable for your new firm, the titles do suggest subject areas in which books would be helpful to your new business.

BUILDING CODES AND STANDARDS

Before completing our examination of office filing systems, we must take a look at the general requirements for filing local, state, and federal building codes. Every consulting engineering firm must be constantly aware of all codes and standards that may apply to its work. The minimum file requirements thus would be a copy of the city or town codes for your immediate area, a copy of the state building code, and a copy of all relevant federal codes.

Municipal codes generally differ slightly from the state and federal requirements. You must have copies of the codes from all municipalities in which you practice. Since some agencies issue the codes in the form of bound handbooks, they can often be shelved in an area separate from the firm's catalog files. Also, you may want to keep a card file listing all codes alphabetically by area, subject, and location in the files.

Local building codes consist of the general requirements for ensuring public safety and often contain sections on plumbing, heating, electrical, and sanitary systems. The state codes generally contain similar items, although the local requirements are often more stringent than those of the state government. Further, a town or city may adopt the state code as its own. But in most cases the local requirements will be the most crucial.

Also, there may be specialized codes that apply to individual structures or types of engineering work, such as bridges, roads, industrial and chemical exhaust, energy conservation, hospital operation, and electronic equipment design. In electrical work, for instance, the consultant must follow not only the *National Electrical Code* but also the *National Electrical Safety Code,* which is more applicable to high-voltage distribution and installation. The *National Electrical Code* does not cover these items in as much detail. All the information must be incorporated into the firm's designs.

A section of the code files should also contain local, state, and federal fire safety requirements. The *National Electrical Code* is currently included in the handbooks issued by the National Fire Protection Association (NFPA). It is a good idea to subscribe to NFPA's updated handbook series to stay abreast of any changes or additions. Copies can be ordered by phone or by writing NFPA at 60 Batterymarch Street, Boston, Massachusetts 02110. The experienced consultant will find her or his acquaintance with the codes invaluable since many of the government's inspectors think of these documents as the "bible" for engineers who work with public and commercial structures.

Incidentally, it is also wise to meet these inspectors or enforcement officers personally, when possible. You should at least try to get acquainted with them by phone and obtain their viewpoints when you have questions about a particular project or code. You benefit by getting to know the inspectors personally since the professional relationship will help build mutual respect and confidence. Perhaps you may even join a committee that develops and interprets various building codes for the enhancement of public safety and improvement of the profession.

Generally, the code enforcement officer is the final authority on all building requirements, except in cases where a problem situation is brought to the board of appeals of a government building agency. Most city and state governments have such a board whose function is to judge special cases based on reasonable interpretations of applicable codes. The requirements for certain consulting engineering projects may be decided by the board. The decisions often result in special exceptions, when necessary. For example, a consultant might be allowed by the board to circumvent the usual requirements for venting floor drains in a dairy room because the location of the room would result in a sanitary or structural problem.

Remember—the codes embody minimum standards for ensuring public safety. Following them does not necessarily mean that you are doing all your engineering work satisfactorily. Sound engineering principles may occasionally require a solution that exceeds government standards. However, parts of some codes may exceed the needs of the appropriate engineering solution. This is especially common with respect to small pipes, wires, parts of foundation walls, and the like. The consulting engineer cannot arbitrarily modify the requirements for every building project to circumvent the codes; but if the firm has copies of all government requirements, studies them, and consults with the enforcement officers on appropriate questions, there should be little difficulty in effecting the proper solution.

Besides developing a filing system for building codes and special government requirements, the managers of the consulting firm must be familiar with all rules concerning the professional "sealing" of drawings. Most municipalities have strict requirements concerning whose seal or stamp must appear on the firm's drawings, not to mention the requirements of state and federal agencies. Basically, the rules state that architects cannot seal engineers' drawings and vice versa. While the seal or stamp can usually be applied by using a rubber stamp and ink pad, many states require that the engineer's signature also appear, either on the stamp or separately. This is done, in part, to ensure that the drawings are properly authorized and are not stolen or damaged. Professional stamps are usually mandatory on work done at the local and state levels. Federal requirements vary from one agency to another. Therefore, it is the consultant's responsibility to make certain that the firm knows the requirements and fulfills them throughout every federal engagement.

EQUIPMENT AND SUPPLIES FOR
SUCCESSFUL CONSULTING

In these days of swiftly advancing electronic devices, engineering consultants are faced with a bewildering array of choices of equipment and supplies for the private practice. While most engineers who were born in the 1930s and 1940s were trained to regard the slide rule as an essential tool of the profession, nowadays consultants everywhere are buying programmable hand-held calculators, office computers, sophisticated copiers—and even electric erasers. To gain a clearer perspective of the materials needed for successful consulting today, we will take a quick look at a variety of useful items.

Drawing Paper

The three types of materials commonly used for engineers' drawings are: paper, linen, and plastic. While linen "paper" seems to be disappearing rapidly from the market—perhaps because of little consumer interest—plastic film is rapidly gaining in popularity. Still, paper that can be used with standard drawing pencils remains the most widely used and least expensive drafting medium.

Most plastic drawing materials can be used with an ordinary drawing pencil or with some of the more recently developed acrylic leads. The acrylic or plastic pencils have much less tendency to smear. Used properly on plastic film, they generally produce a superior-looking drawing with greater sharpness. They provide the user with more flexibility and control over the weight of the lines drawn, and the acrylic lines are easier to erase. Further, the plastic film is less susceptible to accidental tearing than ordinary drawing paper.

There are some drawbacks, however. Not only are the plastic film and pencils more costly but they are also somewhat difficult to get used to at first. The consultant who is accustomed to using ordinary pencils often finds it difficult to avoid breaking the tips of acrylic pencils—at least until he or she learns the best way to hold the pencil to apply proper pressure. Nonetheless, many experienced engineers would rather take the added expense to train their staff to use the plastic materials. Ultimately, personal tastes and economic needs will determine which materials you will choose.

Title Blocks

Consulting firms are more and more frequently deciding to use self-adhesive or pasted title blocks, rather than stamps or handlettering, on their drawings. The prepared blocks save time and money and are more accurate than other types. Many firms have even gone to the extent of printing standard items such as legends, symbol keys, equipment lists, and other

DRAWN BY		DATE
CHECKED BY		CONTRACT NUMBER
APPROVED	DOUGLAS A WILKE	DRAWING NUMBER
SCALE	A R C H I T E C T 3 8 R O O S E V E L T A V E N U E G L E N H E A D, N E W Y O R K. 1 1 5 4 5	

Figure 4-5 Two typical title blocks for working drawings.

details so that they can paste them on quickly. There are obvious advantages to using the prepared blocks and details, but the choice is up to you. Figure 4-5 shows two typical title blocks in use today.

Printing: Blueprints, Sepias, Mylars

The three media that consulting firms most commonly use for design plans are blueprints, paper sepias, and sepia mylars. Blueprints usually consist of white lines on a blue background or blue lines on a white background, although some use a black background with white lines. Sepias are a form of paper transparency that can be used in the original to produce blueprints. Sepia mylars are so named because they consist of a plastic mylar film with sepia print; these generally last longer than other prints and often are used in the firm's permanent record of original tracings.

Mylar is being used more and more frequently, particularly by subconsultants who must submit original tracings to the principal consulting engineers on a specific project. As noted, the mylar has the added advantage of being well suited to permanent storage. Every firm should use the sepia mylars for its records, if possible, since they are highly durable and can be used as evidence of the firm's work if other copies are lost, stolen, damaged, or misused.

The growing firm may eventually consider setting up a small blueprinting operation in its offices. Though expensive, the in-house operation will enable the firm to save time by giving it the freedom to make single or multiple prints without having to wait for delivery to and from an outside printer.

The space required for a small blueprinting operation is usually no more than about 6 ft long by 4 ft wide (2 by 1.2 m). A blueprint machine can thus be placed near the wall in the firm's general drafting area. However, proper ventilation must be provided since most blueprints use ammonia, which could cause a serious health hazard if left unvented.

Photocopying

Many types of copying machines are available and nearly every consultant seems to have her or his own favorite. The large amount of paperwork produced by the average consulting firm does make the copier a helpful—though not always essential—office item.

If you are planning to buy a copier for the first time, it might be worthwhile to rent a few different models for trial periods. A few weeks to a month should be sufficient time to enable you to make a sound choice. The rental fees may be substantial, but they will most likely be cost effective since they will help you find the copier that best suits your purposes. Of course, you should first find out whether the supplier or manufacturer is willing to offer machines on a free trial basis. Generally, the so-called dry copiers are less prone to malfunctions than the older fluid-printing types.

Once you buy a copying machine, you should have some way of controlling its use. Many firms lose substantial amounts of money from indiscriminate use by employees copying papers for personal purposes. The logical way to oversee use of the copier is to note the number of copies produced each month and try to spot unusual variations and their causes. A better method of control is to have all copier users record each of their duplicating jobs in a company log near the machine.

Drafting Equipment

Despite the profusion of sophisticated mechanical drafting equipment on the market today, most successful consulting engineers perform their work with a few highly effective but basic tools. A comfortable drafting stool, adequate lighting, a table, sliding parallel bar, triangles, and throw space are among the items that form the foundation of the private engineering practice. Certain types of engineering do require more sophisticated equipment, but it can be costly and is more of an exception than the rule for most consultants' work.

Every firm should have a light table of some kind. A good light table makes it easier to trace drawings that have become old, faded, smeared, or cluttered. A basic light table consists of a rectangular box approximately the size of the average drafting table, with a milky-white glass top and two or more fluorescent lamps underneath. The light table should be equipped with its own sliding straightedge since it is difficult to copy parts of an original drawing without some way of drawing accurate parallel lines. The illumination provided by the lamps relieves eye strain that may otherwise result from the tracing process.

Another piece of equipment that has become more popular with consulting engineers in recent years is the electric eraser. This is an especially effective tool for erasing small portions of complicated drawings. However, if the eraser is used carelessly, particularly on large areas of a drawing, it can damage the paper or plastic film by digging into the material. When this occurs, it becomes more difficult to letter or draw over the area that has been erased. Employees must be instructed in the proper use of the electric eraser and all other types of equipment.

For the drafting stool, it is advisable to take employee's desires into account in the selection of new ones since an uncomfortable drafting stool may lead to lower morale and reduced efficiency. Generally, drafting stools that do not have backs should be avoided. Further, a padded seat is an important addition. While this might seem unnecessary, one can hardly expect a drafter or designer to be fully satisfied with his or her job while perched on a hard metal seat for 8 hours a day.

Using Calculators in Design

The hand-held electronic calculator is a "must" in the contemporary engineering office. How much you are willing to spend for one is largely determined by the type and frequency of usage you expect. Prices range from as low as $10 for a liquid crystal display unit that will perform the basic functions of multiplication, division, addition, and subtraction, to several hundred dollars for programmable devices that perform hundreds of functions and store information as effectively as many larger computers do.

For the design of virtually any engineering project, the consultant must have some sort of permanent record of the firm's calculations. Records are important since they document the organization's design work and provide a means of reviewing the work for correctness. Hence, it is usually advisable to buy a calculator that can be equipped with a printer, if the unit does not already have one built into it.

While we do not wish to endorse any one model or manufacturer, at the time of this writing two particular calculators are notably popular among consulting firms. These are the Hewlett-Packard HP-97 and the Texas Instruments TI-59. Both are well suited to the needs of most consulting engineers since each can be equipped with printers and manufacturer-supplied programs. Additional programs can be developed using the calculators themselves. Further, the programmable models can be used in conjuction with several excellent books on engineering applications, such as Daryanani, *Building Systems Design with Programmable Calculators,* published by McGraw-Hill.

Figure 4-6 shows how a programmable calculator can be used with a simple program for expansion-tank sizing. The program is titled "Piping System Volume and Expansion Tank Sizing." The engineering task involved is not a major design problem in small installations; but it is far from simple in larger installations using high-temperature hot water. Use of the program makes the job easier. It is from the Daryanani book.

Generally, engineers working in civil and structural areas make more calculations than those involved in the mechanical and electrical fields. Thus, most manufacturers of calculators nowadays are able to supply a larger number of programs for civil and structural work than any other type. In the mechanical and electrical disciplines the primary need for programs is in certain specialized areas involving higher temperatures, high pressures, and unusual expansion characteristics and values. Many handbooks are available to cover fluid flow, both gaseous and liquid; programs to resolve such problems and for setting up standard calculations are of great value to most consultants. Short calculations of all types are required in the electrical field. Chief among them are a variety of tedious lighting calculations and short-circuit analyses.

Office Computers

There was a time when the purchase price of an office computer was far beyond the capacities of even the largest consulting firm. But today, with the ready availability of moderate-priced microcomputers, even small companies are buying their own compact systems.

There is little doubt that an office computer can help most consulting engineering firms with all sorts of calculating, writing, and record-keeping tasks. The question is whether you really need one to perform successful engineering work.

Until a few years ago, for the average consultant, computer costs were prohibitive (you could expect to pay between $7500 and $15,000 for software alone), and there was a shortage of usable programs, especially in the mechanical and electrical fields. Further, you needed an extensive knowledge of programming to use most machines. Most technical programs involved the computer language FORTRAN.

Since then, prices have come down, many more programs are available, and you do not have to spend years learning how to operate most office computers. Programs for the mechanical and electrical fields are now being marketed in novel ways. For example, you can lease programs covering various mechanical systems from the Carrier Corporation, based on the purchase of a Radio Shack TRS-80 computer and printer. The TRS-80 uses the simplest of all the computer languages—BASIC. Electrical equipment manufacturers are preparing similar offers. Some systems can be used to store and print complete project specifications.

Other computer systems available for the consulting engineer's office include a Hewlett-Packard desk-top computer, graphics tablet, printer, and graphic plotter. Using this computer-aided design (CAD) system, the consulting engineer can produce design drawings quickly and easily. Corrections to drawings are made instantly and new drawings printed rapidly. The memory can store mechanical, electrical, and architectural details and plot complete drawings for them in different colors. Smaller drawings can be combined by the computer to produce one large, final

PIPING SYSTEM VOLUME AND EXPANSION TANK SIZING PROGRAM

PROGRAM DESCRIPTION

This program will calculate the volume of a piping system in gallons by determining the volume in each piping branch and adding it up. The volume of the heat exchangers, coils, and other equipment can also be added up.

Given the volume of a piping system, the program can determine the volume of open and closed expansion tanks for hot-water heating systems.

The formula for relative expansion of water has been developed by use of a curve fitting program.

EQUATIONS

$$\text{Pipe area in sq ft } (A) = \frac{\pi \times d^2}{576}$$

Figure 4-6 Calculator programs can speed design calculations in any consulting engineering office. (*Daryanani—Building Systems Design with Programmable Calculators, McGraw-Hill.*)

where
d = Inside diameter in inches

$$\text{Pipe volume in cu ft } (C) = A \times L$$

where
L = Length in feet

$$\text{Pipe volume in gallons} = C \times 7.48$$

$$\text{Expansion tank open } (V_{r0}) = V_s \times E$$

where
V_s = Volume of water in system
E = Relative expansion at operating
   temperature

Expansion tank closed (V_c) =

$$\frac{V_{r0}}{\dfrac{P_0 + P_1}{P_0 + P_2} - \dfrac{P_0 + P_1}{P_0 + P_3}}$$

where
P_0 = Atmospheric pressure at site
P_1 = Gauge pressure in expansion tank when water first enters the tank
P_2 = Initial fill or minimum pressure gauge
P_3 = Operating pressure gauge

Relative expansion of water E

$$= (.97 \ e^{(T_O - 18) \times .000446} - 1)$$

$$- \left[1 - \left(\frac{T_O - 250}{150}\right) \text{abs} \times .02 \right]$$

where
T_O = Operating temperature in degrees F

OPERATING FEATURES

Input for expansion tank calculation can be in any units for volume and pressures. However, input for temperature must be in degrees F.

Input for system volume and operating temperature is combined in decimal format—xxxx.xxx.

Integers represent volume while fractions represent temperature. For example:

1000.099 represents	1,000 gal 99°F
5400.375 represents	5,400 gal 375°F

Figure 4-6 (*Cont.*)

PIPING SYSTEM VOLUME AND EXPANSION TANK SIZING PROGRAM
USER INSTRUCTIONS AND EXAMPLES   Number of Cards: ONE

Step	Procedure	Enter	Press		Print out		Explanation
1.	*Example* Initialize	—	2nd	E'	0.		
2.	Enter run no.	1.02	—	A	1.02	R.NO	
3.	Enter inside pipe diameter, in	4.0	—	B	4.00	DIA	Gal/ft of pipe
					0.65	G/FT	
4.	Enter length of pipe, ft	594	—	C	594.00	L	
					388.	GALS	Gal
					388.	ΣGAL	Total gal
	Repeat steps 2, 3, and 4 as many times as desired	2.03	—	A	2.03	R.NO	
		6	—	B	6.00	DIA	
					1.47	G/FT	
		954	—	C	954.00	L	
					1401.	GALS	Gal
					1789.	ΣGAL	Total gal
5.	Enter equipment volume to add to system volume	700	—	E	700.0	EGAL	Equipment gal
					2489.	ΣGAL	Total gal

No.								
	Continue steps 2, 3, and 4 if desired							
6.	Enter total system volume and operating temperature, °F	2500.375	—	2nd	A′	2500. 375. 335.	GALS TEMP GALS	Volume of open expansion tank, gal
7.	Enter atmospheric pressure at site	15	—	2nd	B′	15.00	P0	
8.	Enter tank pressure while filling	1	—	—	R/S	1.00	P1	
9.	Enter minimum initial fill pressure	2	—	—	R/S	2.00	P2	
10.	Enter operating pressure	150	—	—	R/S	150.00 397.	P3 GALS	Volume of closed expansion tank, gal

Note:

1. *P0, P1, P2,* and *P3*—all four should be in same units
2. *P0*—atmospheric pressure
3. *P1*—pressure in tank when water enters tank gauge pressure—generally 0
4. *P2*—Minimum initial fill gauge pressure
5. *P3*—Operating gauge pressure

Figure 4-6 (*Cont.*)

drawing. Complete sets of drawings can be transferred from one location to another over standard phone lines or by mail on a small floppy disk.

The Trane Company offers a Customer Direct Service (CDS) Network of several levels—basic, intermediate, and a full package. The network brings user-friendly software packages direct from a Trane Company computer to a microcomputer in the consulting engineer's office. At the basic level, seven programs are offered, including a variety of equipment selection and optimization programs, building load estimating, and equal-friction duct design. The intermediate level includes the basic seven programs plus—in most cases—local dial-up usage of the Trane Trace load design and local estimating programs. Also included is a popular static regain duct design program. At the full-package level you have direct access to 10 full Trane Trace program runs per year and 40 runs per year of the TRACE stand-alone economic life-cycle costing program.

If you own or lease a personal computer (such as an Apple or IBM), you can use the TK! Solver Packs to speed a variety of consulting engineering office design calculations. Available from Software Arts, the TK! Solver Packs give you great flexibility in solving mechanical, electrical, and many other design problems. Both the U.S. Customary System (USCS) and the International System of Units (SI) can be used with TK! Solver, which permits you to use the results for both domestic and overseas design assignments.

Still you might wonder, is an office computer necessary for a successful consulting practice? For most small beginning consulting firms, the answer seems to be no. A calculator system consisting of a programmable calculator and recorder is capable of meeting the needs of the beginning small firm. Certainly, any firm concerned about expanding its capabilities should investigate the equipment currently available before spending large amounts of money on elaborate computer hardware.

One other option is to seek an arrangement for subletting or time-sharing. If some of your staff members are skilled in programming and operating computers, you may be able to save much unnecessary expense by organizing with other consultants to acquire a good computing system and use it on the time-sharing basis. A number of consultants have done so successfully. However, others feel that such arrangements are risky because they sometimes make it possible for competitors to gain access to the engineer's confidential information.

With the continued reduction in microcomputer prices and a simultaneous increase in the power of these machines, many business advisors recommend that beginning consultants obtain a microcomputer as early as possible in their career. Having your own micro helps you protect your confidential information because only you and your staff have access to it.

CONSULTANT'S PROFIT KEY 7

Michael T. McGovern, PE, a consulting civil-sanitary engineer in Redwood City, California, owns Paradigm Systems, a software development-marketing firm. Here are his thoughts on software strategies for the modern consulting engineer's office, as reported in *Consulting Engineer* magazine.

As microcomputers grow in sophistication and versatility, more and more are finding their way into engineering offices. The latest generation of microcomputers rival or exceed the power of their minicomputer progenitors while providing advantages in many of the same areas.

Because the microcomputer industry is young, large libraries of commercial software to drive them are not yet available. To compensate, engineers eager to tap the new tool are developing in-house software which, in most cases, is limited to immediate technical needs. Individualized, technical application programs are important and in themselves are enough justification for the purchase of a microcomputer. But when combined with a wealth of general purpose software, the programs serve to make the micro an indispensable item for overall office operations and project management.

GENERAL PURPOSE SOFTWARE

General purpose software—that used for word processing, spreadsheet analysis, database management, and scheduling—is so useful that the microcomputer is rapidly becoming akin to the office copier of a generation ago; you will wonder how you got along without it.

Word processing programs have the ability to make corrections and insertions, to move blocks of text, and to find and replace copy. Contrary to popular belief, however, these programs are not just for the secretarial staff; they should be incorporated into the daily routine of all employees involved in administration.

In the realm of project management, word processing can play a significant role in the preparation of proposals, contract documents, progress reports, and correspondence and specifications. The computerization of construction specifications alone justifies the acquisition of microcomputer and word processing software.

Spreadsheet analysis programs have been instrumental in making the microcomputer industry what it is today. Each program emulates a columnar pad which comprises many rows and columns. The intersection of each row and column—termed a cell—can contain text, numeric data, or a formula which, if changed, can be reflected instantly throughout the spreadsheet.

In project management, spreadsheet analysis commonly is used to estimate costs, and its applications are limited only by one's imagination. Other uses include progress payment estimates, budget, and engineering calculations in addition to the generation of camera-ready tables and reports.

Database management programs are the list keepers and data manipulators, providing a means of storing, retrieving, and rearranging large amounts of similar data. For project management, database systems are most useful for record management and technical analysis, e.g., from maintaining records of thousands of toxic waste barrels and soil samples removed from an illegal disposal site to paying the contractor and reporting to health agencies.

Scheduling programs rest on the borderline between general purpose and specific application software. Although these programs are more project-management oriented than other so-called general purpose programs, they too can be applied to every business and industry.

A scheduling program is an important part of the project manager's arsenal of computer software. At the low end are programs that provide a fairly simplistic approach to schedule development, but they are disproportionately awkward to use. At the high end are one or two programs that offer both sophisticated scheduling and ease of use.

APPLICATIONS SOFTWARE

A microcomputer equipped with the general purpose software described above comprises a powerful system for management of projects of all sizes. Still, areas remain which will not be directly supported by these programs and which must be addressed by specific applications software. One such area is job costing.

Perhaps the most critical of all aspects of project management is the need to maintain costs within budget constraints and to react swiftly to discrepancies. Most firms rely upon their accounting systems to alert them to cost problems. Unfortunately, accounting systems are geared to monthly increments of time while projects are sensitive to events occurring within a much shorter time span. Reports compiled on a monthly basis may alert a project manager to a developing problem, but may be too late to allow effective resolution. And such reports may not provide the information necessary to identify a specific problem or its source at the project level.

Job costing procedures developed as part of an accounting system also may characterize a project after the fact better than they provide in-progress control, particularly when the format of the project budget is inconsistent with the job cost accounting code structure. Relating actual costs to corresponding budget items in this case may be as difficult as maintaining a separate manual system for the project.

Microcomputer applications developed specifically for control of engineering projects allow the manager to define and monitor progress of a project within the context of the individual project. Additionally, they provide various levels of reports when and as they are needed.

Criteria for project programs of this type vary depending on the user, project, and desired end result. Features of a program developed by Paradigm Systems lend insight into establishing a cost management system:

- The program is menu-driven and completely interactive with the user; it is designed for operation by clerical rather than professional staff.

- Projects are "setup" in the system with individually defined tasks appropriate to the project, as opposed to uniformly defined accounting cost codes.

- Direct expense centers are individually defined for each task as needed, to make the task a complete accounting entity.

- As the project progresses, additional tasks and expense items can be added to account for unforeseen conditions or extensions to the original project.

- Budgeted labor rates are defined for each labor category and actual employee labor rates are maintained within the system.

- The type of contract (e.g., cost plus fixed fee, time, and materials) is established for each project so that cost reports can reflect the actual accounting rules for overhead and profit.

- Timesheet and expense data entry is simple, fast, and verified, minimizing errors caused by the operator or by those filling out the timesheet.

- Project records are updated automatically whenever timesheet or expense data are entered into the system. Reports then can be generated upon demand.

- Cost reports show actual performance against budget, both for manhours and dollar costs. They are designed for various levels of management so that only information appropriate to each level is distributed.

- Multiple projects can be consolidated into a number of combinations for either executive review (by individual office or division) or as a single large project with multiple phases.

- Due to the degree of control provided by the program, the cause of a real or potential cost overrun is isolated easily at the individual employee or expense level.

Application programs of this nature are geared specifically to the needs of a particular user, in this case an engineering project manager. They are developed to meet these needs, with external office-wide accounting considerations taking second place. The latter, however, can be integrated with project cost management to yield a complete, company-wide system. Such programs also can be coded to allow the addition of new modules to accommodate specific operations. Often these extensions can be developed in-house, using a language other than that used for the original program.

When coupled with appropriate software, microcomputers can significantly enhance the efficiency and effectiveness of the engineering project manager. General purpose programs can speed up many operations and remove much of the drudgery that long has been part of the job. Specific applications software, particularly project cost management systems, monitors costs and produces information quickly and in a meaningful format.

To derive the greatest benefit from these programs, employees at all levels of the organization must be willing to accept and learn new methods. This often is easier said than done, but can result in reduced costs and increased profit, both at the project and company-wide level.

CONSULTANT'S PROFIT KEY 8

As your consulting firm grows you will certainly want to consider CAD as a way to increase your staff's (and your own) efficiency. William J. Meehan, PE, who is associated with R. G. Vanderweil, Engineers, Boston, Massachusetts, and is a lecturer at both Northeastern University and the University of Massachusetts, shows in the following profit key how to integrate CAD into your office system. His observations first appeared in *Consulting Engineer* magazine.

The computer has played a large role in consulting engineering firms for well over 20 years. Its role, though, has not been directed toward the center of the business, the engineering and design process. Instead, it has been directed toward service, in areas such as project management and control, and a wide range of financial matters. Even in engineering areas that traditionally use the computer, i.e., structural, stress, and electrical system calculations, the program input often is copied manually from a drawing to a computer terminal, and the results are manually incorporated into the design.

Within the next several years, this no longer will be the case as computer technology will move from the perimeter to the center of the engineering and design process. It will be integrated into the very essence of the process itself, replacing the paper and pencil concept forever. The computer will be doing graphics; performing design functions, data management, specifications, communications, and computations; and controlling the interrelation of all of these functions.

COMPUTER-AIDED DESIGN

Perhaps the most significant computer applications for the engineer are those that revolve around computer-aided design. CAD is not new. It has been used successfully throughout the past decade in the design of machine parts, aircraft structures, space-related projects, and mapping systems. But until recently CAD systems were driven by large main-frame or super-mini-computers, and their cost often exceeded $1 million. The integration of low cost microcomputers and microprocessors into CAD systems has and likely will continue to reduce the cost associated with CAD.

What exactly is CAD? Basically, it is a generic term for any hardware/software system using computer graphics as a replacement for the traditional designer/drafter. CAD systems are not alike. At one extreme are those that offer sophisticated graphics with three-dimensional displays. At the other are systems that only draw lines, curves, and letters. Some systems are convenient and efficient to use; others are awkward and inefficient.

THE WORKSTATION

A typical CAD workstation consists of a high-resolution graphics display, a typewriter-like keyboard, and a digitizer with an input device such as a light pen. The digitizer is a tablet with hundreds of sensors spaced on a grid, an arrangement that is intended to simulate an x-y coordinate system. The placement of an input device at any point on the digitizer sends the x-y coordinate point to the system. The location of the point can be displayed or permanently recorded.

Each CAD workstation is connected to a central processor that controls the workstation, provides data storage allocation, and manages the communication to other CAD workstations or other computers. Typically, hard copies of the graphics are produced by electrostatic plotters or pen plotters also controlled by the central processor. A system also can be equipped with a printer and magnetic tape drives.

A typical system equipped for a consulting firm would have:

• One or more central processors.
• A data storage system (usually hard disk storage).
• A backup storage system (usually a tape drive).
• Plotting and printing devices.
• One or more workstations consisting of:

  —Graphics display

  —Typewriter-like keyboard

  —Input device

• CAD software.
• Communications capability.

The components themselves, however, are not enough to make the system complete; they must be coordinated carefully. For example, when several workstations are operating simultaneously, the central processor must be able to service all of the workstations without degradation of response to the operator. Adequate on-line data storage also must be available to allow for sufficient system flexibility and future expansion.

CAD'S CAPABILITIES

A CAD system for the consulting engineer ought to be more than a fancy drafting machine. Although some significant productivity improvements can be obtained just by automating drafting, a complete system has the potential to do much more, including the ability to (in increasing order of complexity):

- Perform complex graphic operations.
- Capture and manipulate graphic and nongraphic data.
- Define and associate attributes to graphic elements.
- Perform design functions, engineering analysis, and simulation and optimization.

The graphic operations can be either software-driven or built into the CAD workstation. When routine graphics are built into the workstation, the central processor is free to perform more complex tasks, improving the overall performance of the system. In either case, CAD should automate the process of making engineering drawings. Points, lines, and surface definition should be straighforward and require a minimum of operations. For example, circle definition should require no more than the definition of the center and one point on the perimeter; line weights and patterns should be readily definable; and character size, angle of text, and font should be specified easily. Since only portions of the graphics can be viewed at any given time, it should be possible to pan and zoom the entire graphics file.

The CAD system also should be able to capture and store graphic entities, such as typical details, standard symbols, and commonly used configurations. It should be able to move the entities, scale them up or down, and manipulate them in a variety of ways while, at the same time, building up a catalog of "graphic parts" to be called upon whenever needed.

The ability of the CAD system to associate data with a graphic entity is termed "attribute definition." To illustrate how this term can be applied to an engineering problem, consider an electrical wiring diagram. An electrical cable shown on the diagram is identified by its conductor size, insulation type, and voltage level. The cable itself is shown as a line which, in turn, is actually a model of a section of the cable and accompanying data. The system should be able to keep track of the line's (cable) attributes to the extent that if the line is moved, the attributes move with it.

An "attribute" often is the key to another associated set of data. One attribute of a pump, for example, may be a specification number which may be the key to a specification file. With appropriate attribute definition, the CAD operator can routinely ask the system for a list of all pumps with a certain specification number. Or, as the case may be, ask to replace all pumps with one specification number with pumps of another specification number. The possibilities are endless.

The ability to perform graphic operations, capture and manipulate graphic entities, and define attributes allows a CAD system actually to perform functions such as routing pipe, selecting proper fittings, and keeping track of the pieces. Its ability to perform complex engineering calculations eliminates the tedious, time consuming, and error-prone operation of collecting the data. Further, if the design is stored in the system and the calculations directly access the data from CAD, the calculations will reflect the current data, thus eliminating the need to change the calculations when a design is altered.

Advanced engineering simulations and design optimizations also can be a part of CAD. So a truly integrated CAD system could produce the picture of the engineering model, count the pieces in the model, move the model around, simulate the model's functions, and even figure out how to make the ultimate product for less cost and more efficiently.

THE RIGHT DIRECTION

Other applications will help round out the computer as the center of the engineering and design process. These include database management (DBMS), word processing (integrated with engineering specifications and standards), and inter-computer communications networks.

• Database management provides a consistent and simple way to sort and sift through complex and large databases, including the graphic data developed by a CAD system.

• Word processing can be integrated into the engineering and design process by coupling the standards, procedures, specifications, and building and fire protection codes to databases developed by CAD.

• Inter-computer communication can be used to transfer computer data between architects and engineers, and engineers and constructors. This will shorten the overall project completion time and minimize error in interpretation and translation of data.

To be effective, these applications must not stand alone. Each should be a part of the whole, an integrated engineering and design application system.

WHAT IT TAKES

The ability to perform interactive design functions requires a microcomputer with remarkable speed, one that is able to access huge amounts of high-speed data storage. Therefore, an 8-bit machine with the standard 64k bytes of memory cannot be expected to drive all but the most simple CAD systems. Even the newest 16-bit machines would be hard pressed to provide significant CAD and design operations simultaneously. The 32-bit microcomputers, however, offer the speed and ability necessary to bring the computer into the center of the design process without, at the same time, becoming an economic liability.

Purchasing the hardware is the obvious first step toward integrating CAD. As challenging as that may be—requiring a good deal of research, bench mark testing, and comparative study—it is far more straightforward than evaluating the engineering and design processes to be integrated. The challenge for consulting engineers is to use creativity, imagination, and ingenuity in applying the new design tools to the old problems.

One way to begin is to wipe the slate clean, to imagine that an old design problem has never been solved before. Then, given only the new tools, imagine how to solve the problem, making sure the solution conceived solves the right problem. Don't be fooled into thinking, for example, that the problem is to produce drawings of a bridge rather than to design a bridge. For the consulting engineer who thinks the problem is to computerize drawings, the solution will be left to the limitations of automated drafting.

In examining the engineering and design process, the process can be viewed in the same terms as building a model. Drawings, specifications, diagrams, bills of quantities, and calculations form a model, which, in turn, must be interpreted by the builder. The key in CAD is to find effective ways to build the model and communicate its details to the builder. Obsolete procedures should not be retained nor should the new tools be forced to conform to traditional practices.

THE PLAN

How does a consulting engineer integrate computers into the engineering and design process? The answer is to develop a plan, one that spans from five to 10 years or more, avoids specific reference to hardware and software, and establishes what functions are required immediately, in the near term, and in the long term. Changing the plan to meet changing needs should be encouraged. Changing the plan to account for specific hardware or software that may be available or acquired should be discouraged.

The plan is driven by needs rather than by what is or will be available. It should serve as an ideal, a guidepost, to keep the consulting engineer on track. Any deviation from it should be documented, but the plan itself should remain intact.

The plan should look at the objectives in a top-down fashion, from the general to the specific. It should be based on the examination of a specific segment of the business and include a recognition and documentation of uncontrollable parameters, both those that are and those that are not subject to change. In developing the plan, one must look at the entire engineering and design process. In this way, the solutions to isolated problems will fit into the overall plan and later will be properly integrated rather than ending up isolated and unrelated.

SECURING PATENTS

It is sometimes said that if you scratch the surface of any engineer, you will find an incipient inventor lurking beneath. Only a handful of PEs, however, actually pursue their inventive urge to the point of producing new products and having them patented. Yet a number find that the patent process offers significant monetary and psychological rewards.

It is beyond the scope of this book to detail the steps required for securing patents on specific engineering designs and processes. The study of patents, inventions, and related laws is so broad that entire college programs are devoted to this one topic alone. For most consultants, the pursuit of a patent may begin with observing a few basic points.

Before you start seeking a patent on a potential invention, you must make a *patent search* to be certain that your invention will not infringe on similar existing patents. Generally, this requires hiring an attorney who is skilled in the patent field. Ideally, the attorney would be someone with connections to the U.S. Patent Office in Washington, D.C., or someone affiliated with a corresponding patent attorney. Unfortunately, many would-be inventors discover that patents already exist for inventions that, to the inventor, may seem totally unique. The complexity of laws and federal regulations governing patents makes the patenting process costly and time-consuming. Thus, it is important to settle questions about your invention as early as possible, preferably during the patent search.

Once you establish that your invention is unique, you must seek some form of legal protection for the product—whether it is a gadget, a design, or even an idea. Legal protection is usually obtainable in the form of a patent, trademark, copyright, license, or franchise.

A *patent* is an exclusionary right which bars others from using or selling your product, regardless of whether or not you intend to use or sell it yourself.

Trademarks include words, names, symbols, or combinations thereof, used to identify products being sold.

Copyrights commonly apply to written items (books, articles, etc.), photographs, works of art, music, or literature. *Licenses* allow the holder limited use of legally protected items. *Franchises* are similar to licenses. For example, a manufacturer of an item may grant a franchise to particular individuals to authorize the individuals to distribute or sell the manufacturer's products.

Once the appropriate legal protection and control has been obtained for your invention, you still must find a way to produce and market the product before competitive items are developed. Many inventors therefore form joint ventures with manufacturers or other organizations that have sufficient resources for fulfilling the patent requirements, producing the product or process, and possibly marketing it as well.

WORKING WITH MANUFACTURERS' AGENTS

As was briefly noted earlier in this Handbook, manufacturers' agents and representatives can be both a blessing and a bane to the consulting engineer. Oftentimes agents can provide invaluable information on the use of their companies' materials, processes, or other products. Indeed, it is sometimes feasible for a consultant to reduce his or her own work load by allowing a manufacturer's agent to perform a large part of a given design for the engineering firm.

Since the agent is usually well acquainted with the manufacturer's products and how they can be integrated into particular projects, he or she may even be able to produce a better design than a less-knowledgeable consultant, and in less time. Unfortunately, this situation can lead to a host of problems.

Not only will the design be oriented toward the materials or equipment supplied by the agent's company, but the project will, in effect, be transferred from the hands of the consultant to those of the manufacturer. It should be easy to see how a conflict of interest may develop from such arrangements. The consultant who permits an agent to perform design work thereby relinquishes some of his or her independence, thus becoming a sort of "property" of the manufacturer. The consultant's specifications consequently may become biased or distorted.

Despite the inherent attractions that such "free engineering" may hold for the busy consulting firm, these situations are best to avoid. Conflicts of interest involving engineers and manufacturer's agents can mar the reputation of an entire firm (not to mention the profession as a whole).

Of course, virtually every consulting engineer needs the expert advice of manufacturers' agents or representatives on some occasions. Few, if any, consultants know as much about a given product as the agent who helps to sell it. The agent concentrates all of his or her energies on learning how to use the manufacturer's products; few engineers have enough time to spare to learn all there is to know about the product's applications. Yet it is the consultant's responsibility to utilize the agent's assistance ethically.

Whenever possible, advice should be solicited from several different manufacturers of similar products. Afterward, contract specifications should be written so as to integrate the consulting firm's engineering know-how with the information provided by various manufacturers. Ideally, all the companies can then be invited to bid on the project together.

So-called free engineering tends to occur in the mechanical and electrical fields more frequently than in other branches of the profession. But the line that divides free engineering from true consulting engineering is not always clear. Certainly, one must know how to distinguish between the two before one can effectively avoid the conflicts of interest that free engineering can engender.

Let's cite a few examples. In one case, the manufacturer of a series of radiators for a heating system designs and sketches the associated piping system for the building in which the radiators will be installed. His agent then gives the design to your office so that you may copy it on your plans. If you adopt the design without question, make no additional design efforts yourself, and are not asked to pay a fee for the design, this would be an instance of free engineering.

Now, imagine that the design requires the application of a complicated temperature-control system. The manufacturer's representative provides a schematic of the controls, reviews the plans with you, and allows you to trace the schematic design onto your plans. Further, the agent writes the complete specification, which you also copy. While you may have exercised discretion with regard to use of the data, at this point you have been only peripherally involved in the design. Hence, this is another case of free engineering and possible conflict of interest. If, however, you did not immediately accept the design and specification of the agent but carefully reviewed them and tailored them to fit your own specifications and plans, you would have acted independently and in the best interests of yourself and your client. The important point to remember is that the agent's job is to provide you with information, which you then use to produce the appropriate design application for the project.

To use the manufacturer-supplied information fairly and effectively, the consulting firm must have expertise in the fields of engineering involved. Free engineering usually occurs when consultants are unfamiliar with the particular type of engagement they have undertaken. Since they may not be fully qualified to do the work required, the consultants allow the manufacturer to perform it in exchange for payment for use of the manufacturer's products. The manufacturer can only profit; but for the engineer, there is substantial risk involved.

The consulting firm that uses free engineering opens itself to accusations of unprofessional behavior, of misrepresenting its services, and of acting as a sales tool for manufacturers. While a number of consultants somehow manage to get away with such actions, others become the focus of crippling legal actions stemming from free engineering.

Clearly, it is the duty of every consulting engineer to exercise discretion over the use of data supplied by manufacturers and to investigate the use of products from several different firms when possible. Once you compare the manufacturers on the basis of the quality of their materials, installation abilities, and general service, you will be in a much better position to complete your engagement to the full satisfaction of your client.

Good manufacturers' representatives are aware of the consulting engineer's concerns; they not only know all the details regarding their products but they also know when their products should be used and when they should not. Also, they can tell you what other materials or equipment should and should not be used in conjunction with their own. They make themselves available for assistance when needed and are readily able to answer most questions concerning their products. Further, they do not exaggerate the importance or applicability of the items they sell.

If a consultant's client actually requests or consents to use designs or other engineering work provided by a manufacturer's agent, the engineer still must check every facet of the design to make certain that it meets the client's needs. Such cases rarely involve conflicts of interest, as long as all parties involved are made aware of the situation as soon as it occurs. At no time should the engineer deliberately attempt to dissuade other manufacturers from bidding since all have the right to compete freely in the marketing of their products.

SUMMARY

The successful organization and management of a consulting engineering office is at once a function of sound business practices and personal style. It requires the establishment of clear-cut objectives, scheduling, and the careful delegation of various daily responsibilities to appropriate staff members.

Managers must establish an atmosphere of respect and mutual rapport with employees. A fundamental tool of the office manager is the managerial accounting report, in all its myriad forms. The reports must be maintained and applied consistently to be effective.

Further, the consulting firm should have a system of supervising and training key employees for leadership positions; at the same time, management must know when and how to use outside experts to provide its services in the manner that is most cost effective and satisfying to clients.

The selection and arrangement of office furniture and other equipment further influences staff effectiveness. While a drafter may be able to work in a small space of 50 ft^2 (4.5 m^2), he or she may need much more space to achieve maximum productivity.

A well-organized filing system is essential to the successful private practice. It helps management save time and reduces the risk of loss or damage to important documents. Careful maintenance and updating of specification files, catalog shelves, and other files can help the beginning consulting firm to build its own library of standard details, specifications, and other vital data. Successful consultants generally keep themselves informed of new developments in building and other safety codes, patents, and inventions in their engineering fields. They stay in touch with reliable agents and representatives of manufacturers but never abuse these relationships in attempts to save time or money.

The Consultant's Reference Shelf[1]

THE BUSINESS OF CONSULTING ENGINEERING

Balachandran, *Directory of Publishing Sources: The Researcher's Guide to Journals in Engineering and Technology,* Wiley.

Balderston, *Modern Management Techniques in Engineering and R&D,* VNR.

Bermont, *How to Become a Successful Consultant in Your Own Field,* Consulting Opportunities Journal.

Cleland & Kocoaglu, *Engineering Management,* McGraw-Hill.

Cohen, *Consulting Engineering Practice Manual,* McGraw-Hill.

Directory of Engineers in Private Practice, National Society of Professional Engineers.

Ellis, *The Financial Side of Industrial Research Management,* Wiley.

Gerwick and Woolery, *Construction and Engineering Marketing for Major Product Services,* Wiley.

Greiner and Metzger, *Consulting to Management,* Prentice-Hall.

Guttman, *The International Consultant,* McGraw-Hill.

Hajek, *Management of Engineering Projects,* McGraw-Hill.

Harkins and Plung, *Guide for Writing Better Technical Papers,* IEEE.

Harvill and Kraft, *Technical Report Standards: How to Prepare and Write Effective Technical Reports,* Banner Books.

Johnson, *Private Consulting,* Prentice-Hall.

Kelley, *Consulting: The Complete Guide to a Profitable Career,* Scribners.

Kemper, *Engineers and Their Profession,* CGS Educational and Professional Publishers.

Kock, *The Creative Engineer: The Art of Inventing,* Plenum Press.

Konold, *What Every Engineer Should Know About Patents,* Dekker.

---

[1]A complete listing of publishers and their addresses can be found at the end of The Consultant's Reference Shelf.

Kuecken, *Starting and Managing Your Own Engineering Practice,* VNR.

Lant, *The Consultant's Kit: Establishing and Operating Your Successful Consulting Business,* Jeffrey Lant Associates.

Meehan, *Getting Sued and Other Tales of the Engineering Life,* The MIT Press.

Miller, *Behavior Management: New Science of Managing People at Work,* Wiley.

Pacey, *Culture of Technology,* MIT.

Roadstrum, *Being Successful as an Engineer,* Engineering Press.

Stanley, *The Consulting Engineer,* Wiley.

Tomezak, *Successful Consulting for Data Processing Professionals,* Wiley.

Holtz, *How to Succeed as an Independent Consultant,* Wiley.

BROAD BASIC REFERENCES FOR CONSULTING ENGINEERS

Aronofsky, *Managerial Planning with Linear Programming in Process Industry Operations,* Wiley.

ASM, *Source Book on Materials Selection,* American Society for Metals.

Babbit, *Plumbing,* McGraw-Hill.

Bailey, *Human Performance Engineering: A Guide for System Designers,* Prentice-Hall.

Baker, *Cooling Tower Performance,* Chemical Publishing.

Baumeister and Marks—*Standard Handbook for Mechanical Engineers,* McGraw-Hill.

Beer and Johnston, *Mechanics of Materials,* McGraw-Hill.

Billington and Allan—*Reliability Evaluation of Engineering Systems: Concepts and Techniques,* Plenum.

Canapathy—*Applied Heat Transfer: A Complete Handbook for Power and Process Engineers,* Penn Well Books.

Cheremisinoff and Regino, *Principles and Applications of Solar Energy,* Ann Arbor Science Publishers.

Chopey and Hicks, *Handbook of Chemical Engineering Calculations,* McGraw-Hill.

Chuse and Eber, *Pressure Vessels—The ASME Code Simplified,* McGraw-Hill.

Considine, *Engineering Technology Handbook,* McGraw-Hill.

Considine, *Process Instruments and Controls Handbook,* McGraw-Hill.

Cox, *International Construction: Marketing, Planning and Execution,* Construction Press.

Crocker and King, *Piping Handbook,* McGraw-Hill.

Croft, *American Electrician's Handbook,* McGraw-Hill.

Faras, *Engineering Evaluation of Energy Systems,* McGraw-Hill.

Ferziger, *Numerical Methods for Engineering Applications,* Wiley.

Fink and Beaty, *Standard Handbook for Electrical Engineers,* McGraw-Hill.

Gartmann, *Delaval Engineering Handbook,* McGraw-Hill.

Gebhart, *Heat Transfer,* McGraw-Hill.

Golde, *Lightning,* Academic Press.

Harris, *Modern Air Conditioning,* McGraw-Hill.

Hicks, *Standard Handbook of Engineering Calculations,* McGraw-Hill.

Hicks and Edwards—*Pump Application Engineering,* McGraw-Hill.

Horikawa, *Coastal Engineering: An Introduction to Ocean Engineering,* Halsted Press.

Isermann, *Digital Control Systems,* Springer-Verlag.

Karassik, *Pump Handbook,* McGraw-Hill.

Karger and Bayha, *Engineered Work Measurement,* Industrial Press.

Kaufman and Seidman, *Handbook of Electronic Calculations for Engineers and Technicians,* McGraw-Hill.

Knief, *Nuclear Energy Technology,* McGraw-Hill.

Kverneland, *World Metric Standards for Engineering,* Industrial Press.

Marter, *Handheld Calculator Programs for Engineering Design,* McGraw-Hill.

McGraw-Hill Dictionary of Science and Engineering, McGraw-Hill.

Merritt, *Standard Handbook for Civil Engineers,* McGraw-Hill.

Mironer, *Engineering Fluid Mechanics,* McGraw-Hill.

Mitchell, *Energy Engineering,* Wiley.

Mueller, *Standard Mechanical and Electrical Details,* McGraw-Hill.

Mueller, *Standard Application of Mechanical Details,* McGraw-Hill.

Mueller, *Standard Application of Electrical Details,* McGraw-Hill.

Nilson, *Design of Prestressed Concrete,* Wiley.

Novak and Cabalka, *Models in Hydraulic Engineering: Physical Principles and Design Applications,* Pitman.

Noyes, *Cogeneration of Steam and Electric Power,* Noyes Data.

Parker, *The Management of Innovation,* Wiley.

Parmley, *Standard Handbook of Fastening and Joining,* McGraw-Hill.

Perry, *Engineering Manual,* McGraw-Hill.

Perry, *Perry's Chemical Engineers' Handbook,* McGraw-Hill.

Probstein, *Synthetic Fuels,* McGraw-Hill.

Reiter, *Industrial and Commercial Heat Recovery Systems,* VNR.

Rice and Hoffman, *Structural Design Guide to ACI Building Code,* VNR.

Roark, *Formulas for Stress and Strain,* McGraw-Hill.

Rosalier and Rice, *Standard Handbook of Plant Engineering,* McGraw-Hill.

Sage, *Systems Engineering: Methodology and Applications,* Wiley.

Schwartz, *Composite Materials Handbook,* McGraw-Hill.

Shinskey, *Process Control Systems,* McGraw-Hill.

Simin and Scanlan, *Wind Effects on Structures: An Introduction to Wind Engineering,* Wiley.

Skeist, *Handbook of Adhesives,* VNR.

Skrotzki and Vopat, *Power Station Engineering and Economics,* McGraw-Hill.

Spiegler, *Salt-Water Purification,* Plenum Press.

Szokolay, *Solar Energy and Building,* Halsted Press.

Taylor, *Alternative Energy Sources for Centralized Generation of Electricity,* Heyden.

Turner, *Energy Management Handbook,* Wiley.

Vernick, *Handbook of Wastewater Treatment Processes,* Marcel Dekker.

Young and Young, *Directory of Special Libraries and Information Centers,* Gale Research.

ENGINEERING AND LAW

Adams, *Legal Guide for your Small Business,* Wiley.

Crosswell, *Legal Aspects of International Business,* Oceana.

Hancock, *Executive's Guide to Business Law,* McGraw-Hill.

Lawrence, *Legal Aspects of Solar Energy,* Lexington Books.

Martson, *Law for Professional Engineers,* McGraw-Hill.

Niederhauser and Meyer, *Legal Rights of Chemists and Engineers,* American Chemical Society.

RADCO Staff, *Legal Problems of Contractors, Architects and Engineers,* Irvington.

Vaughn, *Legal Aspects of Engineering,* Kendall-Hunt.

Wincor, *Law of Contracts,* Oceana.

Wyatt and Wyatt, *Business Law: Principles and Cases,* McGraw-Hill.

CONTRACTS AND THE CONSULTING ENGINEER

Dunham, Young and Bockrath, *Contracts, Specifications and Law for Engineers,* McGraw-Hill.

Lang and De Coursey, *Profitability Accounting and Bidding Strategy for Engineering and Construction Management,* VNR.

Rosenfield, *The Practical Specifier,* McGraw-Hill.

GENERAL BUSINESS PROCEDURES

Amos and Sarchet, *Management for Engineers,* Prentice-Hall.

Barnes, *Communication Skills for the Foreign-Born Professional,* ISI Press.

Davidson and Weil, *Handbook of Cost Accounting,* McGraw-Hill.

Dewhurst, *Business Cost Benefit Analysis,* McGraw-Hill.

Freuhling and Bouchard, *Business Correspondence,* McGraw-Hill.

Mackenzie, *The Time Trap,* AMACOM.

Warson, *Business Calculations,* McGraw-Hill.

ENGINEERING ECONOMY

Barish, *Economic Analysis,* McGraw-Hill.

Chung, *Linear Programming,* Merrill.

Eilon, *Elements of Production Planning and Control,* Crowell-Collier Press.

Ferguson and Sargent, *Linear Programming,* McGraw-Hill.

Ford and Fulkerson, *Flow in Networks,* Princeton.

Grant, *Principles of Engineering Economy,* Ronald Press.

Kurtz, *Engineering Economics for Professional Engineers' Examinations,* McGraw-Hill.

Miller, *Schedule, Cost and Profit Control with PERT,* McGraw-Hill.

Moder and Phillips, *Project Management with CPS and PERT,* Reinhold.

Moore, *Manufacturing Management,* Irwin.

Muth and Thompson, *Industrial Scheduling,* Prentice-Hall.

Riggs, *Economic Decision Models,* McGraw-Hill.

Sasieni and Yaspen, *Operations Research,* Wiley.

Schweyer, *Analytic Models for Managerial and Engineering Economics,* Reinhold.

Schuchman, *Scientific Decision-Making in Business,* Holt.

Starr, *Product Design and Decision Theory,* Prentice-Hall.

Taylor, *Managerial and Engineering Economy,* Van Nostrand.

Thrall, et al. *Decision Processes,* Wiley.

Thuesen-Fabrycky, *Engineering Economy,* Prentice-Hall.

Whittaker, *Economic Analysis,* Wiley.

ENGINEERING ESTIMATING AND COSTING

Bussey, *The Economic Analysis of Industrial Projects,* Prentice-Hall.

Chemical Engineering Magazine Staff, *Modern Cost Estimating,* McGraw-Hill.

Dodge Guide for Estimating Public Works Construction Costs, Dodge Building Cost Services.

Dodge Manual for Building Construction Pricing and Scheduling, Dodge Building Cost Services.

Geddes, *Estimating for Buildings and Civil Engineering Works Incorporating Buildings and Public Works Administration, Estimating and Costing,* Butterworth Publishers.

Kolstad and Kohnert, *Rapid Electrical Estimating and Pricing,* McGraw-Hill.

Komanoff, *Power Plant Cost Escalation,* VNR.

McKee, Berger and Mansueto, *Building Cost File,* VNR.

Means, *Mechanical and Electrical Cost Data,* R. S. Means.

Neil, *Construction Cost Estimating for Project Control,* Prentice-Hall.

Stewart, *Cost Estimating,* Wiley.

Thomson, CM: *Developing, Marketing, and Delivering Construction Management Services,* McGraw-Hill

ETHICS IN ENGINEERING

Alger, *Ethical Problems in Engineering,* Wiley.

Bayles, *Professional Ethics,* Wadsworth.

Bowie, *Business Ethics,* Prentice-Hall.

Firmage, *Modern Engineering Practice,* Garland.

Flores, *Ethical Problems in Engineering,* Rensselaer Polytechnic Institute.

Harding and Canfield, *Legal and Ethical Phases of Engineering,* McGraw-Hill.

Kemper, *The Engineer and His Profession,* Holt.

Kidder, *The Soul of a New Machine,* Avon.

Mantell, *Ethics and Professionalism in Engineering,* Macmillan.

Martin and Schinzinger, *Ethics in Engineering,* McGraw-Hill.

Martin, *Three Mile Island: Prologue or Epilogue,* Ballinger.

Thoyse and Middendorf, *What Every Engineer Should Know about Product Liability,* Marcel Dekker.

COMPUTERS AND THE CONSULTANT

Calmus, *The Business Guide to Small Computers,* McGraw-Hill.

Engineering Databases: Software for On-Line Applications, IEEE.

Isshiki, *Small Business Computers: A Guide to Evaluation and Selection,* Prentice-Hall.

Kolve, *How to Buy (and Survive) Your First Computer,* McGraw-Hill.

Lenk, *Handbook of Microprocessors, Microcomputers, and Minicomputers,* Prentice-Hall.

Long, *Manager's Guide to Computers and Information Systems,* Prentice-Hall.

Walsh, *Understanding Computers,* Wiley.

Wray and Crawford, *What Every Engineer Should Know About Microcomputer Systems Design nad Debugging,* Marcel Dekker.

SOFTWARE FOR THE CONSULTING ENGINEER[2]

General

Column Design

P.E.A.S.
7208 Grand Avenue, Neville Island
Pittsburgh, PA 15225
(412) 264-3553

Developed by: P.E.A.S.

Operating environment: memory requirements depend upon systems configuration.

Source language name and user availability: Basic, source code available.

---

[2]From *datapro/McGraw-Hill Guide to Apple Software,* Datapro Research Corporation. Used with the permission of the publisher. Copyright 1985 by McGraw-Hill, Inc. Companion guides are available for IBM and CP/M software also.

Usage pricing: $125.

Documentation: user's manual provided.

Physical medium: 5¼″ or 8″ diskettes.

Product description: this package calculates safe and actual stresses from concentric and off-center loads. To do the calculations, the Chicago or AISC building codes or user safety factor are utilized. The program makes use of graphics as a visual aid. Automatic selection of AISC beams is also available.

Beams

Software Corp.
1407 Clinton Raod
Jackson, MI 49202
(517) 782-8106

Operating environment: memory requirement depends upon systems configuration.

Source language name and user availability: Basic.

Usage pricing: a purchase price of $115.

Maintenance provisions/pricing: limited warranty included in price, updates available, telephone support.

Documentation: user's manual included.

Physical medium: 5¼″ diskette.

Product description: Beams is designed to calculate the stress and deflection on any beam when given the beam dimensions, length, support conditions and load conditions. The program can also compute the correct American standard I beam when given the length, load conditions, support conditions and maximum allowable deflection. The data input is a menu format so data can be checked and corrected before calculations begin.

Flat Slab Analysis and Design Program

Ecom Associates, Inc.
8634 West Brown Deer Road
Milwuakee, WI 53224
(414) 354-0243

Developed by: Ecom Associates, Inc.

Operating environment: memory requirements depend upon systems configuration.

Source language name and user availability: Basic.

Usage pricing: contact for pricing details.

Maintenance provisions/pricing: maintenance contract is available.

Documentation: user's manual provided.

Physical medium: 5¼" and 8" diskettes.

Product description: this program analyzes and designs flat slab or waffle slab floors in accordance with ACI 318-77. Flat slabs may have drop panels at columns. Columns may have capitals. Either the ultimate strength or the working stress method may be used. Design is for a one-bay wide strip using the equivalent frame analysis method and may be up to 10 spans. The strip may have cantilevers. No beams parrallel to the direction of moments are permitted.

Structural Analysis Software for Micros

Kern International
433 Washington Street, P.O. Box 1029
Duxbury, MA 02331
(617) 934-0445

Operating environment: memory requirement depends upon systems configuration.

Source language name and user availability: Basic.

Usage pricing: a purchase price of $48.

Documentation: User's manual included.

Physical medium: 5¼" or 8" diskette, hard disk.

Product description: this is a package of 11 programs for the analysis of 2D and 3D trusses and frames. The programs will handle linear and non-linear trusses including large deflections and non-linear material properties. They will also handle linear rigid frames. Input consists of forces and moments. Output consists of node deflections and member forces and moments.

TK! Solver Pack for Building Design and Construction

Software Arts, Inc.
27 Mica Lane
Wellesley, MA 02181
(617) 237-4000

Developed by: Software Arts, Inc.

Operating environment: memory requirement depends upon systems configuration.

Usage pricing: a suggested retail price of $100.

Documentation: user's manual provided.

Physical medium: 5¼" and 8" diskettes.

Product description: this package is designed for use with Software Arts' TK! Solver equation processing program. It covers numerous areas in building design and construction. The areas covered that deal with beams include: simple span beams; cantilever beam, uniform and/or point load; simple beam with overhang/ concentrated load; and simple beam with overhang/uniform load. Simple solid columns, which calculate column size based on load, unrestrained length, and proposed least dimensions are also covered in this program. The package allows the user to analyze solar heat gain with respect to heating load per month and per season, as well as heat loss. The package covers estimating for: floor framing, foundations, interior finish, roofing, and exterior wall costs. Section properties, rafter specification and stair calculator are other areas covered by this package.

Surveyor's Aide

Harrisonburg Computer Services
444 Myers Avenue
Harrisonburg, VA 22801
(703) 434-5062

Developed by: Harrisonburg Computer Services

Operating environment: memory requirement depends upon systems configuration.

Source language name and user availability: Microsoft Basic, source code is available.

Usage pricing: a suggested retail price of $500.

Custom modifications: on an individual basis.

Documentation: user's manual provided.

Physical medium: 4¼" and 8" diskettes.

Date first installed: 1982.

No. of packages installed to date: 40.

Product description: Surveyor's Aide is made up of three basic sections. The first is the data entry portion. Using the menu, the user can choose to either load past data from disk, or enter new notes. Once information has been entered, the package calculates and displays the angular error and overall ratio of error. The user can hand adjust any information which had previously been entered to reduce the error for mistakes in entry. When error is low, the traverse can be closed using either the compass, Crandall's or the transit rule. The second section of the program allows the user to pick from the functions on the cogo calculations menu. These functions include: 500+ points capacity, traverse closure with bearings or angles and zenith, all major coordinate geometry calculations are included, graphics capabilities on screen, and plotting capabilities on any printer with or without graphics. The third part of the program includes graphics and plotting. If the user chooses the "Graph Survey" option, selected points from the survey may be plotted on the screen for view of the work. If "Plot Survey" is chosen the user gets a printout of the plot.

Sewer and Drain Pipe Design
Utilities Engineering
213 19th Street, P.O. Box 299
Brigantine, NJ 08203
(609) 266-1774

Developed by: Utilities Engineering

Operating environment: memory requirement depends upon systems configuration.

Source language name and user availability: Applesoft Basic.

Usage pricing: a suggested retail price of $200.

Documentation: user's manual provided.

Physical medium: 5¼" diskettes

Product description: this is a rapid calculation program used to compare various alternate designs. The program will handle 100 manholes and the connecting pipe. The program assumes a "C" value of 100 and calculates for minimum slope if no slope is input. Gallons per minute or cubic feet per second are accepted by the program. Checks flow in pipe sizes chosen.

Miscellaneous Engineering and Scientific

Air Conditioning Requirements
Utilities Engineering
213 19th Street, P.O. Box 299
Brigantine, NJ 08203
(609) 266-1774

Developed by: Utilities Engineering

Operating environment: memory requirement depends upon systems configuration.

Source language name and user availability: Applesoft Basic.

Usage pricing: a suggested retail price of $200.

Documentation: user's manual provided.

Physical medium: 5¼" diskettes.

Product description: this program is written to determine the Tons (BTUH) of Air Conditioning required by a building envelope. These requirements are based on the estimated indoor temperature and the outdoor temperature, the various "R" factors for the building construction elements, the dimensions of the building, windows, and doors. The program then displays the recommended NET Tons of Air Conditioning required. Results can be sent to a printer, and a report is generated.

Interior and Exterior Lighting
Utilities Engineering
213 19th Street, P.O. Box 299
Brigantine, NJ 08203
(609) 266-1774

Developed by: Utilities Engineering

Operating environment: memory requirement depends upon systems configuration.

Source language name and user availability: Applesoft Basic.

Usage pricing: a suggested retail price of $200.

Documentation: user's manual provided.

Physical medium: 5¼" diskettes.

Product description: this program calculates the number of fixtures and lamps required to light a given area to a design requirement of "foot-candles." Lighting choices are built into the program. Some of the choices offered include: Fluorescent, Incandescent, and Sodium lighting.

Fluid Flow Analysis
Kelix Software Systems
11814 Coursey Blvd., Suite 220
Baton Rouge, LA 70816
(504) 769-6785

Developed by: Kelix Software Systems

Operating environment: memory requirement depends upon systems configuration.

Source language name and user availability: Basic, source code available.

Usage pricing: a suggested retail price of $100.

Documentation: user's manual provided.

Physical medium: 5¼" diskettes, tape cartridges.

Product description: this program is designed to solve fluid flow problems. The user may solve interchangeably for flow rate, pressure drop, and pipe diameter. The calculations apply to the laminar, transitional, and turbulent flow regimes. Fittings and valves are also taken into account. The user is able to select from a variety of systems of units, both English and metric. The user is provided with the common values of pipe roughness or may input other values.

Engineering Software for Micros

Kern International
433 Washington Street, P.O. Box 1029
Duxbury, MA 02331
(617) 934-0445

Operating environment: memory requirement depends upon systems configuration.

Source language name and user availability: Basic.

Usage pricing: a purchase price of $21.50.

Documentation: user's manual included.

Physical medium: 5¼" or 8" diskette, hard disk.

Product description: this is a package of 25 programs for computer-aided design and analysis. Emphasis in all programs is on combining computer graphics with engineering problem solving. Programs are included to interactively create engineering drawings, store them on disk, recall, update, merge, and produce 3 dimensional views. Physical properties such as weight, electrical resistance and thermal capacity may be assigned to components of the drawing.

CALFEX Electrical Engineering Application Pack

Interlaken Technology Corporation
6535 Cecilia Circle
Minneapolis, MN 55435
(612) 944-2624

Developed by: Interlaken Technology Corporation

Operating environment: 48KB RAM, DOS 3.3 operating system.

Source language name and user availability: Basic, source code available.

Usage pricing: a suggested retail price of $150.

Maintenance provisions/pricing: telephone support provided.

Custom modifications: custom applications available, price not preset.

Documentation: user's manual provided.

Physical medium: 5¼" diskette.

Date first installed: July 1983.

Product description: this package is made up of a collection of common programs for electrical engineers, for use with CALFEX. The programs deal with common electrical engineering problems. All are written to be as general purpose as is practical. They may be modified or added to by the user. Complete derivation and documentation is provided. Examples or programs included are: Class A Transistor Amplifier Design, Schmitt Trigger Circuit Design, Transformerless Power Supply, Bridge-Rectified Power Supply, Dual-Output Regulated Power Supply, Gaussian Pulse Bandwidth Calculation, Microstrip Design, Strip-Line Design, Lossless Transmission Lines, Single-Wire Transmission Line, Two-Wire Transmission Line, RF Air-Core Inductor Design, Voltage Standing Wave Ratio Calculation, and Decibel Conversion.

Micro-Cap

Spectrum Software
690 W. Fremont Avenue, Suite 11
Sunnyvale, CA 94087
(408) 738-4387

Operating environment: 64KB RAM, DOS 3.3 operating system.

Source language name and user availability: Basic.

Usage pricing: a purchase price of $475.

Documentation: user's manual included.

Physical medium: 5¼″ diskette.

Product description: Micro-Cap is an electronics drawing and analysis package. It enables the user to design and predict the performance of a circuit without having to actually build it. The package also allows the user to: draw an electronic circuit diagram directly on the screen; perform an A/C (Bode) analysis showing gain and phase shift vs. frequency; perform a D/C transfer characteristics analysis showing output voltages vs. input voltage; and run a time domain simulation of circuit responding to user-defined input sources. Circuit diagrams are drawn directly on the screen using the design module. Using the keyboard, the drawing cursor is moved around on the screen adding or deleting components. Each time a component is added the package draws it at the cursor position. Drawing interconnecting lines allows the user to create any type of circuit diagram. Completed circuit diagrams can then be saved in a diskette file.

Digital Filter

Dynacomp, Inc.
1427 Monroe Avenue
Rochester, NY 14618
(716) 442-8960

Operating environment: memory requirement depends upon systems configuration.

Source language name and user availability: MBasic, Basic.

Usage pricing: a purchase price of $50 for cassette, $54 for 5¼″ diskette.

Maintenance provisions/pricing: technical support available.

Documentation: user's manual included.

Physical medium: 5¼″ or 8″ diskettes, cassette.

Product description: the Digital Filter is a data processing system which enables the user to design his or her own filter function or choose from a menu of filter forms. These filter forms are subsequently converted into nonrecursive convolution coefficients. In the explicit design mode the shape of the frequency transfer function is specified by directly entering points along the desired filter curve. In the menu mode, the ideal low pass, high pass and bandpass filter can be approximated to varying degrees according to the number of points used in the calculation. These filters can also be smoothed with a Hanning function. In addition, multistage Butterworth filters can be selected. The system features plotting of data before and after filtering, and display of chosen filter functions.

Wiremaster

Afterthought Engineering
7266 Courtney Drive
San Diego, CA 92111
(619) 279-2868

Developed by: Afterthought Engineering

Operating environment: memory requirement depends upon systems configuration.

Source language name and user availability: C, source code is available for $2500.

Usage pricing: a suggested retail price of $195.

Documentation: user's manual provided.

Physical medium: 5¼″ and 8″ diskettes.

Product description: this program aids in the design, layout, construction, and documentation of electronic hardware. This is designed basically for wire wrap construction techniques, but can be used in layout, error checking, and troubleshooting of PC boards. The inputs to the program are derived from the user's schematic diagram, and are typed into one of the computer's files with a text editor. Wiremaster processes this file, checking for errors and inconsistencies, and produces outputs in the form of disk files, which can be printed on paper, checked over on a CRT terminal, or used to drive numerically controlled wire wrap machines. The information obtained from this can be used for layout checking, wiring error detection, component installation, circuit debugging, and parts list generation.

System Simulation Software

Scientific MicroPrograms
213 Merwin Road
Raleigh, NC 27606
(919) 851-8111

Developed by: Scientific MicroPrograms

Operating environment: 48KB of user memory required.

Source language name and user availability: Basic.

Usage pricing: purchase price lists at $500.

Maintenance provisions/pricing: consulting services concerning software are available.

Custom modifications: custom modifications are available.

Training and installation services: no training is required.

Documentation: complete documentation is included in purchase price.

Physical medium: 5¼″ diskette.

Product description: System Simulation Software is a systems analysis and simulation system, which allows up to 25 variables to be interrelated by differential equations, and will solve, extrapolate, and display the result. This sytem is useful in fields ranging from engineering (the design of mechanical and electrical components and systems, thermodynamic response, etc.) to biology and medicine (description of the response of organisms or populations to stimuli). A few lines of "user program" are all that is needed to run complex simulations including tabulated input data, optimization, and changing initial card boundary conditions.

Robographics

Chessell-Robocom Corporation
111 Pheasant Run
Newtown, PA 18940
(215) 968-4422

Operating environment: 64KB RAM.

Source language name and user availability: Basic.

Usage pricing: a purchase price of $1,095.

Documentation: user's manual included.

Physical medium: 5¼″ diskette.

Product description: installed in any Apple II or IIe computer, Robographics provides a range of technical facilities. The design engineer, architect, draftsperson and illustrator can focus on the creative design and leave the drawing, erasing, and redrawing tasks to the computer. The controller serves as both the drawing instrument and function selector. There are no keystroke commands.

TK! Solver Pack for Mechanical Engineering

Software Arts, Inc.
27 Mica Lane
Wellesley, MA 02181
(617) 237-4000

Developed by: Software Arts, Inc.

Operating enviroment: memory requirement depends upon systems configuration.

Usage pricing: a suggested retail price of $100.

Documentation: User's manual provided.

Physical medium: 5¼" and 8" diskettes.

Product description: this applications package includes 13 models, which are made up of equations, values, and tables to help the user solve problems in mechanical engineering. Some of the problem areas covered include heat transfer, stress/strain, moment of inertia, hydraulics system analysis, fluid flow in pipes, helical spring, and natural frequency of vibration. The models can be used as is or may be modified by the user. Analysis can be done on a two-point supported beam, cantilever beam, and elastic torsion beam. The package also deals with the resolution of axial and tensile stresses using Mohr's circle. This package is designed for use with TK! Solver program.

USEFUL PERIODICALS

Business Week Magazine, 1221 Avenue of the Americas, New York, N.Y. 10020

Consulting Engineer Magazine, The Dun & Bradstreet Corporation, 1301 S. Grove Ave., Barrington, Ill. 60010

Engineering News-Record Magazine, 1221 Avenue of the Americas, New York N.Y. 10020

Harvard Business Review, Soldiers Field, Boston, Mass. 02163

Professional Engineer, National Society of Professional Engineers, 2029 K Street, N.W., Washington, D.C. 20006

Specifying Engineer Magazine, Cahners Publishing Co., 5 S. Wabash Ave., Chicago, Ill. 60603

PUBLISHERS

Academic Press Inc., 111 Fifth Avenue, New York, N.Y. 10003

AMACOM Book Division, 135 W. 50th Street, New York, N.Y. 10020.

American Chemical Society, 1155 16th Street, N.W., Washington, D.C. 20036

American Society for Metals, 9275 Kinsman Road, Metals Park, Ohio 44073

Ann Arbor Science Publishers, Division of Butterworth Publishers, Inc., 10 Tower Office Park, Woburn, Mass. 01801

Avon Books, 959 Eighth Avenue, New York, N.Y. 10019

Ballinger Publishing Co., 54 Church Street, Harvard Square, Cambirdge, Mass. 02138

Banner Press, P.O. Box 6469, Chicago, Ill. 60680

Butterworth Publishers, Inc., 10 Tower Office Park, Woburn, Mass. 01801

CGS Educational and Professional Publishers, 383 Madison Avenue, New York, N.Y. 10017

Chemical Publishing Co., Inc.. 155 W. 19th Street, New York, N.Y. 10011

Construction Publishers, 4552 E. Palomino Road, Phoenix, Ariz. 85018

Consulting Opportunities Journal, 1040 Bayview Drive, Suite 209, Ft. Lauderdale, Fla. 33304.

Crowell-Collier Press, Division of Macmillan Publishing Co., Inc. 866 Third Avenue, New York, N.Y. 10022

Marcel Dekker, Inc., 270 Madison Avenue, New York, N.Y. 10016

Dodge Building Construction Services, McGraw-Hill Information Systems Company, 1221 Avenue of the Americas, New York, N.Y. 10020

Gale Research Company, Book Tower, Detroit, Mich. 48226

Garland Publishing, Inc., 136 Madison Avenue, New York, N.Y. 10016

Halsted Press, Division of John Wiley & Sons, Inc., 605 Third Avenue, New York, N.Y. 10158.

Heyden & Sons, Inc. 247 S. 41st Street, Philadelphia, Pa. 19104

Holt, Rinehart & Winston, Inc., 383 Madison Avenue, New York, N.Y. 10017

IEEE, 345 East 47th Street, New York, N.Y. 10017

Industrial Press, Inc., 200 Madison Avenue, New York, N.Y. 10157

Irvington Publications, 551 Fifth Avenue, New York, N.Y. 10176

Richard D. Irwin, Inc., 1818 Ridge Road, Homewood, Ill. 60430

ISI Press, 3501 Market Street, University City Science Center, Philadelphia, Pa. 19104

Kendall-Hunt Publishing Co., 2460 Kerper Boulevard, Dubuque, Iowa 52001

Jeffrey Lant Assoc., 1040 Bayview Drive, P.O. Box 7514, Ft. Lauderdale, Fl. 33338–7514

Lexington Books, 125 Spring Street, Lexington, Mass. 02173

Macmillan Publishing Co., Inc., 866 Third Avenue, New York, N.Y. 10022

McGraw-Hill, Inc., 1221 Avenue of the Americas, New York, N.Y. 10020

R. S. Means Co., Inc., 100 Construction Plaza, Kingston, Mass. 02364

Charles E. Merrill Publishing Co., 1300 Alum Creek Drive, Columbus, Ohio 43216

MIT Press, 28 Carleton Street, Cambridge, Mass 02142

National Society of Professional Engineers, 2029 K Street, N.W., Washington, D.C. 20006

Noyes Data Corp., Mill Road at Grand Avenue, Park Ridge, N.J. 07656

Oceana Publications, 75 Main Street, Dobbs Ferry, N.Y. 10522

Penn Well Publishing Co., P.O. Box 1260, Tulsa, Okla. 74101

Pitman Publishing Inc., 1020 Plain Street, Marshfield, Mass. 02050

Plenum Publishing Corp., 233 Spring Street, New York, N.Y. 10013

Prentice-Hall, Inc., Englewood Cliffs, N.J. 07632

Princeton Publishing Inc., 221 Nassau Street, Princeton, N.J. 08540

Rensselaer Polytechnic Institute, Troy, N.Y. 11281

Ronald Press, 605 Third Avenue, New York, N.Y. 10158

Charles Scribner's Sons, 597 Fifth Avenue, New York, N.Y. 10017

Springer-Verlag New York, Inc., 175 Fifth Avenue, New York, N.Y. 10010

Van Nostrand Reinhold (VNR), 135 W. 50th Street, New York,. N.Y. 10020.

Wadsworth Publishing Co., 10 Davis Drive, Belmont, Calif. 94002

John Wiley & Sons, Inc. 605 Third Avenue, New York, N.Y. 10158

section 5

Protecting Your Consulting Firm against Liability

In Section 1 we emphasized the importance of maintaining an accurate ongoing appraisal of your consulting engineering firm. By knowing your professional capabilities and limitations, you can make the best use of the essential tools that shape your business. In this section we focus on protecting yourself and your firm from professional liabilities—an area in which your accurate self-evaluation is particularly crucial to achieving and maintaining a successful consulting engineering enterprise.

The question of what to do if you encounter professional or legal problems hinges on several factors. Among them are: your own personal and professional attributes and those of your clients, colleagues, and attorneys; specific requirements of different engineering disciplines; the pros and cons of various types of liability insurance; and the basic dynamics of human psychology.

Here we shall explore each of these factors in depth. We will also discuss some common pitfalls experienced by consulting engineers and provide guidelines on how to avoid those pitfalls. Specifically, we give you details on what steps you might take if you become the subject of client complaints or lawsuits. We also cover some of the latest theories on behaving optimally under stress, plus preventing and resolving conflicts. Several useful checklists are provided.

YOUR PROFESSIONAL IMAGE

An essential ingredient of professionalism is the maintenance of standards of excellence in the performance of the tasks involved in your profession. As a means of safeguarding society against potential harm from the abuse or misapplication of professional skills, many professions—including engineering—have adopted ethical codes and procedures for restricting an individual's capacity to practice. Clearly, the licensing and registration of engineers requires all who wish to join and remain in the profession to fulfill certain responsibilities. Chief among these duties are the upholding of technical standards which support satisfactory relations with clients. Another requisite is the engineer's maintenance of his or her professional image.

Your professional image will help to determine the course of your consulting engineering work and your relationships with others. For example, more than a few engineering lawsuits have their roots in a client's questions about an engineer's ability to produce a design on which the engineer is not known to be experienced. Conversely, some lawsuits originate from projects an engineer may have been involved in many times in the past. Regardless of whether you have produced a particular type of design or other piece of engineering work once or a hundred times, your professional image must support the idea that you undertake all your work responsibly. Further, you must check all your work thoroughly to provide consistently satisfactory results.

Consider, as an example, the project(s) on which you are currently working. If a problem were to develop in the installation of your design and the contractor tried to place the blame on you, how would your client react? Moreover, what is the client's general attitude toward you? If extra costs develop, would your client automatically seek legal redress, or would she or he be amenable to a mutually agreeable settlement out of court? While no consultant can know the innermost workings of a client's mind, we do want to emphasize that your professional image will directly influence your client's responses to any difficulties encountered during the course of an engagement. Generally, an engineering firm will have far fewer claims held against it when its principals are well respected in their field.

How does one gain such acceptance and respect? First, by completing all work competently—on schedule, within budgetary requirements, and in a way that satisfies your clients' needs. But there is more to professionalism than mere competence.

The true professional generally does not take on more work than he or she is capable of completing satisfactorily at any one time. Further, the professional consultant does not attempt to do work for which he or she is not fully qualified. In the actual practice of consulting engineering, this last point may become a source of much consternation since it is often difficult to draw a clear line on the limitations of an engineer's experience. Consultants frequently engage in new areas of design to enlarge the area of their firms' general expertise. But some consultants pursue this expansion to extremes. Occasionally, consulting engineers go so far as to cut themselves off from all but the most experimental and complex type of projects. What often occurs in such cases is a gradual reduction in the number of work prospects for the firm. If the firm is not qualified for the jobs it undertakes, its image will deteriorate and professional and legal difficulties may ensue.

The importance of the engineer's professional image underscores the role that psychology and human relations play in the pursuit of a successful consulting business. Unfortunately, few engineers receive training in the dynamics of human relations. The engineer's difficulties may also be compounded if he or she lacks training in various communications skills. The engineer who is fortunate enough to cultivate skills in these areas will tend to have greater success and fewer complaints from clients since consulting is indeed a people-oriented business.

ELIMINATING TROUBLE SPOTS

Let's face it: Engineers *can* make mistakes. In fact, many engineers tend to repeat the same errors over and over again. Only the strictest attention will help to eliminate trouble spots, which are not confined to technical errors. The most insidious, repetitive mistakes many consultants make are errors that undermine good relations with clients and fellow professionals. For instance, some consulting engineers would rather try to "prove the client wrong" on a point of disagreement than swallow their pride and concede the client may be correct. This is the cause of many lost contracts and complaints about poor consulting work.

It is not unusual for an engineer to try to push a technical point; after all, the engineer is the expert trained to handle engineering problems. But while an aggressive attitude often makes for skillful engineering, the same approach can lead to disasters when it is applied to human relations. Generally, it is far more productive for the engineer to adopt the attitude that the client or fellow professional is correct—at least initially. Then the engineer can begin to explain his or her position in terms that are accepted and easily understood by those who lack the consultant's technical expertise. All successful professionals need to remind themselves of this point periodically.

Thus, one of the first steps you can take to protect yourself and your firm from lawsuits is to learn to recognize trouble spots—i.e., any areas of difficulty occurring repeatedly in your professional image, your communication skills, and your engineering applications. Once you know the trouble spots, you can eliminate them systematically. How? A simple way is to use checklists.

EXAMPLES OF HELPFUL CHECKLISTS

The development and use of appropriate checklists can help you substantially improve your consulting and your income. Helpful checklists can be designed to cover all phases of your firm's day-to-day operations.

Exhibit 5-1 shows one type of checklist used by a firm engaging in electrical work. You can tailor individual lists to the specific nature of your own projects by using Exhibit 5-1 as a model.

Every consulting firm should check all of its specifications thoroughly—not only to verify that all necessary items are included but also to see that the language of the specs is easily comprehended and totally logical. Interestingly, one of the most important characteristics of good specification writing is correct spelling. Engineers often presume that the intent of a specification will be clear even if words are misspelled; hence, they do not pay much attention to checking their spelling. However, if you are ever asked to testify in court about an aspect of your work, a misspelled word could be used as evidence against your version of the specs. Clearly, all drawings, specifications, and other project components must be checked thoroughly by a skillful reviewer if you want to avoid potential problems. Language and spelling may seem like trivial details when you are struggling to meet a deadline, but good checking procedures are one of the most important tools you have for protecting yourself against liabilities.

Even with the tightest deadline, an engineer-reviewer can check completed plans and specifications at the same time that contractors are bidding the work or while the client examines the project. The ideal reviewer is usually a well-organized individual who has not been involved in the preparation of the project. When a detached reviewer is not available, the material must be checked by one or more of the individuals working on the project.

Another area in which a final review proves highly valuable is in cost estimation. If the engineer makes a detailed cost estimate after the project design has been prepared, two benefits are achieved: (1) the cost of the job is determined and (2) the engineer is compelled to give the plans and specifications a thorough, final review.

EXHIBIT 5-1   A typical specification and plans checklist.

Electrical items—specification checklist

1. Work to be done by electrical subcontractor (scope).   _____

2. Work by others—identify others by trades.   _____

3. Costs of connections by power company, telephone company, and fire department.   _____

4. Shop drawings required by specific items.   _____

5. Control wiring requirements for specific items.   _____

6. Circuit breakers (thermal and magnetic).   _____

7. Thermal overloads (all applicable divisions)—three overloads to each three-phase motor.   _____

8. Wire insulation details.   _____

9. Wire and cable splicing details.   _____

10. Conduit and wire sizes.   _____

11. Are specified catalog numbers available?   _____

12. Three vendors named with approval equal for all fixtures, equipment, and systems.   _____

13. Fire alarm system requirements. Flashing red lights and minimum fire station mounting height.   _____

14. Emergency lighting system requirements.   _____

15. Lighting fixtures, standard manufacture.   _____

16. System interruptions—temporary power—detail down-time requirements and limitations.   _____

17. After-installation tests—all systems, primary cable (new only)—secondary wiring, batteries and generators.   _____

18. Specify detailed testing requirements before and after alterations for additions to existing systems.   _____

19. UL approval paragraph.   _____

20. Minimum size of wire—#12 AWG for branch circuits.   _____

21. Minimum concrete over conduit as per ACI Code.   _____

22. Emergency lighting, battery guarantee—10 years.   _____

23. Aluminum conduit embedded in concrete or underground, not allowed.   _____

24. Additional ground wire in all E.M.T.   _____

25. Compatible equipment for existing system.   _____

26. No reference to local codes, or supervision except power and telephone utility companies.   _____

27. Include requirements and recommendations of the latest editions of NEC, state fire safety code, NEMA, IES, IPCEA, ASTM, ASA, ASHRAE, NSPA, NBFU, and all applicable standards.   _____

28. No bidding can be qualified.   _____

29. Are specifications correlated with state of _____ general conditions?   _____

EXHIBIT 5-1   (Cont.)

Electrical items—specification checklist

30. Paragraph requiring bidder to visit site prior to submitting bid.   _____
31. Include the following information:
    A. Is existing service sized for additional loads?   _____
    B. Letter from engineer providing lighting intensities for all areas and con-
       nected and demand loads.   _____
    C. Letters of approvals and cost estimates from all concerned utilities.   _____

Electrical items—plans checklist

 1. Feeder data, wire and conduit sizes—type and source.   _____
 2. Panel details with active and spare breakers and circuit designations.   _____
 3. Complete riser diagrams for all systems with number of wire sizes and race-
    way type and sizes.   _____
 4. Fixture delineation sheet and designation on plan.   _____
 5. Lamps and sizes included.   _____
 6. Emergency lighting and zone control.   _____
 7. Service connections (source and type).   _____
 8. Detail service entrance conduit, wire, and transformer sizes.   _____
 9. Length of cable runs (interior and exterior).   _____
10. Detailed vault requirements including ventilation.   _____
11. Mounting heights of equipment, and fixtures.   _____
12. Grounding details for all outside lighting standards.   _____
13. Detail hazardous area requirements.   _____
14. Generators used for emergency lighting and standby loads must first serve
    emergency lighting use.   _____

Review procedures become more difficult when the plans incorporate new types of procedures, products, or solutions. In this situation, an extra measure of safety can be obtained by asking the manufacturers of the processes or products to inspect the plans to certify that all items have been covered properly by the consulting organization.

Aspects of the project that cannot be clarified in advance present worse problems. For instance, the engineer may be unable to tear down a wall in an existing structure to determine what is behind the wall. This could have a serious effect on work progress since the hidden wall area might contain items such as electrical wiring or flammable materials.

In the past, many engineers attempted to protect themselves from potential liabilities in such cases by including in the specifications the time-honored words, "The contractor shall carefully investigate the existing site and be responsible for all assumptions he may make thereto." This, in effect, put the burden of finding out what was behind the wall onto the contractor's shoulders. At the same time, it enabled the engineer to proceed without fear of difficulties. However, such specification provisions are no longer considered (in most cases) an acceptable means of defending the engineer from liabilities.

Hence, the engineer should discuss the situation with the client to provide the client with a clear understanding that any unusual condition found behind the wall may lead to extra charges. A written agreement should be signed by both parties, stating that the engineer is not to be held responsible for the charges. Alternatively, the client can be asked to absorb the expense of tearing down the wall so the engineer can modify his or her plans according to the findings.

Other problems occur if an engineer utilizes existing plans of a structure and later discovers inaccuracies in the plans' description of the wall's contents. In such a situation, the client may complain that the engineer should have detected the error in advance. To a large extent, the client would be correct. Flawless information must be obtained ahead of time by the engineer—no matter what the cost. When this is not feasible, allowances should be made for the possibility of extra charges arising from conditions unknown at the time the engineering work was prepared.

INSURANCE COVERAGE

Many consultants are unsure of how much insurance to carry to protect themselves and their practices from potential liabilities. In some states, the question may be partially decided for the engineer by the state government; certain state authorities specify the amount of insurance that must be carried for engineering contracts. Where this is not the case, the engineer should consult locally with colleagues and members of voluntary professional organizations. Actual dollar amounts are determined by a complex variety of factors, and requirements are changing constantly.

The number and degree of damage claims against engineers have risen steadily during the last decade, just as they have for physicians and some other professionals. For the consulting engineer, it is even harder to assign values to insurance coverage in structural work since the result of an error in this discipline may not only be property loss but may also include a loss of life.

We cannot properly recommend specific amounts of coverage for your firm without knowing what kinds of work you engage in. However, we can make some general suggestions based on the current practices of consultants around the country. For most small firms with less than 10 members, professional liability insurance valued at $250,000 to $500,000, with a $5000 deductible, seems reasonable. If you do a lot of structural work, it might be worthwhile to choose a higher level of coverage; some firms carry policies worth $1 million or more to have adequate protection against roof problems, for which claims may be enormous.

The premium cost of any errors-and-omissions professional liability insurance will be substantial, ranging from $2000 to $15,000. This could be 5 percent of a small firm's gross income, or about 20 percent of the principal's income.

In view of these high costs, one might reasonably wonder, why buy any insurance at all? A logical answer would be that the consulting engineer needs liability insurance for the same reason a family needs life insurance: The policy provides a measure of protection against harm that could arise from *potential* accidents or loss.

Based on current data, almost every engineering firm will eventually become the target of a potential lawsuit (or series of lawsuits) concerning some aspect of the firm's work. The question of how many lawsuits a firm might encounter—and how much money it might be sued for—depends on many variables. Among them are: the size and type of the consulting organization, the amount of work it handles, the kind of consulting the firm does, the location of the firm's offices, and the characteristics of the client.

The experience of engineers working in a variety of areas shows that most lawsuits cost from $3000 to $15,000 in legal fees. Based on this observation, you might ask, won't an errors-and-omissions policy with a $5000 deductible provide sufficient protection? Provided that the total cost of all legal fees plus damages does not exceed the face value of your policy, your loss will be limited to the deductible plus the value of your time lost from work due to involvement in the lawsuit. The cost of lost time can in some cases amount to a substantial portion of your yearly income.

Clearly, a good professional liability policy is well worth the price. Some firms manage to operate without errors-and-omissions insurance, but they do so at considerable risk.

When looking for professional liability insurance coverage, be sure to check with your professional society. Some are offering such insurance to their members. Here is a description of the professional liability insurance policy offered by the American Society of Mechanical Engineers (ASME) to its members.

Professional liability coverage was introduced when claims were made against medical professionals, and most engineers have only a limited perception of the exposures they face for claims alleging errors, omissions, or negligent acts in their professional services.

Employed engineers may erroneously believe their employers will provide a protective shield and defend them if a claim names the employee for an alleged professional mistake. In fact, though, some engineers have found it necessary to provide their own defense against claims arising out of their professional activities.

The Committee on Group Insurance for ASME Members feels that, in today's legal climate, suits against professionals are increasing, and encourages all members to consider the professional liability program which has been developed to help meet the needs of members of the Society.

The program offers protection for both employed and some consulting engineers against claims of real or alleged errors, omissions, or negligent acts in their professional services. It protects the individual engineer from claims arising out of services rendered by persons for whom he or she is legally responsible. The plan also covers the cost of defense, as well as payments ordered by a court of law or agreed to by settlement or arbitration.

Important highlights of the program are:

- *Occurrence coverage,* which protects against claims arising from services rendered during the policy period, regardless of when the claim is presented. This is much broader than "claims made" coverage, which requires that the alleged error occur and the claim be made while the coverage is in effect.

  This feature is particularly important because an alleged error or omission may take years to manifest itself, or state statutes of limitation may provide many years during which a claim may be made.

  It also eliminates the problem faced by many engineers when they retire because, with occurrence coverage, there is no need to continue the additional expense of carrying a "claims made" policy or purchasing a "discovery endorsement" or a "retired professionals" policy.

- *Coverage for part-time professional services,* which pertains to the employed engineer who provides professional services whether or not they are related to his or her primary employment.

- *Coverage for some consultants,* which applies to individual engineers who work as full-time consultants as long as the engineer remains a sole practitioner with no professional staff.

- *The definition of "engineer,"* which under the program includes graduate engineers and others who, through experience, qualify and hold positions classified for engineers. There is no requirement that the engineer be a registered professional engineer in order to obtain coverage.

- *Limits of liability,* which provide ASME members with a choice of options. Plan I offers limits of $200,000 for each claim and $600,000 for an annual aggregate. Plan II offers a combined single limit of $1-million.

- *No deductible,* which protects the participating engineer against the cost of all defense expenses and any judgment up to the policy limits.

- *Worldwide coverage,* which insures the engineer against the cost of all defense expenses and any judgment up to the policy limits.

- *Professional libel and slander coverage,* which protects the engineer against such claims if they arise out of professional services. Engineers, for example, are frequently asked to review the work of another professional. If this review is unfavorable, a claim alleging libel, slander, or defamation of character could be filed.

For further details, write to the ASME insurance administrator, Smith-Sternau Organization, Inc., 1707 L Street, NW, Suite 700, Washington, D.C. 20036.

THE IMPORTANCE OF LANGUAGE

While engineers primarily deal with facts and figures, they cannot apply their expertise without becoming involved with people and personalities. Prominent engineers have noted that the typical engineering schools devote little time to lessons on psychology, sociology, history, or language. Thus, it comes as no surprise that engineers are often accused of lacking skills needed to communicate and deal with people effectively.

Your skill in handling language will have a direct bearing on your ability to protect yourself and your firm from legal difficulties. There are several approaches to achieving and enhancing good engineering communication. Some of the guidelines that we present will repeat themes that have appeared earlier. However, these are tailored specifically to the topic of protecting your business from legal problems.

The first guideline is that your written and spoken communications should be *concise.* This does not mean that you must limit your engineering discussions to cold, technical terms. It does mean, however, that you should strive to eliminate unnecessary and unclear words from your presentations. A contract or proposal that is brief but thorough generally has much less likelihood of being misinterpreted by the client than one that is long and involved. Hence, a concise communication has less potential for generating legal problems (as long as it is logical, comprehensive, and contains no ambiguous terms).

The second guideline is: Do not use words or phrases that can be interpreted in ways other than the way in which you intend to use them. A simple example of an ambiguous phrase is "Stop the work," which some engineers include in their instructions to contractors. The phrase may sound simple and concise, but it is fraught with potential trouble.

Usually, an engineer will use "Stop the work" to tell a contractor to temporarily hold off on a particular part of a project. While the engineer might reasonably assume that the contractor knows exactly what part of the job the engineer is referring to, the communication can easily backfire—especially when examined in a court of law. For example, if the phrase is not qualified by details on what areas of work you want stopped, the contractor might actually misinterpret "Stop the work" to mean "Stop *all* the work."

Or an unscrupulous contractor wishing to cover up his or her own delays or errors might be able to transfer the blame for a delay to the engineer on the premise that the engineer's instructions had been misleading. Pursued to an extreme, the case might end up with the contractor's successfully suing the engineer for unnecessary expenses allegedly incurred as a result of the command to stop the work. That is a realistic example of how language, improperly handled, can get you into legal difficulties.

Attorneys spend years learning the best ways to use language to protect their clients from legal difficulties. Ultimately, an adept attorney who is well acquainted with the engineering profession will be your best protection against potential lawsuits. However, few consultants can afford to have an attorney examine all their communications.

Nevertheless, *all letters* produced by your office should be reviewed by an individual who possesses superior language skills. It takes a lot of time and hard work to develop a keen eye for trouble spots in engineering communications; but it may save your firm a lot of grief.

In specification writing, opportunities for ambiguity are considerable. Therefore, addenda and clarifications to specs are common. As an example of words that may have to be clarified in the footnotes of an engineering specification or contract, look at the meanings of these words: "furnish," "install," and "provide." Some dictionaries define "install" as to place, establish, settle, or fix in a position. "Furnish" means to supply, provide, or give. "Provide" means to supply, furnish, or make available. Each word has subtle nuances that distinguish it from the rest. Thus, all three words must be used with great care in any formal presentation.

In specifications, engineers often use words that have highly specialized meanings. For example, a piece of equipment may be specified to "have all standard appurtenances required for proper functioning." To a lay jury, these words may be confusing. Therefore, it may be beneficial to clarify the phrase (and others like it) whenever you use them.

Another term that is often used with insufficient clarity in specifications and other engineering communications is "approval." Used by engineers, "approval" generally means limited or conditional permission for a particular item to be used. However, most dictionaries define the word to mean sanction, consent, or favor toward a given action or item.

In actual court cases, lay juries have accepted an engineer's "approval" as implying *unqualified acceptance.* Judges frequently agree with this interpretation.

An engineer's use of the term "approval" in a written agreement has sometimes been regarded as a waiver of the requirements and conditions that the engineer may originally have intended. In view of these facts, it should be easy to see why you must develop strong language skills and review all communications carefully *before* they leave your office.

Not only the words you use but also the tone you convey will have an effect on your audience. Humor, anger, sarcasm, and other attitudes and feelings must be employed with caution so you achieve the effect you desire. If you act brusque or irreverent on the telephone, for example, you may actually jeopardize your working relationships along with your legal standing.

The American Institute of Architects publishes a useful guide called "The Glossary of Construction Industry Terms." The guide provides clear-cut definitions of many terms used in design and construction. All consulting engineers would be well advised to obtain a copy. Further, you can increase and enhance your working vocabulary by using legal dictionaries and glossaries that cover consulting engineering. (See the Consultant's Reference Shelf at the end of Section 4 for more information.)

The following are a few selected terms and definitions that you should be familiar with in your work.

Drafter   The term "drafter" is generally applied to anyone who draws the schematic drawings or plans of a building or other facility, system, or piece of equipment, without rendering an engineering judgment. In insurance, a drafter is an individual who prepares plans and specifications but does not visit the job site. In regard to workers' compensation insurance, a licensed engineer may be called a drafter—even though the profession recognizes her or him as a professional engineer.

Or Equal   The generally accepted definition of "or equal" is that one item can be substituted for a specified item if it has the same general characteristics or is equivalent to the specified item. The term is ambiguous when used in reference to certain construction components since it may be interpreted to mean the exact duplication of one manufactured item by another. Avoid using "or equal" whenever possible.

Project Representative   The term "project representative" is usually defined as an individual who represents the design professionals at a construction site. The project "rep" is someone who is responsible for reviewing the progress of the work to determine whether construction is proceeding in accordance with the design concept. A project representative is not responsible for superintending the contractor's employees, the method of construction, job safety, or other duties connected with implementing the contract. The representative's authority is derived exclusively from the owner and is strictly limited to the capacity of an observer.

Safe Place to Work   The term "safe place to work" is usually defined as a place where the employer or other party chargeable with employee safety has limited the hazardous conditions of the workplace. Failure to provide a safe place to work is often used as a basis for claims by contractors and employees against both owners and design professionals when contractors or employees sustain injuries during construction. Many states have statutes that require a safe place to be provided for workers.

Timely Delivery of Plans and Specifications   Some professional liability policies for design professionals contain exclusions relating to the failure to make "timely delivery of plans and specifications." A plaintiff may allege that failure to complete delivery on time has impeded the work and caused damages. Claims may be made even when the design professional was unable to control factors that created the delay (e.g., inclement weather, labor strikes, or health problems). Contracts imposing deadlines on the delivery of plans and specifications should be qualified so as to ensure that the design professional will not be held liable for conditions beyond his or her control.

There are innumerable other terms that we could cite. But even the preceding short list illustrates how crucial your language skills can be. Words can spell the success—or the failure—of an engineering practice. As a consultant, it is your responsibility to use words carefully.

To enhance your communications systematically, consider recording your phone calls with an electronic tape recorder. By reviewing your telephone performance from the tapes, you should be able to find potential trouble spots. Once found, they can be eliminated.

Few practicing engineers keep elaborate records of all their conversations in the office since doing so would be costly and time-consuming. However, it is important that you try to keep a record—in a telephone diary—of the main points of major conversations affecting your clients and contracts. Your telephone diary notes may prove to be a valuable reference source later on. Of course, any alterations of contracts should be made only with the written consent of the parties involved. Verbal contracts are often difficult to support in a court of law.

Later in this section we will examine specific techniques for successful bidding and contract development. First, let's discuss some general ways to prevent potential problems from turning into crises in your business.

HANDLING STRESS

Virtually every person who has ever owned or operated a consulting firm has had to endure stress at some time. Stress comes in many forms: physical, emotional, psychological, economic, etc. Through experience, most of us learn to adapt to certain types of stress. However, the profession of consulting engineering presents continued challenges. Nearly every project and every meeting leads to new pressure.

Research in human behavior shows that you can learn to handle stress the same way you learn other subjects—by study and practice. An example would be the way you might learn to respond to a fire in your home. If you practice fire drills and learn to use the necessary tools (i.e., fire extinguisher, escape ladder, telephone, etc.) for a fire emergency, you will be better prepared to handle a real emergency in a rational, orderly way. The same principles apply to consulting engineering.

For example, by practicing your sales presentations, as we discussed earlier, you should become well equipped to handle clients' questions and successfully sell your services. Further, you will develop confidence and aplomb.

Stress is a natural response to an imagined or actual threat. Usually, stress develops when two or more parties compete against each other to attain a goal. While you may have to compete against other consulting firms for a particular engagement or for a client's approval, you will rarely, if ever, compete against a client. Unfortunately, many engineers fail to recognize this fact; instead of working with their clients to attain mutual goals, they lay the foundations for disapproval and conflict. Projects proceed much more smoothly when an engineer realizes that the consulting firm and the client form a team along with those who provide the contracting solutions to the client's problems.

The real foundation of a successful engagement is created by your expertise, courtesy, and respect. Thus a proper attitude is frequently a more powerful influence on a project's outcome than are economic factors.

If the possibility of a conflict does arise, you can avoid a great deal of difficulty by promoting an atmosphere of compromise. All members of your firm who participate in contract meetings, conferences, and discussions with clients must be prepared to calmly consider all the relevant facts. As the boxer Joe Louis once said of an opponent in the ring, "He can't run and he can't hide." The same is true of the consulting firm.

Client's questions and problems should be dealt with as quickly as possible. If you do not resolve a problem or complaint immediately, your client may seek a legal resolution, which could cause the question to drag on for years. Further, a court case usually places the resolution in the hands of people who are not trained in the engineering or construction fields. And a court settlement may result in enormous losses to your firm—not only in terms of cash but in terms of your reputation and future job possibilities.

What can you do to resolve conflicts before they can turn into lawsuits? It is a good idea to first seek outside help from a professional mediator or other disinterested party. Mediators generally pursue one of the following paths:

• Allow one party to maximize its outcome

• Allow one party to minimize another's outcome

• Maximize the combined outcome of all parties involved

• Offer solutions that require no concessions by any of the parties

It is important to recognize that a solution which demands no concessions may eventually lead to further demands from one or more parties instead of providing a permanent resolution.

Perhaps the most important steps the professional mediator or solver takes toward ending a conflict are: (1) providing a clear description of what each party expects the other(s) to do; (2) providing a clear description of what each party intends to do or wants to do; (3) analyzing and explaining what will happen if one or more parties does not do what the others desire; and (4) providing suggestions on how to prevent a deadlock and how to achieve a mutually agreeable resolution.

The professional mediator's most important tool is trust. If the mediator can convince the parties involved in a conflict to trust her or him—and especially if the mediator can get the parties to trust each other's intentions—a satisfactory resolution will often become possible. As a participant in a conflict, you can help to create this trust by being receptive to criticism, by acknowledging actual problems, and by demonstrating your willingness to compromise.

If mediation fails, you may experience the beginning of a professional and legal crisis. Lawsuits involving engineering sometimes take as long as 5 years to bring to trial as the parties engage in charges, countercharges, hearings, depositions, and interrogatories. Lawsuits can cost your firm inestimable amounts of time and money. Further, most legal cases involve you in the ordeals of being served a summons or complaint, retaining an attorney, notifying your insurance company, tediously researching and analyzing old project plans, filing answers and cross complaints, and going to court. The only way to avoid this agony is to prevent conflicts from developing in the first place. A good way to start is by properly handling stress as you build and operate your consulting firm. Study the principles of conflict resolution and utilize them in your discussions with clients.

ENGINEERING-RELATED LAWSUITS

Regardless of the defenses of individual engineers who have been the subjects of past lawsuits, few cases are introduced to court unless the claimants, attorneys, and judges decide the cases have some merit. Charges against an engineer that are considered false, frivolous, or malicious rarely get past the initial procedures.

Occasionally, a firm may be able to determine that it will be involved in a lawsuit long before an actual claim has been filed. For example, engineers who designed the civic center in Hartford, Connecticut, probably thought that they would be involved in a lawsuit after the center's roof caved in, causing millions of dollars' worth of damage. But most engineering-related lawsuits begin much less dramatically. Usually, a case begins with the presentation of a subpoena to the consulting firm's principal. A subpoena may also be delivered to a staff member or spouse. Most engineers respond to the subpoena with mixed feelings of anger, questioning, and dread.

If you receive a subpoena, do not rush to phone an attorney or client. Neither should you ignore the matter. Probably the best action to take immediately after being served a subpoena is to calm yourself, try to forget the subpoena for a few hours, then collect your thoughts on what your course of action will be. You certainly will not increase your chances of a favorable outcome if you allow your mind to dwell upon visions of courtroom drama, jail cells, and bankruptcy procedures. The odds are low that you will actually undergo such grief.

Generally, subpoenas are not served without the prior knowledge of the receiver. As noted, most engineering-related lawsuits develop after discussions or mediation attempts with a client fail to bring about a solution that satisfies the client. An out-of-court settlement will usually be much more palatable to all the parties involved, however, since it requires much less time and money to bring about. Seek an out-of-court settlement whenever feasible, even if it may require you to pay a substantial sum to the parties making the complaint. When this is not possible, you should prepare for further legal action.

The first step is to find an attorney who can properly handle your case. It is especially important that you retain an attorney if you receive a subpoena or letter from the client or client's attorney stating that the client intends to sue. Next, inform your client that you are retaining legal counsel and that your attorney will contact the client's counsel shortly.

Finding the right attorney to handle your case can be difficult since few law firms advertise that they are experts in handling engineering lawsuits. However, the person whom you select will have a crucial influence on the future of your case. As far as achieving a favorable outcome is concerned, there can be no substitute for retaining a competent attorney who is experienced in handling engineering lawsuits.

Local chapters of voluntary engineering and consulting organizations may be able to refer you to suitable law firms and legal contacts. If you are already familiar with an attorney whom you consider qualified, you may choose to retain him or her; otherwise it may be necessary for you to interview several prospective attorneys for your case. Most attorneys will be happy to discuss their fees and to provide advice on potential courses of action.

Once you begin to look for an attorney, you should inform your insurance company of your situation. If you do have professional liability insurance, the policy may eventually become your most valuable tool in handling the lawsuit. But the insurance could become useless if you delay notifying the company. Many policies require immediate notification. Write the company a letter explaining the complaint and send the letter by registered mail.

Many professional liability policies specify that the insurance company must select or approve an attorney to prepare the engineer's defense. Occasionally, this may lead to a situation in which the engineer temporarily has two lawyers arguing the same case. To avoid confusion, make sure you know your insurance company's requirements before you select an attorney.

If your liability policy includes a deductible of $5000 or more, you will most likely be required to pay that amount in lawyers' fees (assuming the case proceeds that far). Costs that go above and beyond the deductible should be paid by the insurance company, unless special circumstances prevent the company from acting on your behalf.

Of course, information on your insurance should be kept confidential. Do not inform your client or your client's attorney that your insurance will cover the costs of the lawsuit; doing so could result in the claimant's seeking still higher monetary settlements. Let your attorney handle inquiries about the case.

Engineering lawsuits can take as long as several years to bring to trial. Considerable time may be devoted to questioning by the attorneys from both sides as they establish basic facts about a case. These periods of questioning are usually referred to as "interrogatories." During the interrogatories, the attorneys and their respective clients may provide each other with a good deal of information that will influence the outcome of the case. The engineer may actually be asked to provide a list of people who may be called in as expert witnesses on the engineer's behalf. While the interrogatories may seem never ending, they usually help the participants to determine the exact complaints against the engineer. Further, interrogatories can sometimes provide the basis for a pretrial settlement. At the very least, interrogatories can help the engineer and his or her attorney determine how the court proceedings will be directed.

During this period, the attorney selected by your insurance company may opt for a settlement based on certain facts that the interrogatories bring to light. However, no attorney can legally offer or accept a settlement on your behalf without your written agreement. Instead, the attorney can advise you to accept the settlement or else find a new lawyer. Be sure to know all the terms and conditions of your liability insurance before you reject an attorney's advice; otherwise you may risk losing the protection your policy provides.

Professional liability insurance may prove invaluable even if you win your case in court. The plaintiff in an engineering-related lawsuit is rarely required to pay for losses the engineer may endure as a result of lengthy legal proceedings. Therefore, a good liability policy may be your chief protection against damages.

WORKING WITH LAWYERS

The relationship between the engineer and his or her attorney is of paramount importance before, during, and after an engineering lawsuit. To make the most of the emotionally charged trial period, you must deal effectively not only with facts but with feelings.

Calmly discuss all the details of your case with your lawyer. These details are like pieces of a puzzle that will help the lawyer build a strong argument in your support. Regardless of whether or not all the facts are favorable to you, they are essential to building a defense. Equally crucial is the sense of mutual trust you will create through your cooperation. An engineering lawsuit is usually an uphill battle for the engineer, so you will need all the help you can find.

Utilize photographs and documents whenever possible. Since many of the arguments on both sides of the case will depend on memory and allegations, your defense can be enhanced by incorporating photographs, letters, plans, blueprints, and other materials related to the work in question. People who can demonstrate a superior grasp of the supporting facts usually place themselves at a distinct advantage over the opposing side in pretrial procedures and in court.

A successful defense further depends on your ability to discuss your case in terms that can be easily comprehended by people lacking engineering know-how. This includes lawyers, judges, jurors, and witnesses, in most cases.

While your attorney will most likely exert a lot of energy reminding you to communicate in nontechnical terms, you on the other hand may find it necessary to instruct the attorney on particular technical points. Try not to criticize your attorney's actions. Take a positive approach to advising him or her on issues and questions that could be raised in court to help bring about a favorable disposition of your case. Technical topics can easily become distorted during a trial; thus you and your lawyer must work together for clarity.

Assuming a pretrial settlement cannot be reached, you will probably be required to appear on the witness stand at some point during the trial. When this happens, avoid volunteering more information than is required for answering individual questions. At times it may be appropriate to explain certain technicalities and nuances of engineering to your audience; but long-winded answers may invite more questions than they resolve while doing nothing to advance your case. If a lawyer asks a question that you do not fully understand, you can ask that it be rephrased.

Needless to say, the plaintiff's lawyer has the objective of arguing her or his client's case at your expense. Conversely, your objective as a witness is to support your defense and prevent the prosecuting lawyer from using your answers against you. The prosecutor may use dramatic gestures and speech patterns in an attempt to rouse the emotions of the court. The prosecutor may even ask you questions that imply you are guilty. Do not be swayed by such tactics. Answer questions quietly and succinctly to avoid any implication of guilt.

When the plaintiff takes the witness stand, be sure to follow the questioning carefully. Take notes on salient statements and list relevant questions you may want to discuss with your lawyers.

SURPRISE SETTLEMENTS

It is not unusual for an engineering lawsuit to be settled before the official court proceedings end. Usually, such "surprise settlements" are the result of behind-the-scenes bargaining between the lawyers for the two sides. Surprise settlements require serious consideration by the defendant, especially when they involve a lower settlement than the sum the plaintiff originally sought.

As noted earlier, your insurance company may provide a defense lawyer who wishes to negotiate a particular settlement to end the lawsuit. Regardless of whether you want to continue with court proceedings or not, the wisest choice may be to accept the settlement since doing so will allow you to resume full-time consulting sooner. The insurance company may encourage acceptance of the settlement rather than risk involvement in a protracted and more costly court case.

Not only would a court decision probably cost the insurer more but it would also force you to spend more time away from work. When all the hours spent in a typical court-rendered engineering lawsuit are totaled, an engineer could conceivably have lost 2000 hours' worth of consulting time. At a fee rate of $30 per hour, the engineer would have lost about $60,000 in fees, plus the cost of the deductible on a professional liability insurance policy. The total could easily amount to $65,000!

Even if the engineer were to win the lawsuit, a substantial amount of saleable consulting time would be lost. But a consultant who does not carry or maintain liability insurance would probably fare the worst: he or she would have to pay the costs of lost consulting time, legal fees, court-awarded damage payments, and additional expenditures incurred during the course of the legal proceedings. At that point, the consultant's business could well have been plunged into bankruptcy and the engineer would have to be quite lucky to be able to use the lessons gained from the court experience.

Fortunately, you can learn to avoid all these problems by studying and applying the principles of business protection. Let's examine some more of those principles now.

REDUCING RISKS

To prevent legal difficulties and maximize business success, consulting professionals utilize a variety of safeguards. For example, an experienced consultant carefully researches potential projects to make certain that the consulting firm can accomplish all the work on the project according to scheduling and budgeting requirements. The consultant also investigates the financial capabilities of prospective clients, while evaluating his or her ability to work together with the client and contractors in a way that satisfies the client's needs.

Among the basic questions you should ask before undertaking almost any consulting engagement are these:

1. Does the client have sufficient money to pay both your fee and the cost of building the job?

2. Does the client also have sufficient money to cover extra costs and contingencies that may develop during the course of the engagement?

3. Are you and the client familiar with all the conditions likely to occur during design and construction, as well as with less likely, potential developments?

4. Do you and the client have a clearly written agreement describing all the terms and conditions relative to the engagement, including the three items just listed?

With practice, you can internalize the various business safeguards so that they flow naturally from your daily work activities. Unfortunately, even some experienced engineers overlook these basic techniques in their haste to take up new engagements. Some consultants actually take on greater risks, for instance, by trying to expand their practice to areas in which they have not worked previously. Such risks may be reduced somewhat by hiring an outside consultant who does have the relevant experience. However, responsibility for the success of the project lies ultimately with the consulting firm and its principals.

If problems were to develop during the course of the engagement, the client could possibly bring a claim against the firm *and* the outside consultant. In a court of law, the outside consultant would be held responsible for his or her actions. The consulting firm would also be held responsible for its own actions and for settlement costs that could result from the lawsuit.

The situation could become even more serious for the firm if the outside consultant does not have liability insurance to cover his or her share of the damages. In that case, the consulting firm's insurance might have to cover the entire bill. By neglecting to utilize the appropriate business safeguards, the firm would have created a situation of double jeopardy— i.e., for the outside consultant's liability and its own liability. This example should make it easy to see why safeguards are necessary.

PROJECTS DOOMED TO FAIL

When a consulting engineer performs any professional task which is paid for by a client, a certain element of risk is always involved. No consulting assignment is totally risk-free. But when is a risk so great that it practically dooms a project to failure?

The likelihood of a doomed failure occurs when an engineer attempts to perform a task without any basis of provable calculations in an area with which he or she is unfamiliar and against professional or practical indications that the proposed solution is unworkable.

Such a situation may seem remote. However, it can and does occur. All too often, practicing engineers find themselves in areas of design about which they know too little and they fail to make some sort of calculation to support their position. They then may try to substitute an answer from a different project, presuming that it will work in this case. They may even insist that the work be installed in accordance with their plans and specifications despite the advice of other professionals and contractors against the procedure.

While an occasional job may be successful when done this way, most such jobs are doomed to failure, and the failures can lead to lawsuits. So the experienced and careful engineer will avoid situations in which he or she knows that a lack of knowledge can lead to problems later.

BIDDING FOR WORK

Today most government jobs are done on a lump-sum bid basis. The lump sum is common for a variety of projects for town, city, state, and federal government work. Many governments limit the engineer's fee to 6 percent of the lump-sum bid for design work, with an additional lump sum for supervision. As a practicing professional engineer, you should recognize that there is a direct relationship between professional liability claims and the bidding process.

You will often be told that a number of well-known firms are also bidding on a job you want. You may also be told that the job "will make you famous." A further enticement is the promise of future continuing work from the client if you get the job. Should you, as an engineer, fail to get the job, the possibility of future work for you will be slim.

When the preceding conditions are combined with a lump-sum bid, many problems are created for the professional engineer, the insurer, the client, the contractor, and the courts. The engineer who gives a lump-sum bid is required to work within money restraints. If the engineer is familiar with the type of work being done, the risk involved is minimal. But if the job is one with which the engineer is unfamiliar, a major risk of a future lawsuit exists.

LIMITING LIABILITY

The most common cause of lawsuits, claims, and eventual settlements is when contractors—in their zeal to get work—bid too low for a project. Finding themselves in financial difficulty when the job moves through its various stages, they begin to seek additional sums of money by making claims for all sorts of exceptions. Most such claims are based on alleged defects in the plans and specifications prepared by the engineer. These claims form the basis of a court case against the owner by the contractor (or sometimes against the architect and the engineer directly) based on alleged errors requiring suitable financial compensation.

When a limitation of liability exists, the attorney for the contractor is clearly signaled that the engineer's contribution to any settlement will be severely limited. And, based on the normal rules of settlement, the amount which the engineer will pay will be less than the limit of the liability in the written contract.

Does this liability limitation protect the insuring company holding the professional liability policy for the practicing consulting engineer? This limitation *does* protect the insuring company because the engineer is committed to no more than the amount covered in the policy plus the time involved. However, as noted earlier, the engineer's time may involve many thousands of dollars. Thus, any savings in time which can be made because of the policy will be of direct benefit to the engineer because all the engineer has to sell is time.

THE CONTINUING CLIENT

While it may not seem to apply, the need for liability protection is considerable when an engineer works with continuing clients. Why is this? There are a number of reasons: (1) The client may have been obtained without any liability limitation clause in the agreement; (2) the continuing client may be one with which the engineer does not have a written agreement and has no way of obtaining a liability limitation clause; (3) relations with the client may not be the best, and the engineer feels that it is the wrong time to request a liability clause.

To protect oneself and one's practice in such a situation, the engineer must carefully educate the client. If a client can be made to understand that it is to the client's benefit to have a liability limitation clause in the contract, the engineer who does not have a contract with a client may benefit doubly. The reason for this is that the engineer can effectively combine the limitation of liability clause with a formal written contract with the client which will make the engineer's position even more secure than before.

Most clients already understand the problems of claims which may be made against them by a contractor. Hence, any contract which tends to increase the client's protection by use of a liability limitation clause can be readily presented to the client by an engineer. Frequently the client will quickly accept such a clause.

LIMITATION OF LIABILITY WITH GOVERNMENT AGENCIES

Many state, federal, and local government contracts still contain a "hold-harmless clause," which creates unlimited liability for the practicing professional engineer. Since government contracts usually require acceptance of the lowest responsible bidder, the possibility of lawsuits on a government job is much greater. Therefore, any government agency should have the advantage of the same protection offered by the limitation of liability clause. State professional groups representing practicing engineers have a strong voice with government agencies. These professional groups should work to eliminate the hold-harmless clause from all government contracts.

INTERPROFESSIONAL RELATIONS

An important way to protect your firm and your practice is by hiring capable expert professionals as part of your design team. Besides knowing what you are doing in your design work, you should also be certain that the expert professionals on your design team have the same limitation of liability in their contracts.

The usual problem faced by the practicing professional engineer occurs when he or she is hired by a firm which may or may not have a professional limitation of liability clause in their contract. Since this may put the employer at a disadvantage, you should advise the employer of your proposed limitation of liability and seek the employer's agreement with this limitation.

CONTRACTOR RELATIONSHIPS

The preceding recommendations may sound as though the practicing professional engineer is trying to put all the burden of responsibility on the contractor who does the work. This is not so. The limitation of liability does not make the contractor responsible for your errors, omissions, or negligence. When a contractor signs a contract containing a liability limitation clause in favor of the owner and design professional, the contractor is in no way assuming liability for damage to other parties because of deficiencies in design or specifications.

Many contractors welcome the liability limitation clause because they know that the low bidder frequently bids the project primarily based on the amount of extras which may be secured from actual or alleged deficiencies in the plans and specifications. A liability limitation clause does not adversely affect the contractor who bids accurately and honestly. Once a contractor understands this situation, most complaints will disappear.

UNDERSTANDABLE AGREEMENTS

It is important that every agreement a consulting engineer signs be written clearly and be readily understood by all parties. This concept is vital to properly protect yourself and your firm in its professional practice. What a well-written agreement does is formalize the informal understanding present between consultant and client. A well-written agreement is not intended to favor one side or the other. It is intended to favor both sides equally.

So, in your anxiety to secure consulting work, you should not allow yourself to sign contracts prepared by your client which contain clauses or requirements that are unfair to you. Make it a regular practice to consult with your attorney over the legal implications of every contract before you sign it. Many experienced attorneys are also skillful negotiators and can prove to be a valuable asset in your negotiations with your client.

Further, an attorney can more easily suggest changes that may be readily accepted by your client. Suggestions which you make may be turned down; the same suggestion made by an attorney may be quickly accepted. This preventive sort of legal advice is well worth the investment even if it results in only one instance where possible litigation is avoided.

Take the situation in which an engineer has designed the facilities for a 40,000-ft^2 (3700-m^2) machine tool industrial plant and office building. The engineer—we will say—has designed 14 previous installations of this type. Hence, when the engineer is asked to design the fifteenth job, he or she will have a reasonably accurate knowledge of the labor cost and the overhead involved. He or she can reasonably estimate a lump-sum fee that will pay the engineer a fair profit. Such a situation is fairly common among experienced engineers who have designed a number of different types of facilities, machines, etc.

So the experienced practicing professional engineer will find many situations in which previous experience tells him or her what the cost of a new, similar design will be. Having time cards and accurate records from previous jobs will enable the engineer to prepare an estimate for a new job.

What we are considering here is the engineer's protection of himself or herself against professional liability losses. Should the engineer be required to provide a lump-sum bid at a lower than usual price, both the engineer and the client will know that the bid price is below the normal price for such a job. So the engineer will have to cut corners in his or her office procedures to get the job done at the lump-sum bid price. At the same time, the practicing professional engineer who is knowledgeable will be aware that this cutting of design effort may create personal liability problems which could result in lawsuits.

It is in the business of engineering—especially in protecting one's engineering practice—that the engineer must be particularly alert to the bidding procedure. When engineers deal with the unknown—be they experienced or inexperienced—the possibility of failure exists. And certainly the newly formed firm will have less expertise in handling problems which may arise.

The lump-sum bid is being used more widely than ever. This definitely works to the disadvantage of the newer firm since there is no way for such a firm to gain quickly a backlog of experience in cost analysis of a given design situation. Even a firm which has been in business 10 years or more may have problems because the firm may lack good records of time spent on similar engineering designs previously completed.

Keeping good records of every engineering job is essential for success in consulting engineering. The business side of your engineering practice is one of extreme importance. The engineer simply cannot escape being a good business person and remain in successful practice.

HOW TO SAY NO TO A BID REQUEST

There are times when you should refuse to make a lump-sum bid for a job. When trying to decide whether you should make a lump-sum bid, ask the client, "Do you plan to accept the lowest bid?"

If the answer you receive implies that the price will bear heavily on the selection of the consulting firm, you should quietly, calmly, and firmly decline to offer any sort of bid. This is one way to stay out of lawsuits.

You may sometimes be asked to reduce your bid by a certain amount of money, say, $2000, because you are that much higher than your competitor. Thus you might have bid $103,000 and be asked to reduce it to $101,000 or $95,000. While either reduction may seem to be a small concession, it can lead to future lawsuits. If all consulting engineers were to insist on their bid prices, there would probably be fewer lawsuits because the average job would be performed better. Either your first quotation is correct, or you will suffer some personal loss or some reduction in services will be necessary. If the latter occurs, the possibility of lawsuits rises.

In declining to cut your fee, you should try to show the prospective client that the reduced bid will not be in the client's best interest. Demonstrate to the client how the lower fee would reduce the quantity and quality of services needed for the construction of the project and how it would most likely result in higher construction costs in the end. Increased costs may result from inadequately prepared plans and specifications and consequent errors and change orders. While some clients look only for bargain prices, most will understand the reasons for your original bid once it has been explained.

There is no question that an engineer can secure a large number of jobs of certain kinds if she or he is willing to charge smaller fees than those of competitors. But doing so would not help the engineer to earn a reputation as a first-class professional. Further, fee cutting generally does not attract high-quality clients. Consulting firms that engage in frequent bid reductions often become regarded as second-rate organizations. Since such firms have reputations for engaging in greater risks, insurance companies are sometimes hesitant to provide them with inexpensive protection. Insurers may demand higher premiums from habitual fee cutters or may choose to drop such firms permanently from their clientele.

LIMITATION OF LIABILITY EXPLAINED

Some engineers—by their attitudes and actions—create the impression that no one should question what they do in the course of design and similar work. Yet in the courts of law there are a number of cases being heard because the unexpected has happened on an engineering job.

Frequently such cases arise because the engineer had little or no control over construction or material failures. However, the engineer did control the relationship and understanding with the client prior to the problem's being brought to court. Thus, the question becomes, "What is the professional's limit of liability?"

Professional liability is frequently unlimited where there is no clear statement of its limits. This can occur when there is no written contract containing a specific statement of the limits of liability.

Here is a sample liability limitation statement which is recommended by some professional liability insurance companies:

The owner agrees to limit the design professional's liability to the owner and to all construction contractors and subcontractors on the project, due to the design professional's negligent acts, errors, or omissions, such that the total aggregate liability of each design professional to all those named shall not exceed $50,000 or the total fees paid to the design professional engineer, whichever is greater.

When considering the limitation of liability, keep in mind that courts of law often allow literally anyone connected with a given project to be made part of a lawsuit. For instance, a geotechnical engineer drove his pickup truck to the scene of a landslide to take photographs. His firm's name, observed by a neighbor of the slide victim, was all that was needed to have the engineer included in a lawsuit even though he had performed no work in the slide area. It cost the engineer $15,000 in fees and his own time to get out of the lawsuit. In many courts inclusion of everyone in a suit is an effective legal ploy to secure a small cash contribution from each person to effect a satisfactory settlement for the plaintiff.

DEFINING THE SCOPE OF SERVICES

The usual problem in contractual misunderstandings between an engineer and a client revolves around the understanding the client has of what the engineer will specifically do. The services the client will receive should be covered in every written contract and clearly understood by both the engineer and the client. Further, in the contract the fee that is being charged should be stated, and the work which will be performed should be clearly and completely defined.

Sometimes the description of the services the engineer will render reads, "provide contract plans, specifications, bidding documents, and construction supervision as required to complete the project." This is an all-encompassing phrase which is far from complete and detailed as to exactly what the engineer will do. A number of questions arise instantly. These are:

1. Will project development, programming, scheduling, and planning be done?
2. Will the engineer consult with the owner, the agent, and others?
3. What about cost estimating, economic feasibility and analysis, promotional drawings, and administrative services?
4. What about site studies, master planning, zoning analysis?
5. What about architectural effort, special designs, interior and exterior work, and government agency requirements?
6. What about expert consultants who may be required to complete the project?
7. What bidding and construction documents will specifically be provided?
8. Precisely what services will be provided during construction?
9. What services will be provided after the project is completed?
10. What special services, studies, etc., will be provided to operate the project?

These are just a few of the questions that could be asked about the preceding phrase which an engineer might use to describe his or her services.

Proper answering of all these questions and many others enables the engineer to define exactly what will be done for the owner. Further, the questions may assist in budgeting time and costs required for the project. Answering these questions before a contract is signed will enable an engineer to contract for the work which can be performed within the quoted price. It will prevent embarrassing questions being asked later.

And, answering the questions early allows the quotations to be changed, if necessary. Or, should the engineer decide to do so, the project can be turned down rather than getting involved in something that cannot be financially rewarding.

If an engineer discovers that it is impossible to complete a project within the proposed cost, the chances for design errors increase considerably. Why? Design errors are more likely to occur when an engineer takes shortcuts in a design because the shortcuts may become the source of future legal claims against the engineer.

Defining the scope of services serves many useful purposes:

1. It clearly defines what the engineer will do for the client.
2. It defines what will be done by the client.
3. It will clarify in the engineer's mind what the cost of services will actually be.
4. It will provide a systematic organization of the design effort.
5. It will help the engineer to avoid legal problems in the future.

PROBLEMS IN CONTRACTS

The first question any engineer faces when signing a contract (after the contract has been carefully read) comes when certain words, phrases, or paragraphs are met which are contrary to what the engineer intends to do or what can reasonably be expected of the engineer. The question immediately arises, What should the engineer do about those words, phrases, or paragraphs? Any of them can be a source of future trouble.

The first and foremost consideration every engineer should have when signing a contract is to be certain not to assume responsibilities which are not rightfully part of the engineer's work. This points out the value of a clear scope-of-services clause in every contract. The scope-of-services clause covering the engineer's work should be included in every contract before it is signed.

LEGAL "BOILERPLATE" PROBLEMS

Problems can be created by the so-called legal boilerplate included in almost every contract. Buried in this legal boilerplate are clauses that frequently hold the client harmless and indemnify the client. These clauses may give certain warrantees and guarantees or give defense and indemnity provisions or require certain certifications or certificates of performance. Some of these can be serious in nature. How do you identify these potentially harmful paragraphs? Here are some examples of phrasing that usually precede the troublesome descriptions:

1. The paragraph generally begins with the phrase "The Design Professional agrees to hold harmless. . . . "
2. A warranty or guarantee paragraph generally begins with the phrase "The Design Professional warrants that all work performed. . . . "
3. In defense and indemnity provisions, the paragraph usually begins, "The Design Professional agrees to defend and indemnify the client against liability. . . . "
4. The certification clauses begin generally with the words "The Design Professional will certify that. . . . "

Thus, it is clear that the engineer who reads a contract carefully can quickly spot those paragraphs that may create future problems. Once such paragraphs are found, what should the engineer do about them?

The object of warranty and guarantee clauses is usually to ensure that plans will be free of design defects. Since it is impossible to ensure this with absolute certainty, the engineer should point out to the owner that potential negligence or omissions on the engineer's part may be covered by some form of indemnification. However, ordinary errors can occur in the design of any project and should be covered by a small contingency fund. Public records of building projects indicate that funds in the range of 3 to 5 percent of the job fee are common.

Frequently, if the owner is informed that the hold-harmless clause actually creates a liability for the engineer with regard to potential legal claims from a contractor (e.g., claims concerning the safety of workers, over which the engineer has no control), the owner will be willing to modify the clause to eliminate this sort of provision.

COST ESTIMATES AND SCHEDULES

Cost estimates and construction schedules present another source of potential legal problems for the consulting engineer. Many clients ask for detailed information on these areas when a consultant submits a proposed design. However, a client may fail to understand the many variables that make cost estimates and schedules difficult for the engineer to develop with certainty. Consulting engineers are not contractors and therefore have little or no control over the construction process. Further, many consultants do not gain expertise in cost estimation until they have spent many years in practice.

You can safeguard your business somewhat by prefacing all cost estimates and construction schedules with statements emphasizing that the information reflects careful estimating. Try to make it clear that the estimates are *not* guarantees.

Generally, consultants hope to impress their clients with the idea that their engineering firm can design the most efficient, cost-effective solutions and that these solutions can be built in a minimal amount of time. While the consultants may be quite sincere, they may accidentally underestimate costs and times of completion. However, contractors sometimes produce higher estimates than consulting engineers. An engineer whose estimate differs from the contractor's estimate may suspect that the contractor is padding the job. The contractor may actually set a higher estimate as a precaution against potential delays or additional costs. Regardless, the contractor's estimate usually ends up reflecting the final cost of the project.

Until you build a record of cost estimates and times of completion for various jobs and can compare them to actual construction bid figures and contractor time estimates, you may have difficulty evaluating your expertise in these areas. If you find that your firm's estimates differ from the actual construction estimates by more than 10 percent on approximately five jobs, you should work to improve your organization's estimating procedures. In general, a consulting engineer's estimates are considered good when they consistently fall within 10 percent of the actual construction cost and time value. With practice, you should be able to improve all estimates and thus protect your firm.

MAPPING THE UNKNOWNS

A time when many firms tend to make errors and omissions that could lead to legal hassles is when staff members have to work under extreme pressure to meet close deadlines. Design problems appear more frequently, and the feasibility of making accurate cost estimates and construction schedules decreases.

You can help reduce the risks inherent in these types of situations for your firm by following a few general principles:

1. Provide thorough design supervision to ensure that each of a project's individual tasks is completed in time for the remainder of the work to

proceed without hindrance. (A chief reason why engineers sometimes miss deadlines is that engineers or their designers procrastinate until a deadline draws near. When one task is held up, for valid or invalid reasons, each succeeding task may be slowed until the entire project falls behind schedule.)

2. Maintain a concerted organizational effort in design and all other project areas (e.g., shop-drawing review, cost estimation, blueprinting) to make certain that everyone involved has a clear idea of what they must do to coordinate their work with others and meet deadlines.

3. Try to avoid letting any work pile up to the last minute. While some engineers may claim that they work best under pressure, the proverbial burning of the midnight oil is generally not conducive to success in this demanding, technical field.

CONTROL PROCEDURES

The purpose of checking all work produced by your firm is not only to prevent errors that could interfere with providing excellent services but also to prevent legal claims that could eventually result from problems that your clients may not even be aware of. For example, a design that incorporates an addition to an existing building must not only meet the client's needs for added space; it must also incorporate a study of existing utilities, accessibility, ambulatory and vehicular traffic flow, and many other factors.

Directly related to the plans is the development of complementary specifications. Whether or not the plans take precedence over the specifications (or vice versa) may depend on laws governing engineering in each state. Therefore, it is also important for you to learn the local rules applicable to your practice so you can combine your plans and specifications to provide a complete product. And it is critically important that you check your drawings and specifications completely.

Master specifications based upon a standard format rarely if ever contain all areas required for individual, specialized projects. Therefore, checking specs you write against a master spec does not provide adequate protection against liability claims. The best type of specification checklists are those that break out individual subject areas for each item appearing in the master specification. Ideally, each subject heading should appear on a separate sheet of paper. By checking off your specific project requirements against the items on the separate checklists, you will be able to verify whether or not each requirement has been met properly. Further, you can use a second, more general list of titles to ensure that particular items are covered in both the plans and the specifications. A disinterested third party should review the materials during the final checking process.

Use Exhibit 5-2 as a guide to preparing your own specification checklists. It shows a checklist for supplemental conditions that might follow the general conditions of a project's mechanical specifications. The check-

EXHIBIT 5-2   *Checklist for supplementary conditions.*

1. GENERAL:

2. MODIFICATIONS OF THE GENERAL CONDITIONS:

   Intent:

   Copies furnished:

   Execution, correlation, and intent:

   Labor and materials:

   Taxes:

   Permits, fees, and notices:

   Samples:

   Cleaning up:

   Performance bond and labor and materials payment bond:

   Delays and extension of time:

   Applications for payment:

   Progress payments:

   Contractor's liability insurance:

   Contractor's liability insurance:

   Property insurance:

   Changes in the work:

3. NONDISCRIMINATION:

list shown is simple. It lists the headings of each paragraph that will appear on the specifications under the "Supplementary General Conditions" heading.

At a glance, an engineer using such a checklist can determine whether all the appropriate material has been covered in the written specifications. The checklist further organizes the material in a way that enables the engineer to tell whether certain information belongs or does not belong to a particular section of the specs.

Exhibit 5-2 is intended to be used as a general example, however; each firm must tailor its own checklists to meet the needs of the firm's clients. Other helpful sources of checklists include standard specification formats published by voluntary professional organizations and a variety of commercially available texts listed in the Consultant's Reference Shelf at the end of Section 4.

Many firms have developed their own in-house design manuals to cover the checking of plans and the firms' preferred design procedures. Some of the larger engineering organizations require their staff members to follow the in-house manuals unless a member has received written permission to deviate from the standard steps. While you may not choose to go that far, your firm may improve its efficiency by developing and using an in-house manual. Moreover, such a design manual may help your employees to achieve greater consistency and thus prevent errors and omissions that could lead to lawsuits.

Your design manual should contain solutions that have been tested and proven to be workable, well accepted, and error-free. It should also point to the firm's ability to develop new and better solutions whenever possible. Your manual should provide a brief, general design checklist that can be applied to most projects, plus specialized checklists for particular types of designs (e.g., residential, commercial, institutional, and industrial checklists). Another item most design manuals contain is a list or lists of standard details. Like design solutions, the standard details should not be emphasized to the point that they clutter or confuse a plan or its construction phase. Instead, the details should be applied directly to the areas that a project engineer or project manager has decided must be covered to ensure clarity in development of the design and final construction.

BIDDING DOCUMENTS AND PROCEDURES

Consulting engineers who provide design and supervision services to clients must be skilled in the use of bidding documents and procedures. Their work includes a wider range of responsibilities than that of an engineer whose chief job may be to provide plans to other consultants or to architects, who then incorporate the plans in the bidding documents themselves. In your own consulting practice, you may be asked to offer these additional services, so it is important to know at least the rudiments of the bidding process.

The first item you should become familiar with is the type of advertising used to solicit bids and quotations from contractors for completed project designs. Exhibits 5-3 and 5-4 show samples of typical advertisements.

Both the American Institute of Architects (AIA) and the National Society of Professional Engineers (NSPE) can provide documents that you may use to prepare specifications for bidding by contractors. Materials that these organizations provide include: instructions to bidders, bid forms, performance bonds, labor and material bonds, payment bonds, and other special items. Exhibits 5-5 through 5-8 show examples of some of these items.

EXHIBIT 5-3   A typical invitation-to-bid request.

INVITATION TO BID
CONTRACT NO. A-456
NEWINGTON CHILDREN'S HOSPITAL
NEWINGTON, CONNECTICUT

Sealed bids for the construction of a GAIT LABORATORY will be received by the Newington Children's Hospital, East Cedar Street, Newington, Conn., in the office of the Hospital Director until 2:30 PM EDST on May 21, 19XX

Bids shall be submitted in duplicate on the Contractor's letterhead in the form indicated in the specifications and delivered to John Smith, Director of Purchases.

Bids will be opened and publicly read aloud at that time. Proposals shall be accompanied with a Bid Bond as described in the specifications, payable to the Newington Children's Hospital, in the amount of ten (10) percent of the contract price.

Bidder is required to complete the Qualification Sheet and the supplement thereto and submit it with his bid.

Copies of the bidding documents may be obtained at the offices of Mueller Engineering Corporation, upon deposit of $50 per set payable to the Engineers. Deposits will be refunded on the return of sets in good condition within 14 days after the due date for receipt of bids.

The Owner reserves the right to accept or reject any or all of the bids and to waive any informality in the bidding. No bid may be withdrawn for a period of 30 days after the bid due date. Date: May 2, 19XX.

EXHIBIT 5-4   Federal government invitation-to-bid advertisement.

PHASE II MASTER PLANNING FOR APPROXIMATELY TWELVE MILITARY COMMUNI-
TIES in the Federal Republic of Germany, England and the Netherlands. Firms which
meet the requirements described in this announcement are invited to submit a completed SF 254, "Architect-Engineer and Related Services Questionnaire," and a completed SF 255, "Architect-Engineer and Related Services Questionnaire for Specific
Project," to this office. Firms having a current (less than one year old) SF 254 on file
with this office need not resubmit this form. If not on file with this office, a current SF
254 for each firm shown in block six of SF 255, as "outside key consultants/associates"
is required. Firms as frequently rejected for a lack of or an outdated SF 254. Combined
professional personnel strength of the firms must be at least twenty people to include
architects, civil engineers, landscape architects and planners. Firms must be able to
mobilize personnel to Europe on short notice for field investigations and conferences at
the site with operational users and this office. Firms must be able to perform the work
within the time limitations stated below. The firms must be able to assign their own
personnel to coordinate the project through its entirety for each military community's
Phase II master plan that they are selected to develop. Firms will be evaluated on the
following rating factors: (1) experience within the past five years with Army Master Planning in accordance with AR 210-20. "Master Planning for Army Installations." (2) experience, technical competence and adequate facilities of the firm to prepare drawings
using the overlay composite method in accordance with technical bulletin eng 353 "the
overlay composite method of master plan preparation." and (3) experience, qualifications and organization of the team proposed for assignment to the project, including (A)
technical skills and abilities, (B) project coordination and management, and (C) interest
of company management in the project and expected participation and contribution of
key personnel. It is anticipated to begin the projects in Oct. 82 and complete the projects
in Oct. 83. Only those firms submitting qualifications data will be considered. All qualifications must be received in this office by 4 Jun 82. Approximately eight fixed price
contracts are contemplated. You are advised that mail deliveries from USA to Frankfurt,
Germany, are occasionally slower than anticipated. It is strongly recommended that
those firms responding from the USA to this announcement advice U.S. Army Engineer
Division, Europe (EUD) by Telex or telephone that they have submitted a SF 255. EUD
Telex number is 841-412982, and the telephone number is (0611) 151-7434 (Mr.
Frank Weinstein). Timely receipt of SF 255 and supporting documents is still essential
for consideration and must be received by 4 Jun 82 deadline. No other general notification to firms under consideration will be made, and no further action beyond submission of SF 254 and SF 255 is required or encouraged at this time. This is not a request
for proposals. See note 62. (105).

UNITED STATES ARMY ENGINEER DIVISION, EUROPE, Attn: EUDED-MC, APO 09757

EXHIBIT 5-5   A typical instructions-to-bidders document.

DEFINITIONS

1.1   Bidding documents include the advertisement or invitation to bid, the instruction
to bidders, the bid form, other sample bidding and contract forms, and the proposed
contract documents including any addenda issued prior to receipt of bids. The contract

EXHIBIT 5-5 (Cont.)

documents proposed for the work consist of the owner-contractor agreement, the conditions of the contract (general, supplementary and other conditions), the drawings, the specifications, and all addenda issued prior to and all modifications issued after execution of the contract.

1.2 All definitions set forth in the general conditions of the contract or in other contract documents are applicable to the bidding documents.

1.3 Addenda are written or graphic instruments issued by the engineer prior to execution of the contract which modify or interpret the bidding documents by additions, deletions, clarifications, or corrections.

1.4 A bid is a complete and properly signed proposal to do the work or designated portion thereof for the sums stipulated therein, submitted in accordance with the bidding documents.

1.5 The base bid is the sum stated in the bid for which the bidder offers to perform the work described in the bidding documents as the base, to which work may be added or from which work may be deleted for sums stated in alternate bids.

1.6 An alternate bid (or alternate) is an amount stated in the bid to be added to or deducted from the amount of the base bid if the corresponding change in the work, as described in the bidding documents, is accepted.

1.7 A unit price is an amount stated in the bid as a price per unit of measurement for materials or services as described in the bidding documents or in the proposed contract documents.

1.8 A bidder is a person or entity who submits a bid.

1.9 A sub-bidder is a person or entity who submits a bid to a bidder for materials or labor for a portion of the work.

BIDDER'S REPRESENTATIONS

2.1 Each bidder by making his bid represents that:

2.1.1 He has read and understands the bidding documents and his bid is made in accordance therewith.

2.1.2 He has visited the site, has familiarized himself with the local conditions under which the work is to be performed and has correlated his observations with the requirements of the proposed contract documents.

2.1.3 His bid is based upon the materials, systems, and equipment required by the bidding documents without exception.

BIDDING DOCUMENTS

3.1 COPIES:

3.1.1 Bidders may obtain complete sets of the bidding documents from the issuing office designated in the advertisement or invitation to bid in the number and for the deposit sum, if any, stated therein. The deposit will be refunded to bidders who submit a bona fide bid and return the bidding documents in good condition within 10 days after receipt of bids. The cost of replacement of any missing or damaged documents will be deducted from the deposit. A bidder receiving a contract award may retain the bidding documents and his deposit will be refunded.

EXHIBIT 5-5 (Cont.)

3.1.2   Bidding documents will not be issued directly to sub-bidders or others unless specifically offered in the advertisement or invitation to bid.

3.1.3   Bidders shall use complete sets of bidding documents in preparing bids. Neither the owner nor the engineer assumes any responsibility for errors or misinterpretations resulting from the use of incomplete sets of bidding documents.

3.1.4   The owner or the engineer in making copies of the bidding documents available on the above terms do so only for the purpose of obtaining bids on the work and do not confer a license or grant for any other use.

3.2   INTERPRETATION OR CORRECTION OF BIDDING DOCUMENTS:

3.2.1   Bidders and sub-bidders shall promptly notify the engineer of any ambiguity, inconsistency, or error which they may discover upon examination of the bidding documents or of the site and local conditions.

3.2.2   Bidders and sub-bidders requiring clarification or interpretation of the bidding documents shall make a written request which shall reach the engineer at least seven (7) days prior to the date for receipt of bids.

3.2.3   Any interpretation, correction, or change of the bidding documents will be made by addendum. Interpretations, corrections, or changes of the bidding documents made in any other manner will not be binding, and bidders shall not rely upon such interpretations, corrections, and changes.

3.3   SUBSTITUTIONS:

3.3.1   The materials, products, and equipment described in the bidding documents establish a standard of required function, dimension, appearance, and quality to be met by any proposed substitution.

3.3.2   No substitution wil be considered prior to reciept of bids unless written request for approval has been received by the engineer at least 10 days prior to the date for receipt of bids. Each such request shall include the name of the material or equipment for which it is to be substituted and a complete description of the proposed substitute including drawings, cuts, performance, and test data and any other information necessary for an evaluation. A statement setting forth any changes in other materials, equipment, or other work that incorporation of the substitute would require shall be included. The burden of proof of the merit of the proposed substitute is upon the proposer. The engineer's decision of approval or disapproval of a proposed substitution shall be final.

3.3.3   If the engineer approves any proposed substitution prior to receipt of bids, such approval will be set forth in an addendum. Bidders shall not rely upon approvals made in any other manner.

3.3.4   No substitutions will be considered after the contract award unless specifically provided in the contract documents.

3.4   ADDENDA:

3.4.1   Addenda will be mailed or delivered to all who are known by the engineer to have received a complete set of bidding documents.

3.4.2   Copies of addenda will be made available for inspection wherever bidding documents are on file for that purpose.

3.4.3   No addenda will be issued later than four (4) days prior to the date for receipt of bids, except an addendum withdrawing the request for bids or one which includes postponement of the date for receipt of bids.

EXHIBIT 5-5 (Cont.)

3.4.4   Each bidder shall ascertain prior to submitting his bid that he has received all addenda issued, and he shall acknowledge their receipt in his bid.

BIDDING PROCEDURE

4.1   FORM AND STYLE OF BIDS:

4.1.1   Bids shall be submitted in forms identical to the form included with the bidding documents.

4.1.2   All blanks on the bid form shall be filled in by typewriter or manually in ink.

4.1.3   Where so indicated by the makeup of the bid form, sums shall be expressed in both words and figures; and in case of a discrepancy between the two, the amount written in words shall govern.

4.1.4   Any interlineation, alteration, or erasure must be initialed by the signer of the bid.

4.1.5   All requested alternates shall be bid. If no change in the base bid is required, enter: "NO CHANGE."

4.1.6   Where two or more bids for designated portions of the work have been requested, the bidder may, without forfeiture of his bid security, state his refusal to accept award of less than the combination of bids he so stipulates. The bidder shall make no additional stipulations on the bid form nor qualify his bid in any other manner.

4.1.7   Each copy of the bid shall include the legal name of the bidder and a statement that the bidder is a sole proprietor, a partnership, a corporation, or some other legal entity. Each copy shall be signed by the person or persons legally authorized to bind the bidder to a contract. A bid by a corporation shall further give the state of incorporation and have the corporate seal affixed. A bid submitted by an agent shall have a current power of attorney attached certifying the agent's authority to bind the bidder.

4.2   BID SECURITY:

4.2.1   If so stipulated in the advertisement or invitation to bid, each bid shall be accompanied by a bid security in the form and amount required pledging that the bidder will enter into a contract with the owner on the terms stated in his bid and will, if required, furnish bonds as described hereunder covering the faithful performance of the contract and the payment of all obligations arising thereunder. Should the bidder refuse to enter into such contract or fail to furnish such bonds if required, the amount of the bid security shall be forfeited to the owner as liquidated damages, not as a penalty. The amount of the bid security shall not be forfeited to the owner in the event the owner fails to comply with sub-paragraph 6.2.1.

4.2.2   If a surety bond is required, it shall be written on a standard form, and the attorney-in-fact who executes the bond on behalf of the surety shall affix to the bond a certified and current copy of his power of attorney.

4.2.3   The owner will have the right to retain the bid security of bidders to whom an award is being considered until either: (A) the contract has been executed and bonds, if required, have been furnished, or (B) the specified time has elapsed so that bonds may be withdrawn, or (C) all bids have been rejected.

4.3   SUBMISSION OF BIDS:

4.3.1   All copies of the bid, the bid security, if any, and any other documents required to be submitted with the bid shall be enclosed in a sealed opaque envelope. The enve-

EXHIBIT 5-5 (Cont.)

lope shall be addressed to the party receiving the bids and shall be identified with the project name, the bidder's name and address and, if applicable, the designated portion of the work for which the bid is submitted. If the bid is sent by mail, the sealed envelope shall be enclosed in a separate mailing envelope with the notation "SEALED BID ENCLOSED" on the face thereof.

4.3.2   Bids shall be deposited at the designated location prior to the time and date for receipt of bids indicated in the advertisement or invitation to bid, or any extension thereof made by addendum. Bids received after the time and date for receipt of bids will be returned unopened.

4.3.3   The bidder shall assume full responsibility for timely delivery at the location designated for receipt of bids.

4.3.4   Oral, telephonic, or telegraphic bids are invalid and will not receive consideration.

4.4   MODIFICATION OR WITHDRAWAL OF BID:

4.4.1   A bid may not be modified, withdrawn, or canceled by the bidder during the stipulated time period following the time and date designated for the receipt of bids, and each bidder so agrees in submitting his bid.

4.4.2   Prior to the time and date designated for receipt of bids, any bid submitted may be modified or withdrawn by notice to the party receiving bids at the place designated for receipt of bids. Such notice shall be in writing over the signature of the bidder or by telegram. If by telegram, written confirmation over the signature of the bidder shall be mailed and postmarked on or before the date and time set for receipt of bids, and it shall be so worded as not to reveal the amount of the original bid.

4.4.3   Withdrawn bids may be re-submitted up to the time designated for the receipt of bids provided they are then fully in conformance with these instructions to bidders.

4.4.4   Bid security, if any is required, shall be in an amount sufficient for the bid as modified or re-submitted.

CONSIDERATION OF BIDS

5.1   OPENINGS OF BIDS:

5.1.1   Unless stated otherwise in the advertisement or invitation to bid, the properly identified bids received on time will be opened publicly and will be read aloud. An abstract of the base bids and alternate bids, if any, will be made available to bidders. When it has been stated that bids will be opened privately, an abstract of the same information may, at the discretion of the owner, be made available to the bidders within a reasonable time.

5.2   REJECTION OF BIDS:

5.2.1   The owner shall have the right to reject any or all bids and to reject a bid not accompanied by any required bid security or by other data required by the bidding documents, or to reject a bid which is in any way incomplete or irregular.

5.3   ACCEPTANCE OF BID (AWARD):

5.3.1   It is the intent of the owner to award a contract to the lowest responsible bidder provided the bid has been submitted in accordance with the requirements of the bidding documents and does not exceed the funds available. The owner shall have the right to

EXHIBIT 5-5 (Cont.)

waive any informality or irregularity in any bid or bids received and to accept the bid or bids which, in his judgment, is in his own best interests.

5.3.2   The owner shall have the right to accept alternates in any order or combination and to determine the low bidder on the basis of the sum of the base bid and the alternates accepted.

POST BID INFORMATION

6.1   CONTRACTOR'S QUALIFICATION STATEMENT:

6.1.1   Bidders to whom award of a contract is under consideration shall submit to the engineer, upon request, a properly executed contractor's qualification statement, unless such a statement has been previously required and submitted as a prerequisite to the issuance of bidding documents.

6.2   OWNER'S FINANCIAL CAPABILITY:

6.2.1   The owner shall, at the request of the bidder to whom award of a contract is under consideration and no later than seven (7) days prior to the expiration of the time for withdrawal of bids, furnish to the bidder reasonable evidence that the owner has made financial arrangements to fulfill the contract obligations. Unless such reasonable evidence is furnished, the bidder will not be required to execute the owner-contractor agreement.

6.3   SUBMITTALS:

6.3.1   The bidder shall, within seven (7) days of notification of selection for the award of a contract for the work, submit the following information to the engineer:

1. A designation of the work to be performed by the bidder with his own forces.

2. The proprietary names and supplies of principal items or systems of materials and equipment proposed for the work.

3. A list of names of subcontractors or other persons or entities (including those who are to furnish materials or equipment fabricated to a special design) proposed for the principal portions of the work.

6.3.2   The bidder will be required to establish to the satisfaction of the engineer and the owner the reliability and responsibility of the person or entities proposed to furnish and perform the work described in the bidding documents.

6.3.3   Prior to the award of the contract, the engineer will notify the bidder in writing if either the owner or the engineer, after due investigation, has reasonable objection to any such proposed person or entity. If the owner or engineer has reasonable objection to any such proposed person or entity, the bidder may, at his option:

1. withdraw his bid, or

2. submit an acceptable substitute person or entity with an adjustment in his bid price to cover the difference in cost occasioned by such substitution. The owner, may at his discretion, accept the adjusted bid price or he may disqualify the bidder. In the event of either withdrawal or disqualifiation under this sub-paragraph, bid security will not be forfeited, notwithstanding the provisions of paragraph 4.4.1.

6.3.4   Persons and entities proposed by the bidder and to whom the owner and the engineer have made no reasonable objection under the provisions of sub-paragraph

EXHIBIT 5-5 (Cont.)

6.3.3 must be used on the work for which they were proposed and shall not be changed except with the written consent of the owner and the engineer.

PERFORMANCE BOND AND LABOR AND MATERIAL PAYMENT BOND

7.1 BOND REQUIREMENTS:

7.1.1 Prior to execution of the contract, the bidder shall furnish bonds covering the faithful performance of the contract and the payment of all obligations arising thereunder in such form and amount as the owner may prescribe. Bonds may be secured through the bidder's usual sources. If the furnishing of such bonds is stipulated hereinafter, the cost shall be included in the bid.

7.1.2 If the owner has reserved the right to require that bonds be furnished subsequent to the execution of the contract, the cost shall be adjusted as provided in the contract documents.

7.1.3 If the owner requires that bonds be obtained from other than the bidder's usual source, any change in cost will be adjusted as provided in the contract documents.

7.2 TIME OF DELIVERY AND FORM OF BONDS:

7.2.1 The bidder shall deliver the required bonds to the owner not later than the date of execution of the contract; or if the work is to be commenced prior thereto in response to a letter of intent, the bidder shall, prior to commencement of the work, submit evidence satisfactory to the owner that such bonds will be furnished.

7.2.2 The bonds shall be written on an acceptable performance bond and labor and material payment bond.

7.2.3 The bidder shall require the Attorney-in-Fact who executes the required bonds on behalf of the surety to affix thereto a certified and current copy of his Power of Attorney.

FORM OF AGREEMENT BETWEEN OWNER AND CONTRACTOR

8.1 FORM TO BE USED:

8.1.1 Unless otherwise required in the bidding documents, the agreement for the work will be written on a standard form of agreement between owner and contractor, where the basis of payment is a stipulated sum.

All these documents are of vital importance to the bidding process. Overall, however, the most important written materials a consulting engineer uses in a contract are the standard general-conditions and any supplementary general- or special-conditions clauses relative to a particular project. The standard general-conditions clauses of the contract are the contract's legal basis. Hence they should be documented flawlessly by your firm either with the assistance of an attorney or with the help of your voluntary engineering society. AIA, NSPE, and certain other professional organizations can usually provide preprinted versions that have been checked by attorneys and engineering professionals for accuracy.

BASE BIDS AND ALTERNATIVES

Frequently, contractors will bid a project based on the plans and specifications and then provide additional prices for various alternatives required by the engineers. There is an inherent danger in such alterna-

EXHIBIT 5-6   A typical bid form.

BID OF: _____ BIDDER.

FOR: OPHTHALMOLOGY LAB—NEWINGTON CHILDREN'S HOSPTIAL, NEWINGTON, CONN.

TO: Mr. James Piro
Director of Engineering
Newington Children's Hospital
Newington, Conn.

I (We) have received the Bidding Documents entitled: Ophthalmology Lab—Newington Children's Hospital, Newington, Connecticut, and Addenda numbered and dated as follows:

_____

_____

I (We) have included the provisions of the above Bidding Documents and Addenda #   in my (our) Bid. I (We) have carefully examined the Bidding Documents, the existing Building, and the Site and submit the following Bid.

In submitting this Bid I (We) agree as follows:

1. To hold my (our) Bid open for 30 days after Bid opening.
2. To enter into and execute a Contract if awarded on the basis of this Bid, according to a Standard Form of Agreement Between Owner and Contractor where the basis of payment is a Stipulated Sum.
3. To furnish Guarantee Bonds as set forth in the Bidding Documents.
4. To accomplish the work in accordance with the Contract Documents.
5. To begin work as soon as possible and to complete the work by or before _____.

I (We) will construct the Project for the Lump Sum Price (Base Bid)
OF: _____

_____ ($_____) Dollars

EXHIBIT 5-6   (Cont.)

SIGNED AND SEALED THIS _____ DAY OF _____
19XX

_____
Full Legal Name of Bidder

_____
Signature of Bidder—and Title

_____
Address of Bidder   Telephone No.

_____
City/Town   State   Zip Code

Corporate Seal (If Any)

EXHIBIT 5-7   *Typical performance bond.*

Bond No.

_____

KNOW ALL MEN BY THESE PRESENTS:

That _____
as Principal, hereinafter called Principal, and _____
_____ as Surety, hereinafter called Surety, are held and
firmly bound unto the City of Hartford, Connecticut, as Obligee, hereinafter called
Owner, in the amount of_____
_____Dollars ($_____),
for the payment whereof Principal and Surety bind themselves, their heirs, executors,
administrators, successors, and assigns, jointly and severally, firmly by these presents.

WHEREAS, Principal has by written agreement dated_____
_____entered into a Contract with Owner for_____
_____
_____
which contract is by reference made a part hereof, and is hereinafter referred to as the
Contract.

EXHIBIT 5-8 A typical payment bond.

Bond No. _____

LABOR AND MATERIAL PAYMENT BOND

Note: This bond is issued simultaneously with another bond in favor of the owner conditioned for the full and faithful performance of the contract.

KNOW ALL MEN BY THESE PRESENTS:

That _____
as Principal, hereinafter called Principal, and _____
_____ as Surety, hereinafter called
Surety, are held and firmly bound unto the City of Hartford, Connecticut, as Obligee, hereinafter called Owner, for the use and benefit of claimants as hereinbelow defined, in the amount of _____ Dollars ($), for the payment whereof Principal and Surety bind themselves, their heirs, executors, administrators, successors, and assigns, jointly and severally, firmly by these presents.

WHEREAS, Principal has by written agreement dated _____
_____ entered into a contract with
Owner for _____
_____, which contract is by reference
made a part hereof, and is hereinafter referred to as the Contract.

NOW, THEREFORE, THE CONDITION OF THIS OBLIGATION IS SUCH, that if the said Principal shall pay for all labor and materials furnished to himself or his subcontractors for use in the prosecution of the work, and used therein, then, this obligation to be void; otherwise to remain in full force and effect:

PROVIDED, HOWEVER, that this bond is executed pursuant to the provisions of Sections 49-41, 49-41A, 49-42, and 49-43 of the 1975 Revision of the General Statutes of the State of Connecticut, as amended, and the rights and liabilities hereunder shall be determined and limited by said sections to the same extent as if they were copied at length herein.

Signed and sealed this _____ day of _____,

.D. 19_____

In the presence of:

_____   _____ (SEAL)
                                 (Principal)

_____   BY: _____

---

tives, however. In particular, there is the temptation for the consulting engineer to secure prices for a larger number of alternatives. The profusion of alternatives from different bidders can create confusion in determining who is really the lowest bidder.

In general, the use of more than three alternatives leads to problems in choosing bidders. Further, using more than three alternatives increases the risk of disagreements that could turn into legal arguments. Why? Because a project's owner may have trouble remembering whether or not she or he has accepted each of several alternatives. The owner may change her or his mind about a particular alternative; or worse, a contractor may claim the alternatives have been chosen in a way that precludes her or him from becoming the successful low bidder. Such complications have often emerged in projects that have numerous alternatives. Fortunately you can avoid most of these hassles if you limit your alternatives and provide clear-cut substitutions in your documents.

LUMP-SUM AND UNIT PRICES

Plans and specifications may require a special lump-sum price or unit price for certain aspects of labor, material, or equipment. Unit prices are considered standard in most specifications. They are perfectly acceptable as long as you do not make them too extensive. Unit prices require the contractor to supply unit costs for separate categories of labor that may be required to complete a project.

Certain kinds of jobs (e.g., many federal or state government contracts) are affected by the Davis Bacon Construction Law. In effect, this law requires the use of unionized labor. When a project falls under the law's requirements, certain types of prices, such as lump-sum or unit prices, may be required.

The issue of using unionized or nonunionized labor is a complex matter that is beyond the scope of this book. However, we can note that the strong influence of labor unions in the construction industry led to a situation where most projects now use unionized labor.

RESPONDING TO INQUIRIES

Protecting your firm during bidding requires you to use the utmost care and discretion in dispensing information about your work. Facts and figures must not be handed out casually. The bidding process is a crucial time in the life of any project, and information that is given incorrectly or given to the wrong people can lead easily to legal problems.

To answer queries properly, you must first decide upon these details:

1. Who in your firm should have the authority and responsibility to answer inquiries?
2. What is the safest procedure to follow in communication with bidders?
3. What are your client's requirements relative to the bidding process?
4. How do your answers to the preceding questions affect the special subjects of guarantees and warranties?
5. What consulting office policies have been developed or can be developed to ensure the proper dispensation of information?

Of the five items, probably the most important is, "Who will answer the questions from bidders?" Clearly, confidentiality is essential in fair bidding and in preventing abuses. Exhibit 5-9 shows a typical in-house communication used to instruct employees in the proper handling of phone calls during bidding. The sample may help you to develop and refine your office policy.

EXHIBIT 5-9   In-house communication on answering queries.

ANSWERING QUESTIONS DURING THE
BIDDING PERIOD

1. For any question of clarification of the plans and/or specifications: DO NOT VOLUNTEER YOUR OPINION BEYOND WHAT IS DRAWN OR WRITTEN.
2. Only principals and project managers shall issue clarifications, and these MUST be in writing to ALL bidders.
3. If the caller is not satisfied with your answer, turn the call over to your project manager or principal.
4. If the project manager or principal is not available, carefully record the caller's questions, name, address, and phone number, and give this data to the project manager or principal.
5. All telephone or written requests MUST be answered by the project manager or principal with COPIES TO ALL OF THE BIDDERS AND THE CLIENT.
6. If there is insufficient time to get a written reply to all bidders, the caller must be notified that NO answer can be given and HE MUST ARRIVE AT HIS OWN CONCLUSION BASED ON THE EXISTING DOCUMENTS.

XYZ ENGINEERING CORPORATION

Generally, information should not be dispensed verbally on projects that are in the bidding phase. A good office policy is to make all your answers to questions on the project available to bidders, in writing, even if the client has not made this a requirement. Your client may add specific procedures that he or she wants followed. The client's requirements must be coordinated with your written responses.

GUARANTEE AND WARRANTY

A surprising number of consultants do not fully understand the difference between guarantees and warranties. The difference is made clear by the following illustration.

Equipment manufacturers usually warrant that the materials they provide will be free of defect. However, they do not guarantee that the installation of the items in a particular system will provide satisfactory results. They do warrant that their equipment will provide a specified level of performance.

Here are some good definitions for "guarantee" and "warranty." The definitions are condensed from entries appearing in several respected sources:

Guarantee   An agreement through which one person assumes the duties of assuring payment or fulfillment of another's debts or obligations. Basically a guarantee ensures that a system that has been guaranteed will perform as described.

Warranty   A formal guarantee by the warrantor certifying that the systems of goods sold are as represented. The seller vouches for the security of the title to the item sold and promises to indemnify or make good any defects that arise.

Any unusual requirements affecting the guarantee or warranty in your specifications should be stated in a separate "Notice to the Contractor." Such a notice should at least call attention to these special requirements. If you do not include the notice with the bidding documents, under no circumstances should you change any aspect of the guarantee or warranty that has been expressed or implied in the specification materials.

Remember that the manufacturer of products for your project does not guarantee any results beyond those specified by the manufacturer for particular design conditions. For example, you may specify, and the manufacturer may guarantee, that the product will produce 5 tons (4.5 kW) of refrigeration at a leaving air temperature of 57°F (13.9°C) dry bulb and 56°F (13.3°C) wet bulb when the entering air temperature is 80°F (26.7°C) dry bulb and 67°F (19.4°C) wet bulb. The manufacturer in this case does not guarantee that the system will provide proper air conditioning to the owner or that the system will even work in the given installation, only that it will perform as specified in the conditions stated, i.e., the manufacturer's unit will produce 5 tons of refrigeration effect.

However, a guarantee that appears in the specifications is a guarantee by the *contractor* that the contractor's system will meet all the conditions specified by the engineer. Again, this does not mean that the contractor's installation will necessarily satisfy the owner. It merely means that the contractor's installation will satisfy the requirements of the engineer's design. In both these cases, the usual limitation of guarantee or warranty is 1 year from the date of installation.

An important point to address in the contract is the date when the warranty or guarantee becomes effective. Commonly, equipment is installed many months in advance of the completion of a project. The manufacturer may warrant the equipment from the date of installation. Then the contractor may guarantee the project for a year, and, if the effective date has not been written into the contract, the argument may arise that the contractor's guarantee period started, say, 6 months after the equipment has been installed. This would in effect create a 6-month warranty on the equipment and a 1-year guarantee by the contractor.

Your contracts should carefully delineate the effective date of the manufacturer's and contractor's 1-year warranty periods. Otherwise, claims may emerge whereby the owner argues that the engineer has not satisfactorily upheld the provisions of the contract since the manufacturer can contend that its warranty had expired. Further, the contractor may claim that the specifications created a condition that made it impossible to provide a 1-year guarantee on the entire installation.

Potential problems in this area can usually be avoided if you state in writing that all warranties and guarantees are to begin on the date that you specify in the contract. If possible, you should state that they will become effective on the date that you certify that the project has been substantially completed. The guarantee and warranty period should begin at that time, regardless of previous discussions regarding the effective dates.

SHOP-DRAWING REVIEW

A major service performed by engineers during the construction of any project is the review of shop drawings submitted by the contractor for approval. These drawings cover equipment and materials required by the plans, specifications, and overall contract.

Shop-drawing review requires thorough checking and investigation of materials submitted by contractors. However, the review process should not be allowed to drag on so long that it interferes with timely processing of documents needed for the project to stay on schedule. Time frames will vary depending upon the complexity of the material submitted.

In general, shop drawings should be reviewed and returned to the contractor within 2 weeks after being received. Ideally, a week or 10 days is the upper limit for the review period.

Drawings returned to the contractor must be accompanied by some form of transmittal document noting the results of the review. As mentioned earlier, your firm should become sensitive to the nuances and proper uses of words applied to the review procedures. Most firms stamp the drawings with a standard form; the appropriate comments are then checked to describe the review's findings.

For years, most of these stamps have used the words "approved," "approved as noted," and "disapproved." Unfortunately, the experience of many consulting engineers has shown that these terms—especially "approved"—can take on connotations that are damaging to the professional engineer, especially in a court of law. The reason is that a lay jury may interpret "approval" to imply all sorts of things that an engineer may never have intended. Most engineers intend "approval" to mean that they have accepted the submitted materials as generally meeting the intent of their specifications. Unlike the lay jury, the engineers do not mean to imply that the submitted materials are exactly equal to, or are superior to, materials originally specified by the engineers.

Several engineering firms and voluntary professional organizations have developed approval stamps that get around such problems. Here are some examples of terms and phrases they employ (note that none use the word "approve"):

1. Furnish as submitted
2. No exceptions taken
3. Furnish as corrected
4. Submit a specified item

In addition, some stamps note that a drawing has been "rejected" or instruct the contractor to "revise and resubmit."

Exhibit 5-10 shows a sample of one of the more popular types of stamps used in shop-drawing review. The example is a multipurpose stamp that covers all actions that normally apply to submitted materials. Further, the stamp carries a disclaimer stating precisely what the engineer intends or does not intend in transmitting the shop drawings.

When it is necessary to explain the engineering firm's actions in greater detail than a particular stamp allows, clarifying statements should be provided along with the transmittal memo. Statements should be concise

EXHIBIT 5-10 *Typical shop-drawing stamp used by consulting engineers.*

```
┌──────────────────────────────────────────────┐
│ ☐ NO EXCEPTION TAKEN        ☐ MAKE CORRECTIONS │
│                                NOTED           │
│ ☐ REJECTED                  ☐ REVISE AND RESUBMIT │
│              ☐ SUBMIT SPECIFIED ITEM           │
│                                                │
│  Checking is only for general conformance with the design concept │
│  of the project and general compliance with the information given in │
│  the contract documents. Any action shown is subject to the require- │
│  ments of the plans and specifications. Contractor is responsible for: │
│  dimensions which shall be confirmed and correlated at the job site; │
│  fabrication processes and techniques of construction; coordination │
│  of his work with that of all other trades; and the satisfactory per- │
│  formance of his work.                         │
│                                                │
│                 XYZ, Inc.                      │
│                                                │
│      Date _____  By _____    │
└──────────────────────────────────────────────┘
```

and should define exactly what the engineer intends or implies by the statement of condition indicated on the stamp.

CONTRACT ADMINISTRATION

Still another area in which you can learn to protect your business against legal liabilities is in project supervision. The roles and responsibilities of consulting engineers in supervising a job are complex. When things go well, the field supervisor has one of the more pleasant tasks. Unfortunately, few situations are ideal, and thus the supervisory staff may become burdened with questions such as how to substitute alternative solutions to those originally specified without increasing the risk of complaints from the client.

The person (or persons) who handles supervision for an engineering firm is usually referred to as the "Contract Administrator" or "Contract Administration." This person must be able to thoroughly comprehend the plans and specifications as well as the many nuances and general intent of the design. Frequently, the contract administrator may be required to detect and help resolve minor flaws in the design's intent. It is most important that the administrator handle these situations before they can become an issue with the installing contractor.

The time spent in contract administration usually amounts to about one-third of the total time spent on the entire project. This obviously is a major portion of the firm's total consulting effort.

For large projects, an arrangement may be made by the owner and engineering firm to provide the job with a "clerk of the works" who acts as full-time supervisor for the duration of the construction period. This individual is paid by the owner and may be hired independently of the engineering firm. Thus the relationship between an engineer and an outside supervisor can be delicate. To prevent problems, make certain that the relationships to the owner, engineer, and contractor are carefully defined in a document signed by all the parties concerned.

Further, you should provide written reports of each instance when you or a member of your firm is present at the project site. Conferences and informal meetings should be documented too, regardless of duration or content. Furnish copies of the reports to all parties involved. There is no substitute for a written report, and any engineer who does not take advantage of this protective tool does so at a risk.

Written reports describing conferences should be discussed with other parties who were present at the time of the meeting so that you can determine in advance that the documents do not contain material that will lead to disputes later on. Preferably, the discussions should take place before the final draft is written.

Any questions that could not be resolved during a meeting should be noted in the report. Try to present all sides of a given issue; do not make statements that only favor your firm. Note in the reports that unresolved questions will be dealt with in upcoming meetings; then try to resolve them at the appropriate time. The job meetings and reports essentially provide a basis for resolving potential conflicts before they can begin. If you do not record the actions taken at a meeting and the steps toward resolution, the chances increase that the issues may later return as legal problems.

CHANGE ORDERS

Many engineers fear processing change orders. This concern is understandable since a change order that adds to the cost of a project may reflect adversely on the engineer. The engineer who processes the order may feel that it indicates he or she made a serious error. However, failure to process a change order when necessary is a much worse mistake. Neglecting to put the order through practically guarantees that a lawsuit will result.

Change orders that lead to added costs usually occur because of failure to provide for a contingency fund or other protective devices in a contract. However, no design is perfect, and virtually all contracts have some changes. Change orders should not reflect adversely upon you or your firm's work if you handle them properly.

As an aid to engineers, the AIA, NSPE, and other professional organizations provide standard forms for the proper processing of change orders. Try to make such forms part of your office supplies for use on most projects. Exhibit 5-11 shows a typical standard change form.

In practice, projects rarely have less than five formal change orders. Large projects frequently have 100 or more. Since the basic construction contract is between owner and contractor, proper procedure requires that the engineer submit the change order for approval to the owner, along with a letter explaining why the change is mandated. The letter should also tell how the order will benefit the project.

There are no rules governing the acceptable cost of change orders; but most experienced consulting engineers agree that orders within 2 or 3 percent of the total construction cost for a complete new project are reasonable. For projects that include larger alterations or additions to existing facilities, change order costs can be as much as 5 to 10 percent before they engender complaints from the owner. However, engineers with many years of experience often cite jobs in which change orders ranged from 20 to 50 percent (and higher) of the entire construction cost. Change orders of such magnitude can present serious problems for the consulting engineer. Even when an owner considers the high costs justified, they can lead to complications that could land the engineer in court.

PAYMENT REQUISITIONS

Payment requisitions are documents that are usually submitted monthly by a contractor for payment for work done up to the date of the submittal. While the requisition process may seem like it would be no cause for potential legal hassles for engineers, that assumption can safely be consigned to earlier, less complicated times in engineering practice.

Most contracts contain a clause that provides for the engineer to approve all payment requisitions before they can be processed. For reasons which we shall now outline, it is imperative for engineers to scrutinize every payment requisition to certify that it is fair and just to all parties concerned.

EXHIBIT 5-11   Typical change order form.

PROJECT: (name, address)	CHANGE ORDER NUMBER:

TO (Contractor):

```
┌                              ┐
                                     CONTRACT FOR:


                                     CONTRACT DATE:
└                              ┘
```

You are directed to make the following changes in this Contract:

The original Contract Sum was . . . .$
Net change by previous Change Orders . . . .$
The Contract Sum prior to this Change Order was . . . .$
The Contract Sum will be (increased) (decreased) (unchanged)
by this Change Order . . . .$
The new Contract Sum including this Change Order will be . . . .$
The Contract Time will be (increased) (decreased) (unchanged)   (   )Days
The Date of Completion as of the date of this Change Order therefore is

Engineer	Contractor	Owner
Address	Address	Address
By	By	By
Date	Date	Date

Nowadays many contractors bill in excess of their work. Two reasons account for this practice:

1. Many contracts state that the owner can withhold about 10 percent of payment requisitions as a form of security allowance for final completion of the project. This withholding often represents a substantial por-

tion (or all) of the contractor's profits as the work progresses. Thus the contractor may overbill during the project on the premise that she or he can thereby bring in profits while ensuring that the final billing will represent the actual, final portion of her or his contract.

2. Contractors frequently withhold part of the payment from their subcontractors. This is a practice that can create problems for the engineer. An engineer who approves payment requisitions without thoroughly checking every aspect of the work done can directly or indirectly foster the potential bankruptcy of a subcontractor. If a subcontractor actually does go bankrupt, the situation becomes very serious. For instance, the contractor may not be able to find a substitute subcontractor who is willing to complete the work for the value left in the contract. Then the owner may call in a bonding company. This would in turn lead to an investigation of the cause of the subcontractor's bankruptcy. The engineer would most likely be drawn into a lawsuit as a consequence and would lose time and money even if he or she were to be vindicated.

Probably the only way to prevent such problems is to review all payments requisitions to screen out bills that appear excessive. Be strict in your approval or disapproval of requisition orders. If you have questions about a particular bill, ask the contractor to explain it.

PROJECT COMPLETION

Before a project is finally completed, the engineer is usually required by contract provisions to conduct a final inspection of the job. The engineer must verify that the work has been done according to plan. During any final inspection, you should use a final-condition checklist prepared ahead of time to verify items necessary to meet the client's needs.

Before the project can be handed over to its owner, you should make sure that final operating tests have been conducted and that performance has been demonstrated to be satisfactory in all respects. Furnish any appropriate operating instructions to the owner.

The final inspection report must list and delineate conditions that are incomplete and which are required by the construction contract, plans, and specifications. However, the report should *not* include items that are beyond the scope of the job documents and their requirements. The report must state a time period in which any deficiencies will be eliminated.

Here we encounter a hotbed of potential problems. For example, the owner may claim that some elements of the project are unacceptable. The engineer may feel that the client is thus trying to get more than was paid for. Or the owner may be expecting total perfection—a goal that may be practically unattainable. Worse, the owner may suddenly decide to impose machine loads on the structure that exceed its actual, planned capabilities. Or the owner may want to occupy the building before some deficiency is eliminated.

To prevent such difficulties from arising, most experienced engineers use a "certificate of substantial completion." The certificate informs the owner and the contractor that the project has been substantially completed and that all monies remaining, except a certain fixed sum, will be paid to the contractor. The fixed sum should not be paid until all items listed in the documents are completed, repaired, or otherwise made acceptable according to the requirements of the contract, plans, and specifications.

The period that immediately precedes completion of a project is a sensitive time. The importance of making certain the owner is actually getting everything in accordance with the contract cannot be overemphasized. The client's requirements that had been documented from the beginning of the job should be reviewed thoroughly, along with any new requirements that may have emerged during the course of construction. If you do not address and rectify problems, then it may be too late to prevent a lawsuit.

OWNER RIGHTS

A project's owner has certain rights during construction. For example, an owner who pays for parts of the project as they are completed may choose to use those parts even though the total job is not completed.

But if the owner decides to try to use a part, say, a recently installed boiler, the boiler manufacturer may issue a statement declaring that the warranty will become effective on the date the owner accepts the boiler. Acceptance of the boiler may be predicated on its having been paid for in a recent payment requisition.

An owner who agrees with the manufacturer's statement may find the warranty actually expires before the total project is complete. Further, many other complications can emerge; for instance, the owner may not want the warranty to become effective so early, but may be unable to operate some essential part of the plant without formally accepting the boiler. There is not enough space in this book for us to list all the other sorts of troubles that could result from such an action. But we can emphasize that such situations must be investigated carefully by the engineering firm to determine the respective positions of the owner, contractor, manufacturer, etc., and put those positions in writing.

The use of partially completed facilities is a nebulous topic in almost any construction project. The engineer must explore all the factors related to use of any parts of a project to avoid argument concerning the guarantees and warranties. He or she must establish the precise dates that warranties begin.

Another area that involves owners' rights is the subject of utilities and their costs. In many commercial and institutional projects, the costs of oil, gas, coal, steam, water, and electricity are paid by the contractor. In other situations, these costs may be borne by the owner. An engineer can easily be drawn into disputes concerning such costs. Therefore, it is wise to have someone in the engineering firm document all agreements that relate to utilities services and their costs.

Serious arguments can develop, for example, if a contractor lowers the temperature in a building during the winter to save money. Frequently, an exposed pipe will freeze and burst under such conditions, possibly damaging newly completed work. An investigation usually follows to determine who is liable for the damage; commonly, the engineer becomes implicated. The ideal answer to this problem would be to have the owner pay all utility costs. Then the owner knows at least that the money is being put to good use.

No clear line exists to mark some of the specific rights of an owner of a project. Certain legal rights, such as ownership of a new building after construction and final approval, are basic provisions of nearly every contract. Any well-written contract should define clearly all legal rights as they pertain to each party involved in the job.

Many engineering-related lawsuits in court today are a result of failure to delineate properly the lines of authority between owners and engineers in contractual documents. Whether the fault lies primarily with the owners, the engineers, or other individuals in those cases is irrelevant now. What matters is that you yourself, as a consulting engineer, develop a keen eye for the root causes of such disputes. Once you know how to spot omissions, errors, and troublesome clauses in contracts, you should be able to root them out and replace them with the correct ingredients to protect your business. Accordingly, your contracts should define your firm's relationships with its clients and at the same time clear lines of authority regarding who is responsible for what and under what circumstances they are responsible.

Let's consider the question of supervision. Project supervision is so important that many consultants actually write two separate contracts: one covering plans and specifications and the other covering supervision.

Projects involving private-sector clients allow a large number of different supervisory arrangements, as we noted earlier. For instance, a client may hire a clerk of the works from an independent firm or ask the engineering firm to provide supervision.

When the latter occurs, you must be wary of the use of the term "supervision" in the contract. Experienced engineers prefer to use the term "contract administration" instead. Why? Once again, we have the nuances of words. "Supervision" may imply meanings you never intended. The lay public usually understands "supervision" to mean that the person appointed to supervise a project will cover the job on a full-time, daily basis.

For most engineers, "contract administration" is a much more accurate and much less troublesome term than supervision. "Contract administration" has a narrower definition: It implies that the engineer will provide certain supervisory services during construction and will process shop drawings and other details as defined in the contract. Clearly, you can help to prevent disputes merely by using the correct words in your contracts. Contract administration is just one example. Consult the engineering manuals of voluntary professional organizations for others.

Supervision has a whole different set of parameters in most government-sponsored projects than it has in the private sector. Nearly all federal, state, and municipal agencies act as the final authorities on projects they sponsor. Thus they control the ultimate approval process and most of the construction supervision. This constitutes special problems for the consulting engineer. Government employees handling a given project may do excellent work; however, their agencies often allow little latitude in the reading of plans and specifications provided by consultants.

For example, unlike an experienced consulting engineer, the government employee may follow the plans as designed, without reading anything into their intent. Further, government procedures usually require specifications to follow a performance-based format rather than a materials-based format. This procedure is followed to ensure that no supplier will be discriminated against unfairly. Unfortunately, this practice can create problems for the consulting engineers, who may find that a particular manufacturer's product is inferior from a materials standpoint.

Taken to an extreme, the government employee's rigid reading of plans and specifications can create a situation where change orders are necessitated or work is stopped completely. Rarely will the engineer be required to pay for such problems in terms of cash. Nonetheless, the change order or work stoppage may be counted as a professional liability even if the engineer was not directly responsible. It is conceivable that the government agency would refuse future work to the consulting firm as a result.

The relationship between consulting engineers and government employees who act as construction supervisors is delicate. The engineer should strive constantly to prevent conflict and foster cooperation. To this end, your greatest tool is your ability to translate your professional image into effective consulting.

CONSTRUCTION REVIEW PROBLEMS

When an engineer functions as contract administrator or supervisor of construction, he or she may or may not discover defects in a project during regular visits to the site. Obviously, defects in the manner of installation, in materials, and other areas must be reported promptly. But what happens if, as contract administrator, you fail to detect some defect that is later discovered?

An owner or contractor may try to claim that you are responsible for the results of the defect. In court, the jury's view could easily be influenced by the words you use to describe your function. The jury may be less likely to misinterpret your position if the contract describes your function as "contract administration" rather than "supervision." Likewise, the proper legal interpretation of your duties can be protected by incorporating the term "construction review" into your contracts. This term is relatively new in engineering and is no mere semantic exercise. It is a term with a much more precise meaning than "inspection" and "supervision."

Your responsibility for construction review does not imply that you are to be held liable for a contractor's errors or omissions. A contractor is never legally relieved of the responsibility of discovering his or her own errors and correcting them. But, if your duties are described formally as "inspection" and "supervision," a jury may regard you as having been derelict in your duties for not having discovered a defect for which the contractor may actually have been responsible.

QUALITY CONTROL

A common source of claims against engineers is lack of adequate description of technical functions in the specification of materials and equipment. Usually, the problem stems from failure to research and evaluate available information on the items specified. Some engineers compound this problem by describing the items in language that is too vague to be of practical, contractual value.

This often occurs when an engineer allows her or his education in information-collecting activities to stagnate. As a result, the engineer's plans, specifications, and language fail to keep up to date with recent developments. The engineer may be unaware of the availability of new products that meet a project's needs; or worse, may develop specifications that describe obsolete equipment or equipment which is no longer built to meet the specifications' requirements.

Too many engineers tend to repeat certain elements of design without checking to see if a manufacturer can still supply the necessary items with the same quality and performance ratings they once had. For instance, an engineer may frequently specify a rooftop gas-fired heating unit for certain types of structures. He or she would specify the model number for the unit on various designs. But the heating unit's manufacturer may recently have changed the model number. What once designated an exterior unit may now indicate an interior unit.

If the error is not noticed during the bidding process, the engineer's untimely specification will almost surely become an extra cost, or a reason for a legal claim, during construction. The way to prevent this problem is simple: Make sure you and your firm are kept up to date on the latest developments in products that you may specify, and in your field as a whole. In shop-drawing reviews, carefully investigate all materials to determine whether or not they can meet your specifications.

Experienced consultants do not approve materials that do not fully live up to specification requirements. But what if a particular item meets performance requirements and not material requirements? As we noted earlier, engineers, by and large, are not accepted as product experts in court. Therefore, your opinion that the product is made of unsatisfactory materials may not be sufficient to support a rejection of the product. Unless you have demonstrable evidence, the mere fact that the material seems to be of low quality may not be considered a valid reason for disapproving the submission.

Another problem arises when an engineer designs a process or system that pioneers a new path in installation or design. For such projects, there may be only one manufacturer who can meet the exact requirements of the specifications. On a government-sponsored project, use of such a product may not be allowed. Generally, most publicly funded contracts require three or more responsible suppliers to be sought for engineering specifications. While this is not often a requirement in private work, it is always a good idea to be sure your design can be bid by more than one supplier.

Still another difficulty may emerge if a client wants you to specify items produced by the client's business acquaintances, friends, or colleagues. This situation represents a classic conflict of interest. If it happens to you, it is your responsibility to tell the client what could happen if you accept the terms. Provide the client with documentation to support your position that, for example, using the client-sponsored items will lead to structural instability, change orders, or delays in completing the work. If necessary, disapprove the materials as submitted. Then if the client still demands you use the items, you will at least have taken the proper precautions to protect yourself from a conflict of interest.

BUSINESS PROTECTION FORMS

Besides the sample forms presented earlier in this section, there are a variety of documents that you can use to improve your handling of construction contracts. Many voluntary professional organizations, including AIA and NSPE, will provide such forms on request. The forms include a certificate of project completion (Exhibit 5-12), and an application and certificate for payment (Exhibit 5-13). Exhibit 5-14 shows a typical surety bond covering payment.

COLLECTING FEES

Any book that tells engineers how to get 100 percent of their fees paid, by the date due, would be an automatic best seller. Virtually every consulting firm has difficulty collecting its fees from time to time. In some cases, a client may refuse to pay a bill due to alleged errors or omissions by the engineer.

Imagine, if you can, submitting a bill for $20,000 for services you rendered and receiving a response from the client saying that you have been professionally negligent and will be sued for $200,000 in alleged damages. It happens!

Before submitting any bill, make certain that all work required by the contract up to the date of the bill is completed properly. The next step is to obtain documentation of the client's acceptance of the work. In general, successful firms cover both the design and supervision aspects of a job by writing the client a simple letter stating briefly that the required services have been performed. In the letter, they ask the client to sign an attached sheet to verify the engineering firm's statement.

The approval sheet is usually described as a "Completion Statement." It states briefly that the firm performed the design or supervision services in accordance with the contract, dated as shown, in a manner acceptable to the client. Most completion statements include a space for the client to list any exceptions. There are also lines for the signatures of the engineers and the client, plus the date of signing.

EXHIBIT 5-12   Project completion form.

FORM OF PROJECT COMPLETION NOTICE BY CONTRACTOR

To: _____

_____

_____

Re:

I, the undersigned, certify that on this _____ day of _____
19 _____, I have reviewed all parts of the drawings and Project Manual and all
portions of the work and, in accordance with the Contract Agreement dated the _____
day of _____ 19 _____, state that all punchlist corrections have been
completed and that all work is completed and ready for final inspection and Owner's
acceptance.

Signed: _____
                                            Contractor

For: _____

FORM OF RECOMMENDATION FOR FINAL ACCEPTANCE BY ENGINEER

To: _____

_____

_____

Re:

Gentlemen:

The work of _____, Contractor on the subject project, has
been inspected by this office and our Consultants and has been found to be satisfactorily
completed in accordance with the Contract Documents.

We therefore recommend acceptance of this Contractor's work by the Owner. Work
requiring correction under the guarantee is to be accomplished by this Contractor during
the normal one-year guarantee period, except those specified items which have a longer
period called for in the specifications.

Your action in accepting this recommendation will start the effective date of the
_____ day lien period for release of retention.

Signed: _____
                                            Architect

For: _____

The completion statement will not necessarily prevent a client from instituting a lawsuit if he or she is determined to do so. However, the signed document does represent firm evidence that the client agreed at some point that the proper services were performed.

If a client does not respond to billing at all, send second, third, and fourth bills in the succeeding months. The bills should note the amount of time by which payment is overdue. This may seem like commonsense advice; but any consultants actually wait several months to follow up an unpaid notice. It is better to send monthly reminders since a bill can be lost in the mail or misplaced by the client.

In cases where the client continues refusing to pay, the engineer may opt to use arbitration to settle the matter. However, arbritration does not provide an easy solution to unpaid bills! Even when a contract contains a clause allowing disputes to be taken to arbitration, this may not be an effective way to compel a client to pay. If a client insists on arbitration, you may be compelled to give up your right to conduct "discovery proceedings." In court, however, you can often use discovery proceedings to examine your adversary's files. This is not always possible in arbitration. And when such information is unavailable to you, you will most likely have a harder time defending yourself. Another problem with arbitration is that an arbitrated decision is not subject to appeal in most jurisdictions. The time and cost needed to defend yourself in an arbitrated settlement may force you to lose a considerable portion of your fee.

INSURANCE POLICIES

In this section we have discussed the importance of having professional liability insurance to protect your firm against legal damages. There are many other types of policies that every consulting engineer should consider obtaining. Most of these we discussed in earlier sections. Aside from your own personal life insurance and automobile policies, the policies that can help protect you against claims include: health and accident insurance; unemployment insurance; "key employee" insurance; valuable papers policies; property insurance against loss due to interruption of business; and fire and theft policies.

We've covered many of the situations in which professional liability insurance could be highly beneficial to you as a consulting engineer. But what situations does a professional engineering liability policy not cover? It does *not* insure the following:

1. Liability assumed under any agreement, contract, warranty, or guarantee certificate
2. Liability arising out of ownership of any vehicle

EXHIBIT 5-13 Application and certificate for payment form.

PROJECT:
(name, address)

TO (Owner):

ATTN:

ENGINEER:

CONTRACTOR:

CONTRACT FOR:

APPLICATION DATE: APPLICATION NO:

PERIOD FROM: TO:

Application made for Payment, as shown below, in connection with the Contract, is attached.

The present status of the account for this Contract is as follows:

ORIGINAL CONTRACT SUM .... $ _____

Net change by Change Orders .... $ _____

CONTRACT SUM TO DATE .... $ _____

TOTAL COMPLETED & STORED TO DATE .... $ _____

CHANGE ORDER SUMMARY,

	Additions $	Deductions $
Change Orders approved in previous months by Owner— TOTAL		
Subsequent Change Orders		
Number	Approved (date)	
Totals		

Net change by Change Orders $

State of: County of:

The undersigned Contractor certified that the Work covered by this Application for Payment has been completed in accordance with the Contract Documents, that all amounts have been paid by him for Work for which previous Certificates for Payment were issued and payments received from the Owner, and that the current payment shown herein is now due.

Contractor:

By: Date:

RETAINAGE _____ % .... $ _____

TOTAL EARNED LESS RETAINAGE .... $ _____

LESS PREVIOUS CERTIFICATE FOR PAYMENTS $ _____

CURRENT PAYMENT DUE .... $ _____

Subscribed and sworn to before me this   day of

19

Notary Public:

My Commission expires:

In accordance with the Contract and this Application for Payment the Contractor is entitled to payment in the amount shown above.

OWNER:

ENGINEER:

CONTRACTOR:

Engineer:

By:

This Certificate is not negotiable. It is payable only to the payee named herein and its issuance, payment, and acceptance are without prejudice to any rights of the Owner or Contractor under this Contract.

EXHIBIT 5-13   (Cont.)

Page   of   Pages

APPLICATION AND CERTIFICATE FOR PAYMENT, containing
CONTRACTOR'S signed Certification, is attached.

APPLICATION NUMBER:

ENGINEER'S PROJECT NO.:

Item No.	Description of Work	Value	Work Completed		Stored Materials	Total Completed and Stored to Date	%	Balance to Finish	Retainage 1
			Previous Applications	This Application					
Subtotal or Total									

EXHIBIT 5-14   Typical surety bond covering payment.

NOW, THEREFORE, THE CONDITION OF THIS OBLIGATION is such that, if principal shall promptly and faithfully perform said contract, and shall certify in writing that all wages paid under said contract to any mechanic, laborer, or workman were equal to the rates of wages customary or then prevailing for the same trade or occupation in the City of Hartford; then this obligation shall be null and void; otherwise it shall remain in full force and effect.

The Surety hereby waives notice of any alteration or extension of time made by the Owner.

Whenever Principal shall be, and declared by Owner to be in default under the contract, the Owner having performed Owner's obligations thereunder, the Surety may promptly remedy the default, or shall promptly

1. Complete the Contract in accordance with its terms and conditions, or
2. Obtain a bid or bids for submission to Owner for completing the Contract in accordance with its terms and conditions, and upon determination by Owner and Surety of the lowest responsible bidder, arrange for a contract between such bidder and Owner, and make available as work progresses (even though there should be a default or a succession of defaults under this paragraph) sufficient funds to pay the cost of completion less the balance of the contract price: but not exceeding, including other costs and damages for which the Surety may be liable hereunder, the amount set forth in the first paragraph hereof. The term "balance of the contract price," as used in this paragraph, shall mean the total amount payable by Owner to Principal under the Contract and any amendments thereto, less the amount properly paid by Owner to Principal.

Any suit under this bond must be instituted before the expiration of two (2) years from the date on which final payment under the contract falls due.

No right of action shall accrue on this bond to or for the use of any person or corporation other than the Owner named herein or the heirs, executors, administrators, or successors of Owner.

Signed and sealed this _____ day of _____
A.D. 19 _____

_____ (SEAL)
(Principal)

_____

_____   BY: _____

_____ (SEAL)

_____

_____   BY: _____

3. Liability for death or injury of an employee during the course of the person's employment

4. Workers' compensation

5. Property liability

6. Activities that are not customary or usual in the performance of the professional services designated in the form you sign when taking out your liability policy

7. Liability arising out of some types of cost estimates or quantity surveys

8. Property in the insurance company's care

9. Liability attributable to a failure by the insured to affect or maintain an insurance policy or bond

10. Civil engineering liability arising out of operations in tunnels, bridges, dams, certain types of connections, and other items (except under certain qualified conditions)

11. Any loss caused intentionally or by the direction of the insured

12. Liability arising out of surveying activities, ground testing, and sub-surface soil survey

13. Loss that is the result of insolvency of the insured

Note that not every professional liability policy is the same. When you consider adopting a new insurance policy, be sure to investigate all aspects of the policy's coverage and exclusions before you buy.

In essence, professional liability policies cover you solely as a design professional. They generally do not cover you if you act in another capacity, e.g., as an owner, investor, or contractor. Likewise, such policies cannot be expanded merely on the basis of clauses appearing in a contractual agreement with clients. Agreements which attempt to extend the insured's coverage without notifying the insurance company may fall within the bounds of an exclusion and eventually negate the effect of the coverage.

However, cases do exist where engineers have benefited from so-called piggyback insurance situations. A piggyback situation is one in which the engineer is included in the provisions of a policy taken out by the owners or builders of a project. This type of insurance is commonly referred to as "builder's risk" or "all-risk" coverage. Generally, owners or builders obtain such policies to protect themselves against damages that could arise from "acts of God" (e.g., earthquakes or hurricanes) or certain human-caused events. It is a good idea to check to see that you are included in all-risk policies that have been bought by owners and builders of your projects.

MISCELLANEOUS LIABILITIES

Construction Injury

A growing source of legal claims against consulting engineers stems from construction workers who are injured while working on engineering projects. In such cases, the claimants may argue that the injuries have been sustained as a result of errors or omissions by the engineers who designed, supervised, or inspected the projects.

To protect yourself against damages from such claims, research the laws of your state and the regulations applying to each engagement you accept. The laws should provide you with a firm idea of what you can expect in this regard. Depending on the laws in their states, some consulting engineering firms use these protective measures:

1. Before performing any supervisory services, the engineers make sure that all parties involved in contract negotiations agree that hold-harmless clauses and indemnity clauses in the general conditions extend from the general contractor to the engineer and owner. An important aspect of such an agreement is that its hold-harmless clause usually makes the contractor responsible for the safety of his or her employees and employees of subcontractors. The only time this does not hold true is when an injury is caused by the sole negligence of a design professional or owner.

   Most contractors can obtain this type of coverage easily. Perhaps more importantly, the arrangement serves to remind people that the design professional is not generally responsible for job safety.

2. The engineering firm can ask that the contract include a clause stating that the firm's staff is not responsible to the project owner in cases of worker injury.

3. The engineers can place a disclaimer on each drawing furnished for construction purposes. Such disclaimers usually state that the drawings do not include components necessary for construction safety.

New Partners and Employees

New partners constitute a significant liability risk for engineering firms unless certain formal agreements between prospective partners are reached prior to formation of the partnership. A partner is considered an agent of his or her firm so long as the partner has authority to act for the firm in certain matters. If a new partner independently completes his or her own project, a client may be able to argue that the engineering firm is somehow connected to that project. Likewise, clients may sue a partner for work the partner completed prior to joining the partnership; the suit may consequently name the engineering firm as a defendant. This situation is more frequent when the new partner has worked on projects that are similar to ones on which the consulting firm works.

Another area of risk involves staff members who "moonlight." The danger of the moonlighting employee is twofold:

1. When a moonlighter produces a design or other engineering work as a favor to people outside your firm, you may be held liable for errors and omissions found in the moonlighter's work. A plaintiff may sue the moonlighter, the employers, and the contractor, especially when they have professional liability insurance.

2. An employee who works at a second job does so to earn extra money. However, no benefits accrue to the engineering firm that employs the moonlighter. In fact, a moonlighter may cause overall consulting work quality to decline. The extra burden of a second job may make the employee tired and prone to sloppy work. Worst of all, the consulting firm may be held liable for the moonlighter's mistakes.

To remain successful, consulting engineering organizations must take every precaution in hiring new employees to maintain high-quality staffs. Diligent checking and interviewing is essential to effective hiring practices.

Retroactivity

Many beginning consultants see no need to buy liability insurance since their new firms may have engaged in little or no engineering work. Often a consultant will decide to obtain a liability policy only after working for several years in private practice. When a client lodges claim in regard to a project completed in the early days of the practice, however, many engineers start worrying about insurance protection. Fortunately, some professional liability policies provide coverage on a retroactive basis.

Such insurance policies can protect a firm against damages for work the firm may have completed prior to buying the policy. For the policy to cover past errors or omissions, the claim must have been made during the period in which the policy was effective. More importantly, most policies will provide retroactive coverage only when there has been no prior knowledge of the errors or omissions cited in a claim.

Note that most insurance policies provide coverage only for problems that arise during the period in which the policies are effective. For example, if you retire from practice, a policy may not protect you against claims for problems in past work since the policy will most likely have expired. In some cases, special provisions are written into the policy to extend the coverage beyond the normal effective date.

Another example would be a new joint venture formed for the purpose of engaging in a special project. Since most joint ventures use a special group liability policy, the insurance may be discontinued following the completion of the project and the end of the joint venture. If an alleged error or omission appears after the insurance ends, there may be no coverage for the potential claim. Therefore, it may be advisable to maintain a special group policy beyond the date on which the joint venture dissolves. Alternatively, you can try to obtain some types of extension on the policy when you buy it. Of course, the most important step you can take is to obtain appropriate protection before any problems develop.

Consulting firms can function without having their own professional liability insurance policies. If you can tailor your practice to avoid engaging in work that will require construction and instead confine yourself to producing reports and surveys that do not impinge on public safety, your chances of being exposed to legal claims will be relatively low. Most consulting engineers do engage in projects involving construction and public safety, however. Thus the professional liability policy is invaluable for most consultants.

SUMMARY

The possibility of being sued starts the day the consulting firm obtains its first assignment. So it is the wise consultant who recognizes this possibility and takes steps to be prepared in advance of any lawsuit.

Steps the consultant can take include obtaining the services of a competent attorney or legal firm, being careful in all bids to choose work one is competent to perform, watching the language of contracts and other documents, having adequate insurance, and being especially careful in dealings with contractors.

The wise consultant conducts business in a manner which is alert to the possibility of lawsuits but does dwell unduly on this situation. Being ready for trouble helps one avoid it. Taking heed of the advice in this section of the Handbook should diminish the possibility of being sued.

Expanding a Professional Practice

The freedom of the individual to expand personal and professional interests remains a crucial pillar supporting the vision of success in America. The fruits of this freedom have been evident in U.S. history from the earliest colonial settlements to the industrial revolution and beyond. Many consulting engineering practices today are thriving through the creative use of opportunities that pervade the engineering field.

TECHNIQUES AVAILABLE TO PROFESSIONAL ENGINEERS

The variety of techniques available to the professional engineer for expanding a consulting firm derive from three basic goals of business growth: personal, organizational, and financial. These three kinds of growth sometimes overlap; for example, steps to streamline the organization and increase efficiency of checking drawings, plans, and specifications may indirectly lead to increased profits and personal rewards.

Therefore, you may choose to emphasize one type of growth over others; but it is always worth remembering that changes in one area of your practice may affect other areas, and you should control your firm's expansion in a balanced way. This is our objective in this section: to examine the goals and techniques of growth, in detail, and look at some of the possible avenues you may want to pursue in your own engineering practice.

Personal Growth

Very often, an engineer's desire to fulfill his or her personal ambitions becomes the driving force behind the excellence in the work a firm produces. It is not unusual, for instance, to find a consultant placing high priority on the freedom to be involved in nearly every detail of every project rather than emphasizing efforts to increase the organization's size or gross revenues. Such involvement may sometimes tend to limit the amount of work the consultant will choose to accept. But instead of short-changing the firm's economic and structural resources, it may allow the engineer the personal satisfaction of taking full responsibility for the firm's work while helping to ensure top-quality results.

Professional engineers are inclined by training to pay strict attention not only to details but also to the broad scope of their endeavors. The engineer who focuses on personal growth, therefore, is likely to be one who also attends conferences in various engineering disciplines and maintains contacts with other professionals. As a result, such an individual often gains the personal rewards of heading a small- or medium-sized firm known for outstanding engineering.

Organizational Growth

Almost at opposite poles from personal goals is the pursuit of organizational expansion. Many experienced company managers consider the development of the structural aspects of a firm to be of paramount importance. Strong administrators are often more than happy to delegate responsibilities to individual members of their firm. Management controls the overall plans and direction of the organization, frequently with the aim toward enlarging the staff and diversifying the work that is to be produced.

For the consulting engineer, this sort of emphasis may require considerable financial sacrifice and the devotion of large amounts of time to ensure that the organization will stay healthy, even during periods of industrial difficulties and slumping economies. Typically, the effective organizer grooms her or his organization in ways that relieve management of the burden of constant involvement in production of the engineering work. The firm's managers can consequently devote more energy to such vital functions as marketing, arranging contractual matters, financing, and dealing directly with past, current, and prospective clients.

Financial Growth

It should surprise no one that maintaining and increasing income remains a primary concern of many engineering professionals. There is certainly nothing objectionable about the goal of financial growth. In fact, the growth of a consultant's income is often based directly on the production of excellent work that meets the needs of clients and directly or indirectly benefits the public. Consulting engineers can meet this objective in sundry ways. You may choose to focus on maintaining a small, highly efficient and profitable firm whose specialty is local civil engineering work, for example; or you may pursue increased income by building a complex and diversified practice specializing in large government contracts.

Now let's examine personal, organizational, and financial growth in more concrete terms. Keep in mind that none of these forms of growth are mutually exclusive.

The history of consulting engineering is full of true stories of men and women who learned not only how to gain great personal reward from their professional work but also how to build strong, diverse, and profitable organizations. Some of the techniques for expansion, such as joint ventures, will no doubt sound familiar. However, it would be almost impossible to know how to implement effective changes without first analyzing the reasons and common results of various expansion efforts. These fall into four categories:

1. Concentration on particular kinds of work
2. Expansion of the firm's offices
3. Merger with other organizations
4. Diversification of work

Concentration

While few engineering firms want to be known as being capable of only a narrow range of projects, listings of members of various engineering societies, along with their chosen fields of expertise, clearly indicate that a good many prominent national and even international firms have chosen to concentrate in specific areas of engineering or industry. Some of the largest engineering firms in the United States work chiefly in individual fields such as oil exploration and processing, particular types of power generation, waste-water treatment and pollution control, or even in the single field of suspension bridges. These are only a few of many examples we could cite.

Many firms that choose to concentrate and expand their business in single areas of engineering have done so because they gained a reputation for doing superb work on certain types of projects early in their history. Consequently, the same firms may deliberately solicit similar projects in an ever-widening geographic range. In such firms, marketing efforts are often based primarily on the premise that the organization is a well-known expert in its chosen line of engineering.

Unfortunately, a consultant who is just starting out may feel discouraged by the presence of long-established firms in the local consulting market and decide that there is no room left for her or his new practice to succeed. Such a decision would very likely be a mistake. It is our opinion that room always exists in the market for consulting engineers capable of demonstrating their ability to do high-quality, efficient work in particular areas of specialization.

There is certainly no requirement that specialized expertise be some new, far-out version of professional engineering. The basic requirements are for the firm to be able to do consistently exemplary work, to market the organization's services, and to maintain a high level of quality as the firm increases the geographic area of its projects and hires additional staff. Using these sorts of guidelines, many well-established consulting groups can boast that they are into their second and third generations of happy partners and employees.

The technique of growth by concentration can further be examined in terms of two major approaches: "vertical" concentration and "horizontal" concentration. Most established consulting firms attribute at least part of their success to one or the other of these approaches.

Vertical concentration denotes specialization in a relatively narrow range of clients and engineering services. Firms that focus their efforts on the vertical approach generally try to market their work over a broad geographical area. The horizontal approach, on the other hand, is most often characterized by concentration on a limited locality. Consulting engineers who use the latter technique frequently attempt to serve many kinds of clients by offering services from a variety of engineering disciplines.

Deciding which growth technique best suits your firm requires extensive analysis of the consulting market from several distinct perspectives. Certain categories of engineering work are available to most firms, regardless of location. An examination of these markets may provide the framework for such a market analysis.

The basic markets for consulting engineers include services to the following kinds of clients: residential homeowners, commercial enterprises, industry, private and nonprofit institutions, government agencies (i.e., city, town, state, or federal bodies), and insurance companies and forensic specialists (for investigative purposes). Potential clients in each of these markets have need for a wide range of professional engineering expertise through all phases of project selection, design, construction, operation, and maintenance.

Effective planning for a firm's organizational and promotional activities is grounded in an understanding of the particular markets that the business can serve while restricting overhead and maximizing income. Thus an organization located in a medium-sized city such as Des Moines, Iowa, may prefer to concentrate horizontally rather than vertically in order to help assure a healthy cash flow and a dependable local market.

The alternative for such a firm would be to spend large amounts of money sending salespeople from one city to another. While well-organized marketing efforts aimed at broad geographic areas can sometimes bring substantial rewards for established consultants, the expenses incurred in travel can frequently spell disaster for the small or beginning firm. On the other hand, the well-known warning against placing all your eggs in one basket clearly applies to the professional practice of consulting engineering.

Many successful small and new firms have established an excellent balance between their goals of growth and marketing by concentrating their efforts in the local area at the same time that they diversify their services. In the corporate business realm, this practice is sometimes known as "cutting territory" since it effectively increases competition with similar organizations operating in the same area. Commonly, it is accomplished by opening an additional office to expand the firm's influence over the geographic area. The same effect can be achieved merely by increasing the firm's aggressiveness in marketing its services to local customers. Depending on usage, the idea of cutting territory may sometimes carry a negative connotation. In the present context, however, it describes a workable and proven technique of growth.

Vertical concentration can be just as effective when properly applied. An engineering firm specializing in a narrow range of services—providing consulting to paper producers, for instance—will almost inevitably incur substantial travel expenses. Such an arrangement may prove ideal, depending on the organization's financial resources and on its members' personal and professional orientation. Often, vertical concentration entails a longer period for the business to establish its reputation, generate sales, and build revenues.

The number of areas in which the consultant may choose to concentrate is virtually limitless, despite rumors that certain disciplines are no longer as viable as they once were. For example, many engineering publications refer to electrical power and transmission engineering as a "mature" science. Unfortunately, first-year college students and their parents frequently interpret such statements to mean that electrical engineering is a dead, nongrowth area, and students are consequently discouraged from pursuing such a course of study. Some established engineers hold similar attitudes. While it is true that the markets for various disciplines and subdisciplines fluctuate according to a wide variety of economic, scientific, and cultural influences, it is questionable whether one or more types of engineering can validly be considered too mature to be worth engaging in. Indeed, electrical energy and its distribution are only about 100 years old. Modern versions, including such hot growth areas as semiconductors, are based on designs that began entirely within the past 50 years. If electrical engineering can be considered mature, how might one describe waste-water treatment and water-supply engineering? The latter categories stem from technology as old as civilization, but the need for qualified professionals and technological improvements will continue to increase for many years to come.

WHEN SHOULD YOU ADD NEW OFFICES?

It seems to be getting harder and harder these days to find a well-established engineering firm that has not seriously considered opening an additional office or offices to expand its practice. In some ways, the desire to open new offices can be seen as the engineer's version of the 7-year itch. And the desire to open a new office can strengthen or weaken a firm.

The addition of an office can be used as the key to successful horizontal concentration, though it is not a prerequisite. The new office can help the organization gain the competitive edge over other consulting firms in the region. Though it requires a substantial initial investment, the new office can save the firm money that would otherwise be spent on travel and accommodations to contact prospective new clients. Further, existing clients in the vicinity of the new office can often be served in a more cost-effective manner than before.

However, the opening of permanent new offices does not always benefit the consulting organization. Firms that can afford to rent or own temporary additional offices sometimes do so to avoid the financial and professional commitments that accompany the permanent office establishment. These commitments may include:

1. The need to acquire additional professional engineering registration for members who will work in the new offices

2. Relocation expenses for firm members and their families
3. Revisions to promotional programs and brochures
4. Management reorganization
5. Motivation to start the branch office, make it viable, and keep it running as a profitable venture
6. A big injection of money to get the added location started

Generally, established firms can borrow on their accounts receivable as they nurture the branch office into a healthy adulthood. They can pledge the personal property assets of their owners and principals; or they can transfer some current work load to the new location. This last choice often represents a proper solution to the question of financing. The transferability of a project nearly guarantees a source of revenue for the fledgling office, and the job may even be done more effectively from the branch location. The change in cash flow caused by such a situation may not become a problem as long as the personnel can be relocated to suit the needs of each office.

Perhaps one of the more difficult tasks in setting up a new office is the reorganization of management and staff that it requires. Ideally, you would be able to hire some excellent additional employees for the new office without much difficulty. But what about the assignment of top partners, officers, and principals? Who will work in which office, and how will their responsibilities be affected by the new office?

In many cases, a single licensed proprietor cannot legally or successfully operate one or more offices. But he or she can manage two offices, each operated by properly licensed and trained members, and possibly direct the selling for both branches. However, the marketing needs of a firm with two or more offices differ markedly from those of a one-office organization. For example, the new office does not merely introduce itself as a newcomer; it necessarily expands upon a reputation that already follows the established office. Nonetheless, the branch office is indeed perceived as a newcomer in some ways. Therefore, its sales effort may have to be budgeted at a rate double or triple that of the first office for the first few years of expanded operation.

Expansion and Combination

Growth by expansion of offices can pose special difficulties for the small engineering firm. Large, older organizations can usually better afford the capital expenditures required to support new offices through various stages of the firms' development. Consequently, a growing number of consultants with small firms are opting to expand by combining their practices with other firms or by acquiring small practices. Thus some consulting engineers find that they can quickly add successful new offices and new business territory without much additional capital or new staff.

Becoming acquainted with other engineers with whom you may be able to develop fruitful professional relationships is indeed one of the virtues of keeping active in various voluntary professional organizations. Many successful mergers and acquisitions have begun as a result of casual, friendly discussions at meetings of professional engineering groups.

For example, a successful firm located in Boston may find that it could add some profitable new projects if only it had an office in, say, Philadelphia. If the principal of the Boston firm maintains good contacts with other consultants whom she or he has met at regional meetings of a voluntary professional group, it is quite possible that the principal would be able to strike up a mutually rewarding arrangement with another consultant based in Philadelphia. In this way, engineers have expanded the reach of their services from Boston to Philadelphia to Washington to Chicago to Los Angeles and beyond. Meanwhile, other firms have rejected this sort of expansion because they decided they would not be able to effectively control the operation of additional offices.

Like engineering and construction contracts, proper arrangements to combine or acquire organizations are delicate and often require the assistance of outside professionals, such as an attorney and accountant.

Many successful mergers and acquisitions occur between engineering firms that share similar areas of service or, at least, similar geographic locations. As we stated previously, the effective combination and expansion of offices frequently results in a substantial strengthening of a firm's abilities to compete against larger and older companies in a given area. For many small firms, combination is by far the best way of expanding the practice to promote personal, organizational, and/or financial growth.

But despite the potential advantages, fiercely independent consultants tend to shy away from the idea of becoming involved in professional ventures that they feel might limit their autonomy. Joint ventures, new partnerships, additional offices, and other efforts toward expansion and combination are occasionally viewed as threats to the hard-earned and cherished independence that attracted the engineer to professional consulting in the first place. This is the main reason why combinations of small firms do not occur more frequently.

An objective evaluation of the risks and advantages inherent in starting or expanding almost any business would probably show that expansion, properly planned and executed, is rarely more dangerous than building a firm from scratch. Both efforts may carry hidden areas of weakness for the professionals involved. In either case, the serious consultant needs to address a variety of practical considerations before starting the venture. Among the questions you should ask are these:

- How will the proposed changes in your professional practice affect your ability to promote and sell your services?
- What will be the responses of clients? For example, if you enter into a joint venture with another firm, can you be certain that you will continue to be able to provide full, personalized attention to your clients' needs? Even if you are sure you can do so, will your clients view the new arrangement as a step that is likely to dilute the effectiveness and efficiency of your firm's work?
- Will you be able to organize the new or expanded business in a way that mutually satisfies the personal and professional interests of all its members?
- Are you confident that a prospective partner or firm with which you may combine will maintain or improve the general level of engineering expertise to which you are accustomed?
- Who will be responsible for the various decisions to be made about the new organization's finances, services, marketing, management, and public relations? Whereas a small firm with 10 members may have been successfully managed by one owner-principal, an expanded venture may require several administrators, each responsible for a different area of operation. For instance, you may choose to remain in charge of all projects for a particular client with whom you have a good ongoing relationship, while a new partner may have similar contacts with other clients. Likewise, the principals may choose to designate administrative specialists, such as a vice president for marketing, a vice president for office operations, a vice president for field services, and so on.
- Will you be able to work with the other members of an expanded firm without interference from feelings of competition, professional jealously, and other "ego" conflicts?
- How will the new arrangement influence your use of the engineering office or offices? Would an expanded partnership require acquisition of an additional office, or would you extend the scope of your services to include the production of certain kinds of work from an office currently held by the prospective partner?

Certainly there are other questions to be addressed in any prospective business expansion, not the least of which is the immediate financial impact that such a move may have on your firm. The act of combining one firm with another through amalgamation or joint venture can sometimes result in literally instantaneous growth. Effective combinations have been known to double a firm's revenues within a few months because of the nearly automatic increase in the number of clients and the strengthened marketing abilities. But the many logistical requirements—including legal fees, transfer of ownership, and equipment purchases—that are vital to expansion may entail considerable expense at the outset of the new arrangement.

The complexity of any arrangement requires the aid of a financial expert who is skilled in analysis of the impact of mergers, acquisitions, joint ventures, and other kinds of expansion on consulting firms. Clearly, each proposed new arrangement must be evaluated individually so the firm can decide exactly what kinds of changes may benefit it most.

Generally, engineering consulting organizations do benefit from combinations and new partnerships—provided that the modifications add depth to the management, sales, and engineering expertise of the group. Greater depth in these areas will almost invariably enhance the general profile of the business as it is viewed by new and old clients, particularly at top-management levels.

What does it take to make the combination of two or more firms a successful operation? Simply stated, it requires a willingness on the part of all individuals involved to work toward a common goal. Further, it requires the members to accept the individual successes, as well as the failings, of every other person in the group. So long as everyone is willing to work toward the resolution of problems that develop—and some problems most assuredly will develop—the capability for continued success remains. The alternative is continual conflict and confrontation.

A common problem encountered by many organizations engaging in combinations with other firms is the question of how many offices will be needed and in what locations. Frequently, one or more of the principals may feel that the expanded operation should use only one office, and one of the principals may favor his or her location over that of the firm with which they are considering combining. This point may become the first stumbling block in a merger discussion.

The case for a single office carries substantial weight when the organizations are in the same town or building. As discussed earlier, however, the use of two or more offices located in separate geographic areas can be a boon to most firms. One of the most successful types of combinations is one which involves two firms that have offices in areas that are remote enough to allow each office to serve a separate clientele and close enough to permit cohesive administrative oversight.

In well-populated areas, the individual offices may be able to operate effectively when they are as little as 30 miles apart. Each group entering into the combination may be able to retain its current work sources, and at the same time the central organization may seek projects that had not previously been available to either group alone.

For example, a mechanical and electrical engineering firm in one area and a structural group located 25 miles away might easily merge their resources to form a firm with mechanical, electrical, and structural expertise. This arrangement could be accomplished without abandonment of any offices. The retention of both offices could prove a great benefit as specific project responsibilities and personnel are rotated from one location to another to increase the overall efficiency and capability of the whole group.

Similar results could be achieved, in a slightly different manner, by groups located as far apart as a few hundred miles. In that case, the logical orientation of the combination would be to develop a pool of professionals who could work on projects secured through either office without constant need to travel from one location to another.

The use of additional offices may encourage a client who has projects in distant areas to employ the expanded organization, whereas the client may have previously seen no reason for seeking the services of one of the individual firms. The practice of doing business with the client prior to combination will often convince the client that he or she can best be served by the expanded firm since one of its offices may be near the proposed project.

Before you do enter into such a combination, however, it may be beneficial to discuss the potential changes with your current clients. Arrangements which may alienate a large customer, such as mergers with firms that have done unsatisfactory work for your clients in the past, should be avoided. The same caution applies to arrangements with firms that compete in any way with your current clientele. Some clients will be delighted with the idea of your firm's expansion, however.

As mentioned, particular members of the expanded organization may consider themselves responsible for all the work for certain clients. This is fine as long as the scope of each individual's responsibility is clearly delineated during the contractual process. Conflicts within an engineering firm may develop if professional limits and duties are not understood by everyone concerned. The apportionment of revenues to executive consultants, in particular, requires clearly written, legally binding documentation.

Equally important during the process of combining and expanding practices is an evaluation of the legal profile and potential liabilities of the individuals involved. All members of a prospective merger, joint venture, partnership, or other combination should reveal openly any potential problem areas before the papers are signed. As noted earlier in our discussion of professional liabilities, the continual involvement of engineers in work affecting public safety makes lawsuits relatively frequent occurences. If a potential partner or firm is the focus of a civil suit, the feasibility of a successful merger is not totally negated; but extreme caution and careful planning, along with expert legal advice, is essential.

Pursued with care, an expanded practice offers many advantages to the practicing consultant, especially to an owner of a small firm. Arrangement for a larger organization provides increased continuity in the management and engineering of the group and greater freedom for its owners and principals. These considerations are valid since even the most dedicated professional consultants will occasionally want to have some time to themselves. Further, the expanded practice provides a vital buffer against the economic and personal losses that could be sustained were one or more members to resign, retire, or die unexpectedly. Small firms sometimes collapse under such shocks. Firms that develop a strong managerial and supportive foundation through growth techniques, on the other hand, can maintain their standing in a competitive market despite a substantial unexpected loss.

Diversification

Closely allied with the goal of combining practices is the idea of diversifying the company's engineering services. This can be accomplished by any one of several methods. Aside from merging with or acquiring other firms, for example, you may choose to increase your staff by adding experts qualified to practice in fields in which your firm had not previously excelled. One or more experts in a given discipline may be attracted by a proposition to become a partner or responsible stockholder in your firm, thus helping you to achieve a personal goal of increased prestige and financial success.

For small firms, the addition of a partner—or, for that matter, a merger with another firm—is chiefly an organizational change geared toward promoting growth. However, medium-sized and large engineering organizations can often add new disciplines and new partners or staff without substantially altering the structure of their organizations.

Historically, the firms that seem to succeed best at diversifying are those that are careful not to accept too much additional work in the new discipline until the entire organization has been "shaken out." Generally, problems in this type of expansion occur when the firm or its clients expect an exaggerated level of proficiency at the start of the diversification program. Before taking on a new discipline fully, the firm's officers should be certain that at least a few projects in the additional areas have been accomplished successfully.

In theory, a diversified, complex organization may seem easy to build. However, it takes time, patience, and perseverance in actual practice. The technique of growth by adding engineering disciplines is nonetheless used by many consulting firms.

Those firms that have the greatest success in their diversification efforts tend to share some common traits from which we can glean valuable information.

Let's start with the case of a firm that hires a new member, or promotes an existing employee, to be placed in charge of a particular engineering discipline in which the firm has not previously practiced. What is the major activity the company must pursue to get projects in its new area of expertise?

The answer is promotion. Without considerable efforts toward marketing the added discipline, it would be practically impossible for the organization to sell its diversified services. However, marketing in this context has special meaning. Unless the individual who has been put in charge of the new discipline has a well-established reputation for doing high-quality work in that speciality, the firm will need to build confidence between prospective clients and the consultant. This requires an extended period of public relations effort during which the clients may gradually get to know the engineer and build sufficient rapport to provide a sound basis for solid consultant-client relations.

Another factor that must be considered prior to attempts to diversify your services is continuity. In other words, you want to be able to add disciplines, staff, and clients without significantly disrupting the current operation of your practice. For some, continuity is best achieved through mergers or partnerships with other established organizations rather than by adding staff members.

Regardless of the specific route you choose, the best way to evaluate your ideas for diversification is to watch the responses of clients. For instance, you might send out a news release or other form of announcement to let your current and prospective clients know about a change in staffing or organization to add an engineering discipline to the firm. The first set of announcements could be followed a few weeks later by advertisements, a revised company brochure, or telephone calls to selected customers.

By engaging in this sort of a promotional campaign, you would, in effect, be testing the market for your diversification efforts. Such a campaign can be accomplished at nominal expense to the consulting organization. More importantly, it reduces the risk inherent in any change in company orientation.

Two areas of engineering that many small- and medium-sized consulting firms have added to existing operations with high rates of success are forensic engineering and construction management, also known as "project management." Often one can add these services to a business without additional personnel or organizational expansion. The National Society of Professional Engineers (NSPE) offers instruction in forensic engineering, covering everything from application of engineering theory and science to accident investigation and courtroom participation. The society also occasionally gives courses in construction claims management, the core concern in project management.

JOINT VENTURES

Perhaps the type of consulting firm that most thoroughly integrates the four types of business growth is the joint venture. It is also one of the most misunderstood. While the joint venture may or may not concentrate in a particular discipline, it inevitably involves expansion, combination, and diversification. An examination of the advantages and disadvantages accompanying most joint ventures may help dispel some confusion.

Advantages to the Client

Most clients who have experience dealing with the various groups involved in a large engineering project—consultants, contractors, vendors of materials, equipment, and the like—are all too familiar with the potential haggling and attendant delays that may accompany such jobs. Consequently, many begin looking for engineering consultants who can provide the total job within a more cohesive framework. More to the point, perhaps, is the fact that such experienced clients prefer an arrangement in which they can hold one person ultimately responsible for the entire project.

This preference accounts largely for the increasing popularity of the design-build concept of consulting engineering discussed earlier. At its furthest extreme, the engineering profession has sought to satisfy the sophisticated customer's desire with organizations that do the total job. In international contracts particularly, a consulting firm may take the form of a multifaceted corporation providing design, construction, manufacturing, financial, and management services. But there are disadvantages, too, to the single-source concept, of which most knowledgeable clients are aware. For example, projects completed by a sole-source firm frequently carry sundry hidden handling and coordination costs capable of increasing the final costs by anywhere from 20 to 60 percent over original estimates.

So why do so many experienced clients still prefer to deal with a single consulting group on each project? The answer lies in the following advantages that such situations present to the average client. Joint ventures and other strong, diverse organizations can offer:

• Unified administration of the overall effort
• Centralized management of personnel
• More effective professional and financial responsibility
• More economical results through coordinated design
• Greater contractual oversight
• Clearer relations between engineer and client

Advantages to the Consulting Firm

To the group(s) involved in a given joint venture, both tangible and less obvious advantages accrue. Among the most important are the following:

• Ability of the joint venture to secure clients who could not be attracted by any of the member firms operating individually
• Increased profit potential
• Increased professional experience and exposure for each firm involved
• Opportunity for staff members to increase their skills in working within, and being responsible for, a team
• Lasting increases in sales, profit, and net worth capabilities for members of the joint venture

Disadvantages

The major disadvantage to the members of a joint venture is the greater number of individuals on which the professional status and ultimate success of the larger group depend. Thus, the chance of errors may increase. While the joint venture will be only as effective as its weakest components, however, the larger group offers increased checking and verification during the course of the job.

Purpose

As noted, nearly all joint ventures start with the aim of carrying out a specific project or consulting service. If representatives of another firm approach you with the idea of forming a joint venture without a particular job in mind, the offer should be subjected to considerable scrutiny. Exactly why is the other firm making such an offer? What is their aim?

Since all loss and liability ultimately must be shared in a joint venture, it is rarely advisable to start one without a well-planned, carefully delineated, finite consulting job as the primary goal. Not only should all parties be concentrating on the proposed project and the joint venture's ability to handle it, but they should also reach agreement on performance standards, ethical principles, business methods, and numerous other components needed to complete the job on schedule and at the right cost.

Contracts

For the purposes of the joint venture, the only acceptable form of agreement is one made in writing which has been properly reviewed by a skilled attorney. This is what nearly every client wants.

Potential liabilities in a joint venture are not only professional but legal and financial. Each member must be willing to accept full responsibility for its partners in the agreement.

Most joint ventures use stationery and a drawing title block developed explicitly for the job at hand. This is no mere piece of trivia: Such technicalities can constitute a legally binding acknowledgment of contractually defined responsibilities.

Management

For an established consulting organization, it is usually not a problem getting people with the proper engineering skills needed to make a project a success. The place where difficulties occur is in eliciting and coordinating such performance within the newly assembled team. Thus management becomes the most critical task facing a well-organized joint venture.

A number of approaches have been taken to minimize the difficulties inherent in managing joint ventures. The two most common are:

• Creation of a separate management group
• Appointment of one of the participating firms as the managing body

The first option generally works best for large or long-term projects. A separate management team concentrates on the administrative requirements of the entire job so it is not bogged down in smaller details of specialized tasks. Usually, all member firms in the joint venture are represented on the roster of the management group, allowing proper acknowledgment and credit while enhancing the structure as a whole.

For projects which involve only two or three firms in work lasting less than a year, the second type of management is often better. When one group handles the administration of the job, decisions can often be made more swiftly and with greater consistency. This type of framework provides the clear-cut essentials for the project's success, provided that all parties have agreed to it in the contract.

How to Start

How can you decide if and when you should get involved in joint ventures? Begin by weighing the positive and negative characteristics of the concept for your firm, in relation to specific projects. Exhibit 6-1 shows a "weighted joint-venture evaluation" form that almost any firm can use. You can tailor such a list to more ideally suit the characteristics of your firm, if you wish.

Rate the prospect of a joint venture with a known organization for each category on a scale of 1 to 5. A top rating for all 20 categories would yield a total score of 100. Scores of less than 100 can be reduced to a rough indicator of the odds for success. No scientific claims are made for this method, of course, but you can weigh the odds. For example, a total score of 75 would mean a ratio of 75 to 25, or odds of 3 to 1 for success. Generally, the odds should be at least 2 to 1 in favor of success for a joint venture to be seriously considered.

The final test for consideration of a joint venture is not shown on your list. It can best be summed up in the words, "If you could have all the principals of the potential member firms as your equal partners, would you choose to do so?" This is the crux of the issue. The consultant entering into a joint venture must be able to work effectively with heads of other firms for an extended period of time. More often than not, this requires compromise and negotiation.

Additional Growth Areas: Government and Overseas

Further means of expanding your professional practice include finding new ways to secure government contracts and international work. Since we have already examined these topics in this book, we shall not repeat points made about them earlier. It is important to note, however, that

EXHIBIT 6-1 *Weighted joint venture evaluation.*

A. *Services offered*

Efficiency   _____

Reliability   _____

Convenience   _____

Uniqueness   _____

Attractiveness   _____

Economy   _____

B. *Venture as formulated*

Selling   _____

Management   _____

Performance   _____

Personnel   _____

Future work   _____

Facilities   _____

Finances   _____

Expertise   _____

C. *Clients available*

Number   _____

Size   _____

Growth possibility   _____

New areas   _____

Distant markets   _____

Other   _____

---

tremendous potential exists in both these areas for new contracts and for increasing your firm's experience and professional renown.

Before you begin soliciting work in international areas, be sure to contact your professional organizations or societies for suggestions. The American Consulting Engineers Council (ACEC) and the International Federation of Consulting Engineers (IFCE) can provide a variety of useful references, registration forms, and other documents for international projects.

Frequently consultants find their best contacts for overseas contracts begin with U.S. government agencies and departments. For example, the Army Corps of Engineers handles a large amount of work done in the oil exporting countries. An important source of funds for many underdeveloped and developing nations is the Export-Import Bank (also called Eximbank), which has offices in Washington, D.C.

As noted earlier, the international model form of agreement often serves as the basis for successful negotiation of contracts with overseas clients. Essentially, this widely used model differs little from carefully drawn contracts prepared for domestic use. Its provisions include:

1. Commencement and project completion dates, material setting forth the conditions of alteration or termination of the agreement

2. Rights, duties, and liabilities of the consulting organization (These sections contain some variations from the ordinary domestic contract, some of which relate to liabilities for patents, copyrights, regular and third-party insurance, and potential damages.)

3. A section on obligations of the client which includes international concerns such as compensation for taxes, duties, and other levies

4. Settlement of disputes (This item is not included in domestic contracts nearly as frequently as in those in many countries in Europe and certain other foreign nations, suggesting cultural differences in approaches to conflict resolution.)

Additional sections deal with specific personnel assignments, payments, and other logistical considerations.

The international model form of agreement should be studied carefully if you contemplate doing work overseas since it contains many details with which you may not have become familiar in your consulting work to date.

Another source of lucrative contracts and references is government in the United States. It is a well-known fact that many government contracts are secured through discussion and negotiation with government representatives, regardless of whether one is dealing with local, municipal, state, or federal work. The advantages and disadvantages of developing and nurturing relationships with people in the political arena were discussed in depth earlier.

Many government contracts are obtained by engineers who are personally acquainted with government representatives and politicians. Most of these contracts are secured in a wholly ethical and fair manner. As noted, virtually all government agencies and departments are required by law to solicit bids from a number of competing consulting firms, thereby assuring equal treatment to all involved. Clearly, however, consultants who have the most knowledge of engineering needs in each area of government will be the first to know about prospective projects. Keeping abreast of the latest developments and needs—through contacts with government officials and by monitoring source documents such as *Commerce Business Daily* and the *Federal Register*—provides a strong competitive edge for the progressive consulting engineer. The U.S. Chamber of Commerce and its branch offices, as well as city and local chambers of commerce, can sometimes provide valuable information on potential consulting jobs. Many times all it takes is a telephone call to one of their offices to start moving toward your next prospective contract.

The one role you cannot expect your government contact to fulfill is that of your sales agent or personal representative. Manage your relationships with government thoughtfully, and the rewards can be tremendous. Perhaps we can sum up the principle of good consultant-government relations like this: *Government contacts cannot sell, but if you make your services available, they can and will tell.*

Consulting for Manufacturers

Have you ever noticed the large number of advertisements placed in daily newspapers by manufacturing firms seeking engineering consulting services? If you have not, you may be missing a vast array of potential jobs. New projects mean growth, and for many consulting engineers, the manufacturing field remains an untapped treasure trove of growth ideas.

Consulting for manufacturers is not to be confused with the practice of joining a manufacturing firm or selling one's practice to a large building concern. Here we are referring to the interesting and often lucrative prospect of providing consulting services, usually on a temporary or part-time basis, to aid a manufacturer in drafting and design.

Consider the manufacturer's options. He or she can either pay a handsome finder's fee to some technical employment agency for temporary workers who lack the professional backing and expertise provided by an experienced consulting firm or go directly to professional engineers best equipped to handle his or her problems. If you were the manufacturer, which would you choose? In most cases the professional consulting firm will be able to supply a core of personnel qualified in a greater variety of fields and skills relating to work within industrial plants. Clearly a market exists for the practicing engineer who is willing to accept the challenge of consulting for manufacturers.

Generally, manufacturers are not nearly as concerned with the size of the engineering firm as they are with its capacity to furnish the appropriate personnel. Once this question has been resolved and the engineering staff is properly assigned, the next issue frequently relates to where the work is to be done. Some consultants may not want, or be able to, perform the work at the location of the manufacturing concern. Sometimes this may hinder its ability to conclude the contract negotiation. More often, however, a mutually agreeable decision can be made whereby most or all of the project is completed at the consultant's office.

The key to securing this type of job is to convince the industrial client that you are fully equipped to supply the required services while under the direction of the manufacturer's industrial engineers. These people in effect become the project engineers for the manufacturer.

Areas of vital importance in this field are as follows:

- The consulting engineer frequently comes into contact with proprietary information; therefore, she or he must be able to present clear evidence or guarantees that such data will not be "leaked" out or used to solicit other clients.
- The engineer must assure the industrial client that the consulting services will be supplied consistently and responsibly; i.e., the engineer cannot attempt to switch personnel on or off the project once individuals have been assigned to particular tasks.
- Deadlines are crucial! Since many manufacturers and other industrial clients hire consulting engineers precisely because they themselves lack sufficient staff to meet a deadline, the consultant's ability to finish a job on schedule is essential to undertaking the project in the first place. When in doubt, do not even negotiate the undertaking.

Two major areas of consulting for manufacturers are process design and industrial control. The emergence and continued growth of high technology in everything from agriculture to urban zoning assures engineers with knowledge in these fields prosperous work for many, many years to come. Computerized and electronic controls for industry are particularly burgeoning areas which every consultant can learn about.

The problem for many manufacturers is they know what they want for their product or plant, and the computer suppliers and electronic instrument suppliers know what they can sell; but neither seems to be able to apply this information to achieve the desired results. That is where you fit in. Frequently, all an engineering firm needs to do to obtain projects with manufacturers is to devote a portion of its overall marketing efforts toward industrial concerns in the firm's area.

Is "Design, Build, and Manage" the Future?

In recent years, the most popular way to grow as an engineering consultant appeared to many to be to mold the professional practice around the design, build, and manage concept, which we mentioned frequently in earlier sections. This concept generally means that the consulting firm not only does all engineering work on a given project but also designs the architecture, performs construction work, and manages both the construction and design functions, including specialized purchasing of materials and equipment. This is indeed an excellent way for many firms to grow, and we can expect to see more consultants becoming involved in design, build, and manage teams in the future.

The range and complexity of such work requires strong organizational capabilities to deliver the finished project in a way that satisfies and benefits all parties involved. The basic supervisory roles in design, build, and manage projects are: project manager, design manager, and construction manager.

The *project manager* functions as general administrator, overseeing the functions of both the design and construction managers. The project manager may perform feasibility studies, real estate analysis, site selection, financing, accounting, and other duties such as leasing and even tenant acquisition. Each of these subspecialities may require the services of additional personnel, of course, whom the project manager would supervise.

The *design manager* is in charge of architectural and engineering services, while the *construction manager* oversees all aspects of construction. All three individuals may share information and duties related to the design process, cost control, scope of construction, equipment purchase contracts, scheduling and expediting, contract documents, and related matters.

Most professional consulting engineers have had experience in project management and management of design services. The particular orientation they acquire through years of work in special engineering fields can occasionally lead to problems in the functioning of the design, build, and manage team, however.

A project manager may bring a decided "bent" toward construction and the practical aspects of the job, for example. That is fine since this approach is needed. But it is only one of many perspectives that can be taken on any given project. The result can sometimes be a neglect of the main design considerations and other abstract concepts needed to build the structure in the first place.

Clearly, administrators of design, build, and manage teams have a special responsibility. They must pay utmost attention to integrating the various needs of each project and to balancing their respective concerns to achieve a common goal.

Another potential source of frustration, especially for the design engineer, is the so-called fast-track method of building. In fast-track projects a municipality, industrial complex, or other group awards construction and design contracts simultaneously. This is done to accelerate completion by ensuring that construction will proceed on the heels of design. Little time lag is allowed. Consequently, substantial pressures may be placed on the design engineers to expedite their work in the most "practical," construction-oriented manner possible.

Fast-track projects sometimes work to the benefit of everyone involved; however, they can also backfire when the rush causes insufficient time and attention to be focused on investigation and formulation of design solutions.

Despite such risks, contracts which involve the consulting firm in project management and construction management roles offer great promise for organizations wishing to improve their competitive position in a given market.

Before you begin looking for design, build, and manage work or fast-track types of contracts, be sure to review their implications for your insurance status with properly skilled legal professionals. General project management services are unlikely to increase your firm's exposure to potential liability as long as they primarily involve managing the project and planning design and construction aspects as the owner's agent. Liability potential does increase when construction management duties bring the consulting firm into the actual construction phase. It is important to remember that providing direct construction assistance may not be covered by your professional liability insurance. In some cases, it may be necessary to form a separate corporation or to become affiliated with a general contractor to modify your insurance profile properly.

How to Get Involved

As a small or relatively new consulting organization, you can expand your work in management areas in various ways. Here are a few of the services you can offer to a potential client:

- Analysis of equipment and structural additions to the client's existing plant or building
- Feasibility studies of development options
- Enumeration of details and their implementation
- Finding and retaining additional consultants needed to complete the project
- Receiving bids and purchase orders in the owner's name on long-lead items
- Negotiating bids with prime contractors
- Maintaining cost accounting and control systems
- Inspection of construction

- Work scheduling assistance for prime contractors
- Assisting owners and contractors in start-up and test activities
- Training owner's personnel in operation of the new facility

Many of the preceding items are well within the capabilities of most consulting engineers. They are also applicable to a wide range of projects, regardless of size. It is the responsibility of each consultant to study his or her market to determine existing opportunities. Ask yourself, What are my clients' needs? Do these needs indicate potential for project management and construction management services? Do the same for prospective clients.

Other questions you should address are: What changes, if any, will be mandated in order for you to get involved in the expanded areas? For instance, you may need to revise your insurance coverage to include certain aspects of construction work. Also, what costs will you incur as a result of these changes?

The particular capabilities necessary for involvement in project and construction management are these:

1. Scheduling, particularly the ability to use such "network" scheduling aids as computer programs, with specialized concepts like "CPM/PERT"
2. Cost control, emphasizing quantity and estimation procedures
3. Inspection—the ability to handle fabrication and on-site quality assurance
4. Safety program recommendations
5. Value analysis
6. Construction procedure planning and sequencing

Charging Fees

The ways of charging for such services run the gamut of fee arrangements, including:

- Lump-sum fees—where the scope of services can be clearly defined
- Percent-of-cost fees
- Hourly consulting fees—where this fits in with client requirements
- Combined lump-sum and hourly fees
- Basic cost fixed fee—usually defined as actual payroll cost and overhead reimbursible expense plus profit

In reviewing the market for project and construction management services, we want to emphasize that utilization of consultants in this capacity is increasing. At the same time, the duties of project managers, design managers, and construction managers are becoming more clearly defined by the accumulation of experience and improvements being transferred to colleagues in voluntary professional groups. In many cases engineers and clients are learning that the client's interests can best be served by design, build, and manage teams or by firms that oversee general aspects of a project and its construction.

Project and construction management services can provide the advantages of the turnkey approach, offering greater client satisfaction, while avoiding the disadvantages. Still, you will have to deal with the competition of general contractors whenever you offer construction management services.

Computers Provide Special Aids for Expansion

The increasing availability of low-cost, high-quality computer systems for small- and medium-sized businesses is making these silicon-brained electronic servants a highly praised tool for expansion. Properly selected, a good microcomputer can help your firm operate more efficiently, increase accuracy, and improve correspondence and promotional capacities immeasurably.

You can use a microcomputer to assist in design, perform payroll tasks, store bookkeeping and account records, write brochures, print letters and mailing labels, and even check your spelling. Hence, you can increase your output and your income at the same time.

Two operations that particularly benefit from the use of a small computer system are the preparation of monthly billing statements and maintenance of project time records. Most computer dealers carry program software specifically designed to handle these tasks. Of course, time will be required for you to learn to use your system and keep it functioning properly. In most cases we know of, consultants have been able to retrieve their initial investment within a year or two. Further, many states allow substantial tax deductions for businesses investing in their own computer systems. Similar arrangements are available at the federal level.

A microcomputer valued anywhere between $1800 and $8000 (including a video display, printer, keyboard, disk drives, and software) will enable you to perform the tasks we mentioned and also allow you to write, store, and retrieve specifications, details, and standards. Often all it takes for a computer to correct a mathematical or typographical error is a couple of keystrokes. You can "call up" data from previous projects and examine a list of standardized details with the same amount of ease.

The use of computers in structural, civil, and sanitary engineering is by no means new. A tremendous variety of programs are available for small and large computers to assist in these areas. Some of these programs are designed for programmable calculators. Calculations for the heating, ventilating, and air conditioning fields—such as piping and ductwork design—are being performed on computers with increasing frequency by consulting engineers.

Computers cannot take the place of top-flight engineering personnel, of course. But they can help two people to do the work of three or more. A microcomputer can speed your production of studies, reports, and estimates while reducing clerical expense. To the client, this can make a big difference.

Consulting firms that can afford larger systems are beginning to get into the exciting fields of computer-aided design and computer graphics. The larger computers allow you to produce, correct, alter, and finalize your design on the screen, press a button, and get a finished print.

CONSULTANT'S PROFIT KEY 9

Minicomputers and microcomputers hold tremendous promise for the future of consulting engineering. They will become the "great equalizer" in professional engineering and market competition as we approach the twenty-first century. Thomas G. Roberts, Information Systems Support Manager for Bechtel Petroleum, Inc., writing in *Chemical Engineering* magazine, gives the following useful pointers for using computers competitively to improve work and cut costs:

As technical decision-makers, we evaluate a proposed technology by weighing its tangible benefits against its costs. This goes for computer technology too. Engineers cannot all be system-design experts, but we should be "smart users," able to identify a potential application, set realistic objectives, and analyze matters with an eye on the bottom line.

This article looks at this front end of computer-system design, and tells where to seek applications and how to weigh benefits against costs. But first, let us examine some of the motivations, and management goals, for computer technology.

WHY COMPUTERS?

The most widely touted advantage of computerization is improved productivity, achieved through greater speed, quality and accuracy of engineering calculations. Less widely acknowledged are the improvements brought about through more effective organization of engineering workflow—e.g., transcribing engineering data only once, rather than over and over again, on drawings, specifications and material requisitions. And there is also the human side of the question—e.g., computer technology may facilitate working in small, multidisciplinary groups.

Perhaps the greatest payoff of computer technology is in minimizing the cost of building and operating a plant. This is management's top goal, and thus the point where computer technology can be applied with the greatest justification and economic leverage.

Altogether, applying computer technology offers a company a way to improve its competitive position in the world marketplace. But which applications can be justified? Detailed analysis compares costs vs. benefits, but as a start one can screen applications to see whether they are attainable, and reasonably likely to pay off.

IS IT GOING TO WORK?

The slow adaptation of computer technology is often attributed to individual footdragging, or fear, but there are more important barriers. One is that some people have a hard time thinking through a computer application. Then, if the system should fail, they may be unwilling to risk another failure the next time around.

How do we break out of this morass? By approaching the computer project systematically, as we do in regular engineering work. This means clearly defining the problem, determining how a computer system can solve it, and analyzing the payoff. The kinds of engineering tasks that computers can handle, and that can be cost-justified, include:

- Eliminating repetitive tasks. Computers can do mindless, boring work with great speed and accuracy. An example is tracking workhours and costs, and assigning them to various accounts. More-complex examples are budgeting and forecasting.

- Handling complex calculations. Such applications are characterized by a small amount of input, a large amount of output, and an ability to change parameters to test the sensitivity of the calculation. Examples include structural analysis, mass and energy balances, and critical-path scheduling of large projects. Computers can do in a short time what it would take a person hours or even years to do.

- Controlling masses of data. A computerized database can organize large amounts of related data, prevent input errors, allow for easy retrieval, and facilitate the sharing of data. An example is a material-control system that tracks requisitions, purchase orders, traffic, logistics, receiving, warehousing and installation.

- Providing sophisticated output. This includes business graphics, and engineering drawings. Computer-aided design/drafting is an example. While it may be hard to prove a cost saving, such an application may be justified on the quality and consistency of the output.

Still, to assure a payoff requires a detailed evaluation. And it is the engineers who have to apply the good business sense. Otherwise, users may get disillusioned when computer projects end up costing too much; programs that are developed and maintained may be seldom used; and work done on the computer system may cost more than the same work done manually.

EVALUATING APPLICATIONS

Effective evaluation means considering computer decisions as part of the business strategy, and *not* delegating authority to computer experts. The evaluation should consider the total system—hardware, software, procedures and training—and this system should be defined in some detail. To do so requires a clear statement of the system's objectives.

These objectives should say, clearly and concisely, what the system is supposed to do. A few sentences are usually enough. They should be written down, so as to avoid lapses of memory later, and agreed on by those who will use the system and those who will supply it. Everyone involved in the project should know what the objectives are; this avoids confusion and crises.

The objectives must be reasonable. This often means limiting them to the easy ninety percent of the problem, since trying to solve the last ten percent can double the cost. For example, reasonable objectives for an estimating system are to: provide a central file of prices and estimates; make the file available to all estimators; provide rapid response to queries; and perform the necessary calculations. An unreasonable objective would be to provide simultaneous access to all records, as this could lead to a conflict among users.

As another example, in some cases it may be better to develop "throwaway" software, for one job only, than to try to develop a multipurpose program. And the cost of a desired system may force one to accept a simpler or smaller version and plan for upgrading or expansion.

In general, a reasonable objective is technically attainable—i.e., something a computer can handle. It is affordable. It provides the degree of precision and timeliness that the work actually requires. And it is cost-effective. But how is cost-effectiveness measured?

TANGIBLE COSTS AND BENEFITS

Cost/benefit analysis means thorough evaluation of the tangible costs and savings associated with an application. Because computers are so complex, and so often argued about in subjective terms, rigorous cost justification is all the more important. And, in order to compare proposals, a consistent approach should be employed.

A cost/benefit analysis begins with analyzing the current method of performing the work, in enough detail so that all the tangible costs are considered. These include the costs of salaries, benefits, equipment, facilities, supplies and services, as in Exhibit 6-2. The analysis should consider the present workload, and allow for increases or decreases in future years.

The next step is to cost out the current mode of operation, and project the costs over three to five years—accounting for planned changes such as lease increases, additional staff, and such. In Exhibit 6-3, this projection is shown as the "cost without automation."

Then, estimate how much it will cost to automate—the one-time investment—and how much it will cost to operate the automated system, on the same basis and for the same time period. Five years is the computer-depreciation schedule typically followed in the U.S.

EXHIBIT 6-2 Tangible and intangible costs and benefits.

Tangibles	Intangibles
Workhours and staff salaries	Efficiency or productivity
Equipment	Operating environment
Facilities (floorspace, security, etc.)	Image among clients or employees
Computer hardware	Communication speed
Computer software	Access to data
Hardware and software maintenance	Quality of service
Communications services	Speed of response
Miscellaneous supplies and services	

EXHIBIT 6-3 Example analysis shows positive, tangible benefits.

	Summary by year ($000)					
	1	2	3	4	5	Total
Tangible factors						
Cost without automation	4,375	4,800	5,300	5,800	6,400	26,675
Cost with automation Operation	4,400	4,450	4,750	5,000	5,300	23,900
Development & installation	225	50				275
Total	4,625	4,500	4,750	5,000	5,300	24,175
Tangible benefit	(250)	300	550	800	1,100	
Discounted at 15%	(218)	227	362	458	547	1,376
Cumulative benefit	(218)	9	371	829	1,376 =	NPV

EXHIBIT 6-3   (Cont.)

		Summary by year ($000)				
	1	2	3	4	5	Total
Intangible factors						
Productivity saving						
Managerial	(10)	(10)	20	20	20	40
Engineering	(20)	(30)	10	20	30	10
Clerical		(10)	40	40	50	120
Other intangibles	(10)	(10)		10	20	10
Intangible benefit	(40)	(60)	70	90	120	180
Discounted at 15%	(35)	(45)	46	51	60	77
Cumulative benefit	(35)	(80)	(34)	17	77 =	NPV
Tangibles and						
** intangibles**						
Total benefit	(253)	182	408	509	607	1,453
Cumulative benefit	(253)	(71)	337	846	1,453 =	NPV

The one-time investment should account for the actual purchases, plus development and installation costs. The operating cost should include all the factors considered in the projection of current costs, plus the new factors associated with the computer system:

- Software maintenance may run three or four times the cost of the original development work over the life of the system.
- Computer supplies (e.g., forms, disks, printwheels and ribbons) can be expensive.
- Extra floorspace may be required, and this space may call for special wiring, air-conditioning, fire protection, and security.
- Communications services and facilities are another big-ticket recurring charge.

To accurately assess all the costs of the computerized system requires a functional specification, as outlined in Exhibit 6-4. This describes, in some detail, how the system will look to the user ("externals"). And it covers computations, logic, file and data characteristics, software tools, hardware, documentation, and user training. Such a specification takes a lot of effort, and part of it is the user's effort. But users need to be involved in order to make sure they get what they want.

Once the tangible costs are noted, it is time to apply a discounted-cash-flow analysis. In Exhibit 6-2, the discount rate (or internal cost of capital, or hurdle rate) is set at 15%, by way of example, and taxes are ignored. The

---

EXHIBIT 6-4   Functional-specification checklist.

---

1. **Application summary**

2. **External specifications (user view)**
   Application, inputs and outputs, performance, future expansion

3. **Software tools**
   Operating system, languages, utilities, database languages, report writers, security features

4. **Hardware, vendor selection**
   Characteristics, quality, performance

5. **Internal specifications**
   Computations, logic, flowcharts, files, data editing, backup and unloading, testing plan

6. **Protective measures**
   Backup, physical security, fire protection, procedures

7. **Documentation, standards**
   Project workbook, user manuals, program documentation, operations documentation

8. **User training**

9. **Plans and schedules**
   Design reviews, acceptance authority, workhour and capital budgets, progress reporting

---

final result in Exhibit 6-3 is a net present value (NPV); other common measures are internal rate of return and the payback period. Whatever the measure, the system is justified in tangible terms if the benefits are greater than the costs over the period in question.

INTANGIBLE FACTORS

Let us now consider intangible measures of productivity and value. Here, productivity improvements are like the tangible performance improvements—e.g., workhours—but they are less certain. For example, the "productivity saving" in Exhibit 6-3 may be the estimated reduction (or increase) in: managerial hours devoted to review and followup; engineering hours devoted to material takeoffs; clerical hours devoted to transcription. One could argue that productivity is a tangible item, but productivity improvements often do not actually reduce expenses—workhours may be redirected rather than actually reduced.

Other intangibles should be identified and quantified (if possible), and their value should be explained. These might include:

- Improved competitive position due to speedier response to business opportunities and demands.

- Better operational environment.
- Clearer and faster communication.
- Faster, timelier and more extensive access to needed information.
- Better service to clients.
- Improved image in the eyes of clients, employees and potential employees.

The analysis of Exhibit 6-3 puts a net dollar value on these factors for every year, and discounts them at 15%. The final result is a measure of the total value of the computer project—here, a net present value of some $1.5 million. But the tangibles and intangibles may be considered separately.

THE BOTTOM LINE

If the system is too costly, or the benefits not great enough, it may be possible to go back, revise the objectives, and work out a system that is more cost-effective. But if the system does appear to be a good investment, the next step is a formal proposal—including the functional specification, the cost justification, and the critical issues and uncertainties associated with developing, implementing and operating the system.

It is difficult to project potential savings far into the future, and easy to disparage claims of great cost savings resulting from computer technology. However, we should not be afraid to try something new, for while failures are inevitable, we cannot get ahead if we do not attempt to.

The first application of computer technology may have low or negative returns, and it will probably increase workload—because one is starting from scratch. But there are indeed opportunities for reducing tangible costs and improving productivity through computer technology, and a provable, positive net benefit is a compelling argument for moving ahead.

REFERENCES

1. Boehm, Barry W., "Software Engineering Economics," Prentice-Hall, Englewood Cliffs, N.J., 1981.
2. Covvey, H. Dominic, and McAlister, Neil H., "Computer Choices," Addison-Wesley, Reading, Mass., 1982.
3. Gerstein, Marc, and Reisman, Heather, "Creating Competitive Advantage with Computer Technology," *Journal of Business Strategy,* 1982.
4. Roberts, T. G., Jenkins, R. D., and Rowse, J. L., "An Executive's Guide to Computer System Analysis and Design," *Software Age,* December 1969.

CONSULTANT'S PROFIT KEY 10

The following profit key appeared in *Consulting Engineer* magazine and was written by Paul E. Pritzker, PE, while president of the National Academy of Forensic

Engineering, president-elect of NSPE and CEO, and senior engineer of George Slack & Pritzker Forensic Engineering Consultants, Quincy, Massachusetts. In his closing, Pritzker points out, "Forensic engineering is an interesting profession, but the special skills required and the adversary relationship when experts disagree preclude this type of practice for every design professional. Instead, design professionals can benefit from engaging a forensic engineering firm to work as part of their team in addressing a client's problems." So in this section we suggest that you consider—as detailed in Section 1—becoming a forensic engineering firm if your skills and experience qualify you for this work. Here are Paul Pritzker's thoughts on investigative techniques.

In the practice of forensic engineering, there are many sources of clients and diverse types of assignments. The common denominator is that all clients call when they are in trouble.

A forensic engineer first and foremost is a professional engineer. His client may be in an adversary relationship, but the engineer should not play an adversarial role. He fulfills the mandate of his commitment when he identifies facts forthrightly.

The *client* may be an owner, insurance company, attorney, surety, or design professional with a direct or tangential interest that requires response. The *incident* may be a building collapse, building system failure, fire/explosion, product liability allegation, or construction related dispute.

Figure 6-1 illustrates the path of a typical assignment and provides a condensed overview of factors germane to field investigation techniques.

Field investigations are critical to every forensic engineering assignment because most answers are found at the site. This is the area in which forensic engineering differs significantly from traditional design services. For the latter, the engineer's work is executed in the privacy of his office. He delivers his final work product to his client in the form of written construction documents consisting of plans and specifications. He seldom is required or requested to share his technical calculations.

By contrast, the forensic engineer practices in an open field. On some assignments there are so many technical experts that one almost needs a scorecard to keep track of the players. As a general rule, each engineer will serve an interest that is inimical to the efforts of others on the scene. His work product is not accepted at face value; he must be prepared to defend his conclusions in a variety of forums including, on occasion, a court of law. Peer critique is comprehensive and can go to the jugular. Calling a horse a cow does not make a horse a cow.

TECHNIQUES IN THE FIELD

Field investigation techniques require a modicum of common sense; they do not require heroics. The engineer should not enter the site until adequate safety precautions have been taken. Access to many parts of a structure, from the subbasement to the penthouse, may require removal of debris and

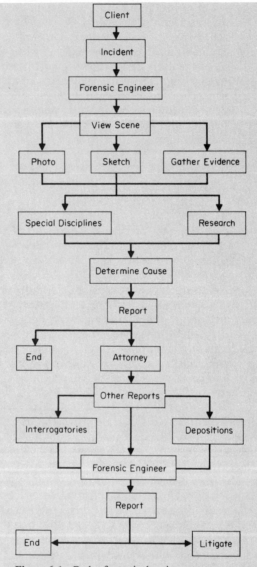

Figure 6-1 Path of a typical assignment.

the shoring of floors and walls to assure safe passage. Good lighting is important. Proper apparel to protect face, eyes, hands, legs, and clothing should be worn at all times.

Special tools and instruments, including dictating and photographic equipment, should be included in the field kit taken to the scene. Although there are assignments that mandate the services of a professional photographer, the investigator should make his own photographic record to document exterior and interior conditions. Frequently, photos will reveal important features that were not readily apparent.

Using 35mm color film provides quality images that can be enlarged; contrasts are diminished on black and white film. However, Polaroid pictures can be effective for interim documentation. We have found the VCR 7lb. low-light video camera with the 7lb. companion recorder useful. When required, the camera can be augmented with a 12v quartz light which can be operated from batteries if no power is available.

Written and graphic documentation of field observations are critical. Graph paper can be used to construct a rough sketch that can be redrawn to scale at a later date. These sketches may be invaluable in clarifying an oblique point or fact. Accurate notes are the cornerstone of an efficient investigation. We use a portable tape recorder and a 6″ × 9″ spiral bound steno pad with blue and red pencils to emphasize details. Yellow highlighting on important documents further separates the vital from the mundane.

The engineer may be contacted to continue with this phase of the assignment months or even years after his initial report was prepared. His notes, photographs, report, and memory may be his only resources at that time, particularly if the structure in question has been demolished.

Physical evidence must be treated with respect. Care must be taken to avoid disturbing or modifying evidence. Physical evidence should be photographed just as it was first observed and notes taken to document each article. When authorized to remove evidence, special care must be taken to preserve and protect it from loss or harm. Clean plastic containers can be used for large pieces; small articles can be placed in glass jars with airtight covers. The evidence should be tagged with name, date, time, and location and placed in secure storage with limited access.

We include copies of photographs and drawings to illustrate the technical report. All field notes, negatives, photographs, and special reference material are retained for our records in a secure jacket.

An example of how this basic level of service can provide maximum benefit to a client was a tragic hotel fire in which 85 people perished. Our client was a prominent West Coast attorney representing a subcontractor's insurance carrier. After reviewing the prodigious documents and conducting a field trip to the site, we were able to conclude that the subcontractor, who was working at the hotel at the time of the fire, did not exacerbate the fire's spread or contribute to the loss of life or property.

UNDERSTANDING THE LEGAL SYSTEM

Generally, field investigations lead to conclusions relating to negligence on the part of one or more parties. A client is entitled to recover from the negligent party the costs incurred as a result of a particular incident. An insurance company that paid a claim, for example, will commence subrogation proceedings against a third party when convinced the third party was negligent.

Forensic engineers should be familiar with the rules of negligence and concepts of foreseeability and connection, including the importance of an intervening cause. Frequently, field investigations provide evidence of negligence on the part of one or more parties. On negligence investigations we work the algebraic equation postulated by Associate Justice Learned Hand: $P \times G < B$ = negligence. We address ourselves to the *probability* of harm, analyze the *gravity* of the risk, and use professional value judgment to determine if the *burden* of adequate precaution created a scenario of negligence. The dispassionate analysis of facts by an experienced forensic engineer should provide a conclusion that is supportable and unequivocal.

Forensic engineers must be able to recognize the factors that have a bearing on negligence. They should be familiar with the terms used by trial lawyers when they commence litigation, as each profession has its own lexicon. An attorney will use his initial pleading to establish the parameters of his case. He will utilize discovery documents: interrogatories (questions on specific issues that are followed with cogent answers); and depositions (statements taken under oath, including adversary cross examination). Supplemental documents may be provided that will include reports by other engineers and parties of interest.

Once the engineer is positive that a conclusion is based on merit, he tries to identify negative factors that preclude other causes. Fire investigators use the technique called "diagnosis of exclusion" to confirm probable cause determinations.

Investigating a particular nightclub fire some years ago, we determined that arson was the cause. Accelerants had been used to "torch" the property. We inspected the electrical and HVAC systems and the general housekeeping to make certain that our opinion was valid based on the analysis of evidence. Sherlock Holmes once observed that "When you have eliminated the impossible, whatever is left, however improbable, must be true."

Another example is one in which we were retained by an attorney representing the interest of a group of cooperative owners who complained about latent defects in property that were making life unbearable for over 1000 residents.

The team approach required the blending of multiple disciplines directed by a registered professional engineer experienced in forensic engineering techniques. The team included structural/civil, mechanical/HVAC, mechanical/sanitation, electrical, and metallurgical engineers, supported by two testing laboratories, and a drilling company retained to obtain borings.

The chemist who studied the soil conditions reported that "the test data show the soil in this area to be alkaline with a relatively high dissolved salt content. These conditions would be corrosive to exposed metals and accelerated corrosion would be expected. Metallic materials exposed to these conditions should be protected with corrosion-resistant coatings or corrosion-resisting materials should have been used in this installation."

The full report addressed the major latent defects by discipline:

• *Roofs.* The roof systems do not meet the specified four-ply construction. The material was improperly installed. The workmanship was poor and sloppy. Samples and laboratory reports conclude that some of the roofs are two ply and others are three ply. The roofs require replacement.

- *Heating system.*   Boiler and distribution pumps are undersized. Boiler room ventilation is inadequate. Radiation is deficient. Underground piping is unprotected from corrosion.

- *Hot water underground distribution.*   Insulation was punctured and was not suitable for particular field application. Metallography of eight samples provides evidence to conclude that the steel piping was inadequately protected. Rate of corrosion exacerbated by soil conditions.

- *Wall surfaces.*   Rain water migrates through brick joints at locations where flashing is missing. Rain water travels via vertical joints in the original brick work. This condition allows water to enter the interior walls and damage apartments. This condition is due to improper and incomplete "buttering" and "tooling" of joints required to assure proper bond. The condition is the result of poor construction procedures.

- *Electrical system.*   Switches and receptacles were utilized that were not rated or approved for connection to aluminum conductors.

- *Plumbing.*   The insulation specified was not provided by the contractor. There was evidence of severe electrolysis.

The case did not go to trial. The client accepted an offer of $1.6 million to remedy the latent defects. The funds were obtained on the basis of an exhaustive field investigation that documented with both narrative and photographs the litany of observations taken by qualified persons.

FOLLOW-THROUGH PROCEDURES

The field trip represents only the initial phase of the forensic engineer's responsibility. Information gleaned from that experience is used to establish requisite follow-through procedures which may entail retaining consultants with special skills to evaluate physical evidence and schedule tests when issues are in doubt. These tests, destructive or nondestructive, should be conducted only after consultation with the client and notification to all interested parties. As part of this procedure, laboratory technicians should be briefed thoroughly as to the results requested from tests.

There often is a time lapse between the initial site visit and the report of associate technical consultants, making record keeping vital to track each assignment as it progresses. Clients should be notified on a two-week cycle of the status of each case; this allows them to adhere to various regulatory procedures. When the analysis is completed, the initial preliminary opinion is rendered to the client or his designated attorney.

RESEARCHING THE PROBLEM

Forensic engineering requires dedication to researching problems. Some of what we are taught as engineering students quickly becomes obsolete but the procedures for thinking and for analyzing problems never change. For example, in a building failure it is necessary to determine if nontraditional design or construction was involved; if there was a curtailment, even partial, of on-site construction phase services by the design professionals; or if there was a post-construction lack of maintenance.

Design changes are a fertile ground for breeding potential building system failures. Field investigation should include review of design and as-built drawings and comparison of these with field observation. Was there an unconventional use of materials or unusual construction erection techniques? Was there a potential for weather-related factors to cause aging of materials?

There are a host of initial and secondary factors that can be ascertained with sufficient construction field experience. U.S. Weather Bureau records can be used to log the weather during the period being investigated. Expansion and contraction due to heat or cold, for example, pose potential dangers. Snow drifts, ponded water, vibration, and air infiltration influence a structure's capacity to remain intact. Frequently, lack of structural redundancy leads to catastrophic damage.

There are many safety problems in building construction. Field investigations should focus on visual clues which suggest that workers can come in *contact* with hazardous conditions that could include electricity, gas, fumes, water, steam, heat, cold, or radiation. Can workers be *struck* by sharp or jagged edges, protruding objects, or stationary or moving objects? Can a worker be *caught* in, on, or between pinch points, protruding objects, or moving or stationary objects? Can the worker *slip, trip,* or *fall* on the same or a lower level? Can physical injury result from lifting, pulling, or pushing? The type of mind set that allows the engineer to search out these factors will help him identify the basic types of hazards that lead to personal injury and property damage.

Forensic engineering is an interesting profession, but the special skills required and the adversary relationship when experts disagree preclude this type of practice for every design professional. Instead, design professionals can benefit from engaging a forensic engineering firm to work as part of their team in addressing a client's problems. Architects have a tradition of retaining engineers with diverse engineering disciplines. Engineers in construction, government, industry, and private practice might consider emulating that practice. The sometimes conflicting objectives of various parties on a construction project are part of the technical problems that nourish disputes. The forensic engineer is in a unique position to search for the truth objectively.

SUMMARY

The consulting engineer is by nature an independent person. But a consultant should not let his or her drive for independence inhibit the growth of a consulting organization. Careful planning and administration afford you the potential for successful growth through:

- Joint ventures
- Concentration
- Additional offices
- Combining practices with others
- Adding disciplines
- International work
- Government contracts
- Manufacturing design services
- Design, build, and manage teams
- Project management and construction management

Each technique and type of expansion requires consulting engineers to modify their authority over their business. There is a price to pay for everything, and we feel that successful growth of the consulting firm is well worth the price you must pay. You may have to relinquish some of your cherished autonomy, but your expertise, profits, and your contribution to society will increase.

To expand properly, a firm's consultants must be able to study the alternatives and likely results of various growth options. They must plan for the firm's growth, administer to its changing needs, supervise the group, and finance the expansion. Particular attention should be focused on marketing and promoting the newly added services through the most effective channels available. Some firms devote considerable effort to researching the prospective market before they undertake a program of expansion. This step can save time and money that could otherwise be wasted.

The professional practicing engineer is made, not born. There is no typical consultant, although most share the desire to be professionally independent. Further, most successful consulting engineers are adventurous enough to speculate on their ability to start new, independent efforts. Risks are inherent in such undertakings, but no one ever succeeded without taking a chance. Perhaps too many salaried professionals remain in unrewarding jobs because they accept the notion that engineers lack business and verbal acumen. We know this notion to be false.

Certain traits common to most established consultants can be discerned. They possess the will and tools to let others know about their available expertise, and they do so by promoting their firms. They decide what kind of assistance they need—whether from a staff or a computer—

and they obtain it. Perhaps most importantly, the successful consultant uses and adopts certain standards of professional ethics, motivation, and interpersonal behavior. In short, the engineer desiring to reap the rewards of a professional consulting practice plans his or her work and works that plan. The rewards are within your grasp. All you need to do is reach!

Index